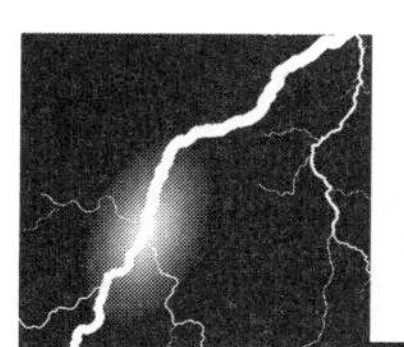

[AODS5] Notes to Reader

Guide to Army Operations & the Six Warfighting Functions

An **operation** is a sequence of tactical actions with a common purpose or unifying theme. Army forces, as part of the joint force, contribute to the joint fight through the conduct of unified land operations.

Unified land operations are simultaneous offensive, defensive, and stability or defense support of civil authorities tasks to seize, retain, and exploit the initiative and consolidate gains to prevent conflict, shape the operational environment, and win our Nation's wars as part of unified action.

Decisive action is the continuous, simultaneous combinations of offensive, defensive, and stability or defense support of civil authorities tasks. Operations conducted outside the United States and its territories simultaneously combine three elements—offense, defense, and stability. Within the United States and its territories, decisive action combines the elements of defense support of civil authorities and, as required, offense and defense to support homeland defense.

Army forces conduct **multi-domain battle**, as part of a joint force, to seize, retain, and exploit control over enemy forces. Army forces deter adversaries, restrict enemy freedom of action, and ensure freedom of maneuver and action in multiple domains for the joint force commander.

Combat power is the total means of destructive, constructive, and information capabilities that a military unit or formation can apply at a given time. To an Army commander, Army forces generate combat power by converting potential into effective action. Combat power has eight elements: leadership, information, mission command, movement and maneuver, intelligence, fires, sustainment, and protection.

The Army collectively describes the last six elements as the **warfighting functions**. Commanders apply combat power through the warfighting functions using leadership and information.

SMARTbooks - DIME is our DOMAIN!

SMARTbooks: Reference Essentials for the Instruments of National Power (D-I-M-E: Diplomatic, Informational, Military, Economic)! Recognized as a "whole of government" doctrinal reference standard by military, national security and government professionals around the world, SMARTbooks comprise a comprehensive professional library designed with all levels of Soldiers, Sailors, Airmen, Marines and Civilians in mind.

SMARTbooks can be used as quick reference guides during actual operations, as study guides at education and professional development courses, and as lesson plans and checklists in support of training. Visit **www.TheLightningPress.com**!

Army "Doctrine 2015" Overview

Under the previous doctrine management program, the Army maintained 625 publications on the Army Publishing Directorate website and the Reimer Digital Library. Many of these manuals remained unchanged for years. In 2009, the Commander, U.S. Army Training and Doctrine Command (TRADOC), directed a reengineering of doctrine. The primary goals of the reengineering project were to reduce the number of field manuals (FM), standardize the content of manuals to less than 200 pages, and establish a more efficient doctrine management program.

In the past, the average life cycle of a doctrine publication was about three to five years. Once proponent authors begin revising it, the revision process takes from three to 24 months to complete, depending on the needs of the field. The current cycle has come a long way in adjusting to the needs in theater; however, when a rapid change is required, the system requires significant time to update a manual. The current method is viewed by many as cumbersome, slow, and unable to keep up with rapidly changing unified land operations. The primary focus of Doctrine 2015 is to produce a body of knowledge related to the conduct of operations that uses technology to leverage and incorporate leader input, especially on mission essential tasks. Doctrine development will become faster and the system will create fewer publications which will be shorter, clearer, and more digitally accessible than the current system.

Doctrine 2015 will have four categories of operational knowledge: Army doctrine publications (ADPs), Army doctrine reference publications (ADRPs), field manuals (FMs), Army techniques publications (ATPs) and digital applications (APPs).

As the window on real-world operations and actual combat knowledge starts to close, the drive to capture the lessons from over a decade of persistent conflict is strong. Doctrine 2015 will be the vehicle for gaining and capturing that knowledge and transmitting it to the Army of the future. By breaking up doctrine into its basic components, the Army will be able to make revisions faster, retain enduring concepts, and gain lessons from battlefield experienced warriors.

Doctrine 2015 is a significant departure from the way doctrine has been developed in the past. Changing times, technical advances, demands from the field and the ever changing battlefield environment prompted these significant and necessary changes. The Army's need to teach both enduring lessons and new concepts remains constant. It will be how the Army obtains and delivers information that must change. The Doctrine 2015 system will allow this change to happen.

The Army Operations & Doctrine SMARTbook

Guide to Army Operations & the Six Warfighting Functions

AODS5 is the fifth revised edition of The Army Operations & Doctrine SMARTbook, with complete discussion of the fundamentals, principles and tenets of Army operations and organization (ADP/ADRP 3-0 Operations, 2016); chapters on each of the six warfighting functions: mission command (ADP/ADRP 6-0), movement and maneuver (ADPs 3-90, 3-07, 3-28, 3-05), intelligence (ADP/ADRP 2-0), fires (ADP/ADRP 3-09), sustainment (ADP/ADRP 4-0), and protection (ADP/ADRP 3-37).

** Change 1 to AODS5 (Apr 2017) incorporates minor text edits from ADRP 3-0 (Nov 2016), FM 6-0 (Chg 2, Apr 2016), FM 3-52 (Oct 2016) and FM 3-34 (Apr 2016). An asterisk marks changed pages.*

Foundations of Doctrine

Unified land operations *describes how the Army seizes, retains, and exploits the initiative to gain and maintain a position of relative advantage in sustained land operations through simultaneous offensive, defensive, and stability operations in order to prevent or deter conflict, prevail in war, and create the conditions for favorable conflict resolution. Unified land operations is the Army's operational concept and the Army's contribution to* ***unified action****.*

Chapter 1

"TLS5: The Leader's SMARTbook"

Decisive Action

Editor's note: For the purposes of The Army Operations & Doctrine SMARTbook, an overview of the following elements of decisive action are represented as they relate to the movement and maneuver warfighting function (including Special Operations):

Chapter 3 - (Movement & Maneuver)

"BSS5: The Battle Staff SMARTbook"

The Six Warfighting Functions

A warfighting function is a group of tasks and systems (people, organizations, information, and processes) united by a common purpose that commanders use to accomplish missions and training objectives. All warfighting functions possess scalable capabilities to mass lethal and nonlethal effects. The Army's warfighting functions link directly to the joint functions.

Chapter 2 *Chapter 3* *Chapter 4* *Chapter 5* *Chapter 6* *Chapter 7*

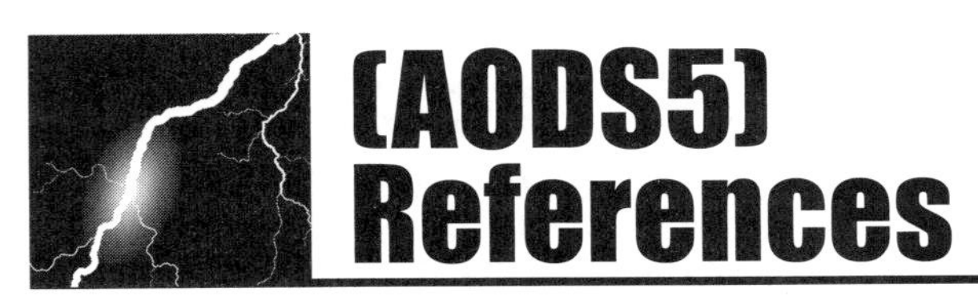

[AODS5] References

The following references were used to compile The Army Operations & Doctrine SMARTbook. All references are available to the general public and designated as "approved for public release; distribution is unlimited." The Army Operations & Doctrine SMARTbook does not contain classified or sensitive material restricted from public release.

Army Doctrinal Publications (ADPs) and Army Doctrinal Reference Publications (ADRPs)

ADRP 1-02	Feb 2015	Operational Terms and Military Symbols
ADP/ADRP 2-0	Aug 2012	Intelligence
ADP 3-0*	Nov 2016	Operations
ADRP 3-0*	Nov 2016	Operations
ADP 3-05	Aug 2012	Special Operations
ADP 3-07	Aug 2012	Stability
ADP/ADRP 3-09	Aug 2012	Fires
ADP 3-28	Jul 2012	Defense Support of Civil Authorities
ADP/ADRP 3-37	Aug 2012	Protection
ADP/ADRP 3-90	Aug 2012	Offense and Defense
ADP/ADRP 4-0	Jul 2012	Sustainment
ADP/ADRP 5-0	May 2012	The Operations Process
ADP/ADRP 6-0	May 2012	Mission Command (with Chg 1, Sept 2012)

Field Manuals (FMs)

FM 3-34*	Apr 2014	Engineer Operations
FM 3-35	Apr 2010	Army Deployment and Redeployment
FM 3-52*	Oct 2016	Airspace Control
FM 6-0*	Apr 2016	Commander and Staff Organization and Operations (w/change 2)

Joint Publications (JPs)

JP 1	Mar 2013	Doctrine for the Armed Forces of the United States
JP 3-0	Aug 2011	Joint Operations

** New/updated since last publication.*

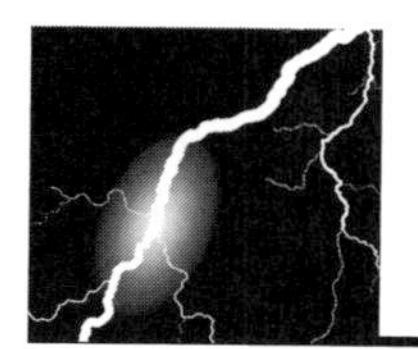

(AODS5) Table of Contents

Chap 1 Operations (ADRP 3-0, Nov 2016)

Chap 2

Mission Command Warfighting Function

Chap 3

Movement & Maneuver Warfighting Function

Chap 4

Intelligence Warfighting Function

Chap 5

Fires Warfighting Function

Chap 6

Sustainment Warfighting Function

Chap 7

Protection Warfighting Function

Chap 1 I. Military Operations (ADRP 3-0, Nov '16)

Ref: ADP 3-0, Operations (Nov '16) and ADRP 3-0, Operations (Nov '16), chap. 1.

An **operation** is a sequence of tactical actions with a common purpose or unifying theme (JP 1). Army forces, as part of the joint force, contribute to the joint fight through the conduct of unified land operations. **Unified land operations** are simultaneous offensive, defensive, and stability or defense support of civil authorities tasks to seize, retain, and exploit the initiative and consolidate gains to prevent conflict, shape the operational environment, and win our Nation's wars as part of unified action (ADRP 3-0). ADP 3-0 is the Army's basic warfighting doctrine and is the Army's contribution to **unified action**.

I. An Operational Environment

An operational environment is a composite of the conditions, circumstances, and influences that affect the employment of capabilities and bear on the decisions of the commander (JP 3-0). Commanders at all levels have their own operational environments for their particular operations. An operational environment for any specific operation is not just isolated conditions of interacting variables that exist within a specific area of operations. It also involves interconnected influences from the global or regional perspective (for example, politics and economics) that impact on conditions and operations there. Thus, each commander's operational environment is part of a higher commander's operational environment.

Operational environments include considerations at the strategic, operational, and tactical levels of warfare. At the strategic level, leaders develop an idea or set of ideas for employing the instruments of national power (diplomatic, informational, military, and economic) in a synchronized and integrated fashion to achieve theater, national, and multinational objectives. The operational level links the tactical employment of forces to national and military strategic objectives, with the focus being on the design, planning, and execution of operations using operational art. Finally, the tactical level of warfare involves the employment and ordered arrangement of forces in relation to each other. The levels of warfare assist commanders in visualizing a logical arrangement of forces, in allocating resources, and in assigning tasks based on conditions within their operational environment.

See pp. 1-21 to 1-32 for a discussion of operational art.

Important trends such as globalization, urbanization, technological advances, and failed or failing states can affect land operations. These trends can drive instability in an operational environment as well as a continuing state of persistent conflict. Persistent conflict is the protracted confrontation among state, nonstate, and individual actors who are willing to use violence to achieve their political and ideological ends. In such an operational environment, commanders must seek opportunities for exploiting success. Opportunities may include greater cooperation among the local population of a town or perhaps the ability to advance forces along a previously unsecured route. To exploit opportunities successfully, commanders must thoroughly understand the changing nature of an operational environment. In understanding an evolving operational environment, commanders must identify how previous experience within the current or a similar operational environment has changed or is no longer applicable and can actually detract from mission success.

Enemies are developing the capability to mass effects from multiple domains at a speed that will impact ongoing operations. Operations in the information environment and cyberspace will attempt to influence U.S. decision makers and disrupt any

ADRP 3-0 (2016): Major Changes (from 2012)

Ref: ADRP 3-0, Operations (Nov '16), preface and introduction.

ADRP 3-0 augments the land operations doctrine established in ADP 3-0, Operations. This publication expands the discussion of the overarching guidance on unified land operations. It accounts for the uncertain and ever-changing nature of operations and recognizes that a military operation is foremost a human undertaking. It constitutes the Army's view of how to conduct prompt and sustained operations on land and sets the foundation for developing the other principles, tactics, techniques, and procedures detailed in subordinate doctrine publications. Combined with ADP 3-0, the doctrine in ADRP 3-0 provides the foundation for the Army's operational concept of unified land operations. This publication also forms the foundation for training and Army education system curricula on unified land operations.

An **operation** is a sequence of tactical actions with a common purpose or unifying theme. ADRP 3-0 discusses operations by expanding on the foundations, tenets, and doctrine of unified land operations found in ADP 3-0. Combined with ADP 1, The Army, and ADRP 1, The Army Profession, ADRP 3-0 provides a common perspective on the nature of warfare and a common reference for solving military problems. Whereas ADP 1 describes the missions, purpose, roles, and core competencies of the Army, ADRP 3-0 describes how the Army conducts operations as part of a joint team working with unified action partners. ADRP 3-0 does this by establishing the Army's operational concept—a fundamental statement that frames how Army forces, operating as part of a joint force, conduct operations (ADP 1-01). Previous operational concepts included airland battle (1986) and full spectrum operations (2001). The Army's operational concept of unified land operations (including its principles, tenets, and operational structure) serves as the basic framework for all operations across the range of military operations. It is the core of Army doctrine that guides how Army forces contribute to unified action. Today, as with each previous version of Operations, ADRP 3-0 shapes all Army doctrine and influences the Army's organization, training, material, leadership, education, and Soldier concerns.

See introductory figure on p. 1-5 for the ADRP 3-0 logic chart.

This version of ADRP 3-0 makes numerous changes from the 2012 version. The most significant changes are the updated version of tenets of unified land operations and the addition of principles of unified land operations. ADRP 3-0 modifies the definition of unified land operations to account for defense support of civil authorities.

ADRP 3-0 retains unified land operations as the Army's operational concept. Additional changes in this version of ADRP 3-0 from the 2012 version include a discussion of the characteristics that Army forces need to display in land combat and a modification of one staff task to account for Army capabilities in cyberspace. Also, the discussion of core competencies that was in the 2012 version of ADP 3-0 and ADRP 3-0 has moved to ADP 1.

ADRP 3-0 employs categories and lists of information, such as principles and tenets, as a means of highlighting key aspects of doctrine. Where categories or lists are employed, a narrative discussion follows to provide details of the subject. It is important to remember that in doctrine, categories or lists serve as guidelines or tools for a Soldier to more easily remember important doctrinal terms. However, Soldiers need to study doctrine in detail and consider how terms are applied to Army operations.

ADRP 3-0 modifies key topics and updates terminology and concepts as necessary. These topics include the discussion of an operational environment and the operational and mission variables, as well as discussions of unified action, law of land warfare, and combat power. As in the 2012 version of ADRP 3-0, mission command remains both a philosophy of command and a warfighting function. ADRP 3-0 maintains combined arms as the application of arms that multiplies Army forces' effectiveness in all operations. However, ADRP 3-0 expands combined arms to include joint and multinational assets as integral to combined arms and discusses how the Army conducts these operations across multiple domains.

New, Modified, and Rescinded Army Terms

Based on current doctrinal changes, certain terms for which ADRP 3-0 is the proponent have been added, rescinded, or modified. The glossary contains acronyms and defined terms. (See introductory table 1 for new, modified, and rescinded terms. See introductory table 2 for modified and rescinded acronyms.)

Term	*Reasoning*
close area	Modifies the definition
combined arms	Modifies the definition
combined arms maneuver	No longer a defined term
consolidate gains	New term and definition
culminating point	Adopts the joint definition
cyber electromagnetic activities	No longer a defined term
cyberspace electromagnetic activities	New term and definition
deep area	Modifies the definition
depth	New term and definition
disintegrate	New term and definition
dislocate	New term and definition
flexibility	New term and definition
hybrid threat	Modifies the definition
individual initiative	No longer a defined term
inform and influence activities	Rescinded
intelligence warfighting function	Modifies the definition
line of effort	Modifies the definition
movement and maneuver warfighting function	Modifies the definition
neutral	Modifies the definition
operational initiative	No longer a defined term
position of relative advantage	New term and definition
simultaneity	New term and definition
support area	Modifies the definition
task-organizing	Modifies the definition
unified land operations	ADRP 3-0 becomes proponent, modifies definition
warfighting function	Modifies the definition
wide area security	No longer a defined term

Ref: ADRP 3-0 (2016), introductory table 1. New, modified, and rescinded Army terms.

Modified and Rescinded Army Acronyms

Acronym	*Full Form*	*Reasoning*
CEMA	cyber electromagnetic activities	Modifies the acronym
IIA	inform and influence activities	Rescinded

Ref: ADRP 3-0 (2016), introductory table 2. Modified and rescinded Army acronyms.

force deployment activities. Land-based threats will attempt to impede joint force freedom of movement and action across all domains, disrupt the electromagnetic spectrum, hinder the information environment, and challenge human perceptions. Just as the enemy will attempt to present multiple dilemmas to land forces from the other domains, Army commanders must seize opportunities across multiple domains to enable their own land operations, as well as the operations of our unified actions partners in the other domains.

Modern information technology makes the information environment, inclusive of cyberspace and the electromagnetic spectrum, indispensable for human interaction, including military operations and political competition. The information environment is the aggregate of individuals, organizations, and systems that collect, process, disseminate, or act on information. (JP 3-13) This environment inherently impacts an operational environment, and that environment will be simultaneously congested and contested during operations. All actors—enemy, friendly, or neutral—remain vulnerable to attack by physical, psychological, cyber, or electronic means, or a combination thereof. Additionally, actions in and through cyberspace and the electromagnetic spectrum can affect other actors.

Refer to JP 3-12 (R) for more information on cyberspace operations and the electromagnetic spectrum.

An operational environment consists of many interrelated variables and subvariables, as well as the relationships and interactions among those variables and subvariables. How the many entities and conditions behave and interact with each other within an operational environment is difficult to discern and always results in differing circumstances. Different actor or audience types do not interpret a single message in the same way. Therefore, no two operational environments are the same.

In addition, an operational environment continually evolves. This evolution results from humans interacting within an operational environment as well as from their ability to learn and adapt. As people take action in an operational environment, they change that environment. Other variables may also change an operational environment. Some changes are anticipated, while others are not. Some changes are immediate and apparent, while other changes evolve over time or are extremely difficult to detect. For example, an enemy force adjusting its geographic position may be easy to detect, whereas changes in a population's demographics or political views may be more subtle and may take longer to understand.

The complex and dynamic nature of an operational environment may make determining the relationship between cause and effect difficult and may contribute to the uncertainty of military operations. Commanders must continually assess and reassess their operational environments. They seek a greater understanding of how the changing nature of threats and other variables affect not only their forces but other actors as well. To do this, commanders and their staffs may use the Army design methodology, operational variables, and mission variables to analyze an operational environment in support of the operations process.

See p. 1-22 for a discussion of the Army design methodology.

A. Operational and Mission Variables

An operational environment for each operation differs and evolves as each operation progresses. Army leaders use operational variables to analyze and understand a specific operational environment. They use mission variables to focus on specific elements of an operational environment during mission analysis.

Operational Variables (PMESII-PT)

Army planners describe conditions of an operational environment in terms of operational variables. Operational variables are those aspects of an operational environment, both military and nonmilitary, that may differ from one operational area to another and affect operations. Operational variables describe not only the

Operations (Unified Logic Chart)

Ref: ADRP 3-0, Operations (Nov '16), introductory figure. ADRP 3-0 unified logic chart.

An operation is a sequence of tactical actions with a common purpose or unifying theme (JP 1). Army forces, as part of the joint force, contribute to the joint fight through the conduct of unified land operations. Unified land operations are simultaneous offensive, defensive, and stability or defense support of civil authorities tasks to seize, retain, and exploit the initiative and consolidate gains to prevent conflict, shape the operational environment, and win our Nation's wars as part of unified action (ADRP 3-0). ADP 3-0 is the Army's basic warfighting doctrine and is the Army's contribution to unified action.

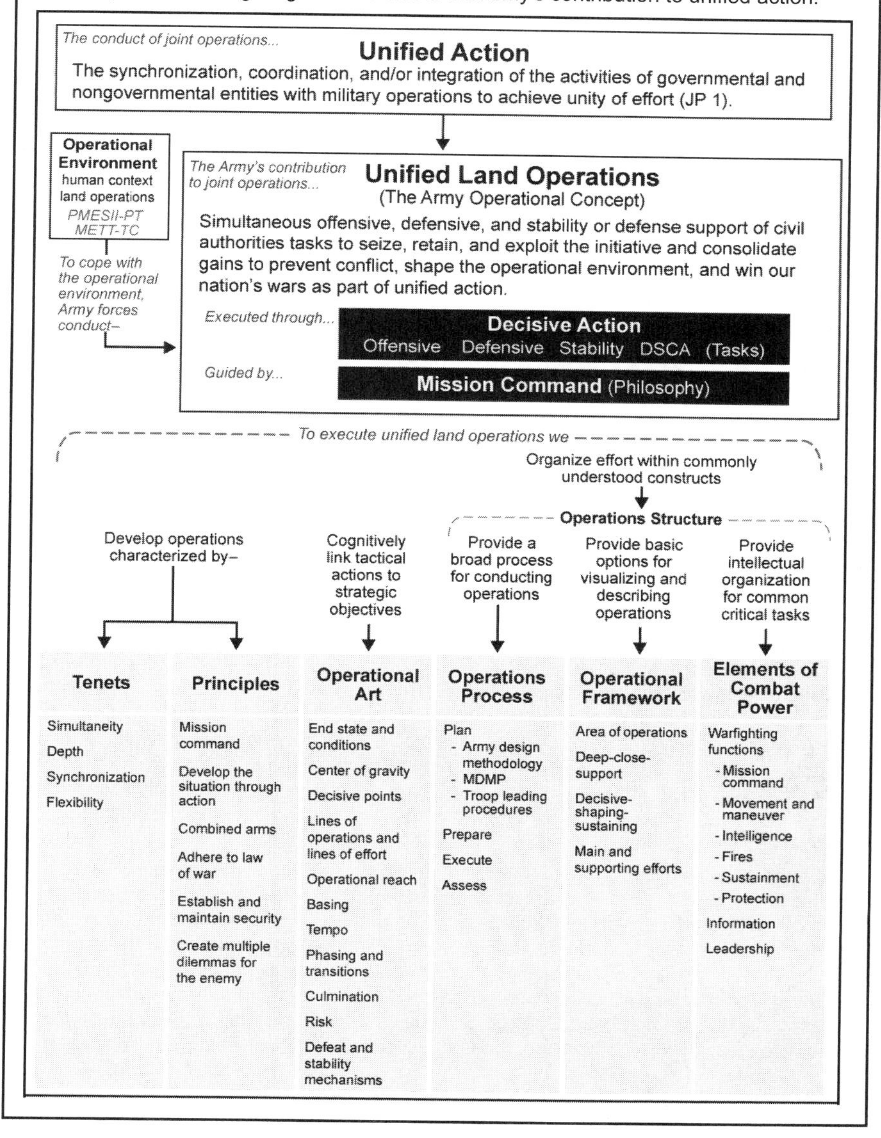

B. Threats and Hazards

Ref: ADRP 3-0, Operations (Nov '16), pp. 1-3 to 1-4.

Threats are a fundamental part of an overall operational environment for any operation. A threat is any combination of actors, entities, or forces that have the capability and intent to harm United States forces, United States national interests, or the homeland. Threats may include individuals, groups of individuals (organized or not organized), paramilitary or military forces, nation-states, or national alliances. Commanders and staffs must understand how current and potential threats organize, equip, train, employ, and control their forces. They must continually identify, monitor, and assess threats as they adapt and change over time. In general, the various actors in any operational area can qualify as an enemy, an adversary, a neutral, or a friend.

Enemy

An enemy is a party identified as hostile against which the use of force is authorized. An enemy is also called a combatant and is treated as such under the law of war. Enemies will apply advanced technologies (such a cyberspace attack) as well as simple and dual-use technologies (such as improvised explosive devices). Enemies avoid U.S. strengths (such as long-range surveillance and precision strike) through traditional countermeasures (such as dispersion, concealment, and intermingling with civilian populations).

Adversary

An adversary is a party acknowledged as potentially hostile to a friendly party and against which the use of force may be envisaged (JP 3-0).

Neutral

A neutral is a party identified as neither supporting nor opposing friendly, adversary, or enemy forces.

Friendly

Finally, a friendly is an individual or group that is perceived to be supportive of U.S. efforts. Land operations often prove complex because an enemy, an adversary, a neutral, or a friend intermix, often with no easy means to distinguish one from another.

Hybrid Threat

The term hybrid threat captures the seemingly increased complexity of operations, the multiplicity of actors involved, and the blurring between traditional elements of conflict. A hybrid threat is the diverse and dynamic combination of regular forces, irregular forces, terrorist forces, or criminal elements unified to achieve mutually benefitting threat effects. Hybrid threats combine traditional forces governed by law, military tradition, and custom with unregulated forces that act with no restrictions on violence or target selection. These may involve nation-state actors, possibly using proxy forces to coerce and intimidate, or nonstate actors such as criminal and terrorist organizations that employ protracted forms of warfare using operational concepts and high-end capabilities traditionally associated with states. Such varied forces and capabilities enable hybrid threats to capitalize on perceived vulnerabilities, making them particularly effective.

Although not strictly viewed as a threat, a hazard is a condition with the potential to cause injury, illness, or death of personnel; damage to or loss of equipment or property; or mission degradation (JP 3-33). Hazardous conditions or natural phenomena are able to damage or destroy life, vital resources, and institutions, or prevent mission accomplishment. Understanding hazards and their effects on operations allows the commander to better understand the terrain, weather, and various other factors that best support the mission. Understanding hazards also helps the commander visualize potential impacts on operations. Successful interpretation of the environment aids in correctly opposing threat courses of action within a given geographical region. Hazards include disease, extreme weather phenomena, solar flares, and areas contaminated by toxic materials.

Hostile state actors may attempt to overwhelm defense systems and impose a high cost on the United States to intervene in a contingency or crisis. State and nonstate actors attempt to apply technology to disrupt the U.S. advantages in communications, long-range precision-guided munitions, movement and maneuver, and surveillance. Enemy actions seek to reduce the U.S. ability to achieve dominance in the land, air, maritime, space, and cyberspace domains. Army forces cannot always depend on an advantage in technology, communications, and information collection. They must account for what adversaries and enemies know about friendly capabilities and how Army forces operate. Army forces must anticipate how enemies will adapt their operations and use their capabilities to struggle for superiority in important portions of the land, air, maritime, space, and cyberspace domains. Additionally, to accomplish political objectives, enemy organizations may expand operations to the United States. Enemies and adversaries may operate beyond physical battlegrounds. Enemies often subvert friendly efforts by infiltrating U.S. and partner forces (acting as insider threats) and by using cyberspace attacks, while using propaganda and disinformation through social media to affect public perception.

Enemies and adversaries may pursue anti-access and area-denial capabilities. Such efforts make U.S. power projection increasingly risky and enable near-peer competitors and regional powers to extend their coercive strength well beyond their borders. In the most challenging scenarios, the United States may be unable to employ forces the way it has in the past. For example, the ability of U.S. forces to build up combat power in an area, perform detailed rehearsals and integration activities, and then conduct operations when and where desired may be significantly challenged. Additionally, enemies may employ cyberspace attack capabilities (such as disruptive and destructive malware), battlefield jammers, and space capabilities (such as anti-satellite weapons) to disrupt U.S. communications; positioning, navigation, and timing; synchronization; and freedom of maneuver. Finally, enemies may attempt to strike at homeland installations to disrupt or delay deployment of forces. These types of threats are not specific to any single theater of operations, and they create problematic consequences for international security. Such an environment can induce instability or erode the credibility of U.S. deterrence, and it may weaken U.S. international alliances, including associated trade, economic, and diplomatic agreements.

Concurrent with state challenges, violent extremist organizations work to undermine transregional security in areas such as the Middle East and North Africa. Such groups are dedicated to radicalizing populations, spreading violence, and leveraging terror to impose their visions of societal organization. They are strongest where governments are weakest, exploiting people trapped in fragile or failed states. In many locations, violent extremist organizations coexist with transnational criminal organizations, and both organizations conduct illicit trade and spread corruption, further undermining security and stability. Also, actions by computer hacking and political extremist groups create havoc, undermine security, and increase challenges to stability.

For Army forces, the dynamic relationships among friendly forces, enemy forces, and populations make land operations dynamic and complicated. Regardless of the location or threat, Army forces must synchronize actions across multiple domains to achieve unity of effort that ensures mission accomplishment. Commanders and staffs must be prepared to adapt and thrive in environments where problems bind actors together rather than formal authorities.

II. War as a Human Endeavor

Ref: ADP 3-0, Operations (Nov '16), pp. 1-4 to 1-5.

War is a human endeavor—a fundamentally human clash of wills often fought among populations. It is not a mechanical process that can be controlled precisely, or even mostly, by machines, statistics, or laws that cover operations in carefully controlled and predictable environments. Fundamentally, all war is about changing human behavior. It is both a contest of wills and a contest of intellect between two or more sides in a conflict, with each trying to alter the behavior of the other side. Success in operations is often determined by a leader's ability to outthink an opponent to gain and maintain the initiative. The side that forecasts better, learns and adapts more rapidly, thinks more clearly, decides and acts more quickly, and is comfortable operating with uncertainty stands the greatest chance to seize, retain, and exploit the initiative in order to succeed over an opponent.

War is chaotic, lethal, and inherently human. The ability to prevail in ground combat becomes a decisive factor in breaking the enemy's will. To break the will of the enemy requires commanders and Soldiers alike to understand the human context that reinforces the enemy's will (for example, societal norms, history, and religion). Commanders must not presume that superior lethality necessarily equates to causing the desired effects on the enemy. Commanders must continually assess through the various means at their disposal, whether their operations are influencing enemies, as well as populations, in the ways that they intend; this same assessment must occur for nonlethal application of force as well.

Throughout history, people have been innately tied to the land upon which they live. This reality places the U.S. Army, as part of a joint force, in a unique position to best influence foreign populations in accordance with U.S. national objectives. Insurgents often engage in warfare by blending into the local population, requiring commanders to understand the human context of the insurgency and the local population to ensure the effectiveness of operations. When unified land operations occur among civilian groups, these groups' actions influence and are influenced by military operations. The results of these interactions are often unpredictable—and perhaps uncontrollable. Commanders must seek the support of local populations and, when necessary, be able to influence their behaviors.

Human context concerns play a critical role in shaping operations as well. Soldiers interacting with partner units and local security forces garner trust when they engage these forces with respect and cultural understanding. Commanders should possess a historical understanding of operations of partner forces and how those forces have integrated with other nations.

The scope of operations associated with war reaches to the lowest echelons and the Soldiers operating at those echelons. These Soldiers number in the thousands for a brigade combat team commander. Soldiers receive orders passed through multiple echelons of command. Each Soldier must understand the limits within which to exercise disciplined initiative. In addition, subordinate commanders must understand the higher-level commander's intent, the capabilities and limitations of their unit and subordinate leaders, and the effects of their actions on the operations of the entire force. To be effective, commanders must communicate and receive information in an understandable form to and from the lowest echelons to ensure shared understanding.

U.S. military forces seek to achieve the goals and objectives given to them by the President and the Congress. Normally, this calls for establishing or re-establishing conditions favorable to U.S. interests. Setting these conditions is the role of unified action, and Army forces are a vital partner in unified actions.

military aspects of an operational environment, but also the population's influence on it. Using Army design methodology, as applicable, Army planners analyze an operational environment in terms of eight interrelated operational variables: political, military, economic, social, information, infrastructure, physical environment, and time (PMESII-PT). As soon as a commander and staff have an indication of where their unit will conduct operations, they begin analyzing the operational variables associated with that location. They continue to refine and update that analysis even after receiving a specific mission and throughout the course of the ensuing operation.

Mission Variables (METT-TC)

Upon receipt of a warning order or mission, Army leaders filter relevant information categorized by the operational variables into the categories of the mission variables used during mission analysis. They use the mission variables to refine their understanding of the situation. The mission variables consist of mission, enemy, terrain and weather, troops and support available, time available, and civil considerations (METT-TC). Incorporating the analysis of the operational variables with METT-TC helps to ensure that Army leaders consider the best available relevant information about conditions that pertain to the mission.

III. Unified Action

Unified action is the synchronization, coordination, and/or integration of the activities of governmental and nongovernmental entities with military operations to achieve unity of effort (JP 1). Unity of effort is coordination and cooperation toward common objectives, even if the participants are not necessarily part of the same command or organization, which is the product of successful unified action (JP 1). As military forces synchronize actions, they achieve unity of effort. Unified action includes actions of Army, joint, and multinational forces synchronized or coordinated with activities of other government agencies, nongovernmental and intergovernmental organizations, and the private sector. Through engagement, military forces play a key role in unified action before, during, and after operations. The Army's contribution to unified action is unified land operations. Army forces are uniquely suited to shape operational environments through their forward presence and sustained engagements with unified action partners and local civilian population.

Army forces remain the preeminent fighting force in the land domain. However, Army forces both depend on and support joint forces across multiple domains (land, air, maritime, space, and cyberspace). This integration across multiple domains, as well as both the contributions that Army forces provide and the benefits that Army forces derive from operating in multiple domains, is multi-domain battle. The Army depends on the other Services for strategic and operational mobility, joint fires, and other key enabling capabilities. The Army supports other Services, combatant commands, and unified action partners with foundational capabilities such as ground-based indirect fires and ballistic missile defense, defensive cyberspace operations, electronic protection, communications, intelligence, rotary-wing aircraft, logistics, and engineering. Unified action partners are those military forces, governmental and nongovernmental organizations, and elements of the private sector with whom Army forces plan, coordinate, synchronize, and integrate during the conduct of operations.

The Army's ability to set and sustain the theater of operations is essential to allowing the joint force to seize the initiative while restricting the enemy's options. The Army possesses capacities to establish, maintain, and defend vital infrastructure. It also provides to the joint force commander unique capabilities, such as port and airfield opening; logistics; chemical defense; and reception, staging, and onward movement, and integration.

Interagency coordination is inherent in unified action. Interagency coordination is, within the context of Department of Defense involvement, the coordination that occurs between elements of Department of Defense, and engaged United States Government agencies and departments for the purpose of achieving an objective (JP 3-0). Army forces conduct and participate in interagency coordination using established liaison, Soldier and leader engagement, and planning processes.

Combatant commanders play a pivotal role in unified action. However, subordinate commanders also integrate and synchronize their operations directly with the activities and operations of other military forces and nonmilitary organizations in their areas of operations. Additionally, commanders should consider the activities of the host nation and its local population. Unified action may require interorganizational coordination to build the capacity of unified action partners. Interorganizational coordination is interaction that occurs among elements of the Department of Defense; engaged United States Government agencies; state, territorial, local, and tribal agencies; foreign military forces and government agencies; intergovernmental organizations; nongovernmental organizations; and the private sector (JP 3-08). Building partner capacity helps to secure populations, protects infrastructure, and strengthens institutions as a means of protecting common security interests. Building partner capacity results from comprehensive interorganizational activities, programs, and military-to-military engagements that work together. As a group of partners, they enhance their ability to establish security, governance, economic development, essential services, rule of law, and other critical government functions. The Army integrates capabilities of the operating and the institutional Army to support interorganizational capacity-building efforts, primarily through security cooperation interactions.

See facing page for discussion of security cooperation activities.

A. Cooperation with Civilian Organizations

Commanders understand the respective roles and capabilities of civilian organizations and contractors in unified action. Other government agencies work with the military and are part of a national chain of command under the President of the United States.

When directed, Army forces provide sustainment and security for civilian organizations, since many of these organizations lack these capabilities. Within the context of interagency coordination, other government agencies are non-Department of Defense agencies of the U.S. Government. Other government agencies include, but are not limited to, Departments of State, Justice, Transportation, and Agriculture.

Another civilian organization is an intergovernmental organization. An intergovernmental organization is an organization created by a formal agreement between two or more governments on a global, regional, or functional basis to protect and promote national interests shared by member states (JP 3-08). Intergovernmental organizations may be established on a global, regional, or functional basis for wide-ranging or narrowly defined purposes. Examples include the United Nations and the European Union.

Finally, a nongovernmental organization is a private, self-governing, not-for-profit organization dedicated to alleviating human suffering; and/or promoting education, health care, economic development, environmental protection, human rights, and conflict resolution; and/or encouraging the establishment of democratic institutions and civil society (JP 3-08). Nongovernmental organizations are independent, diverse, and flexible organizations focused on grassroots aid that ranges from providing primary relief and development to supporting human rights, civil society, and conflict resolution organizations. Their mission is often one of a humanitarian nature and not one of assisting the military in accomplishing its objectives. In some circumstances, nongovernmental organizations may provide humanitarian aid simultaneously to friendlies and enemies or adversaries.

Security Cooperation & Military Engagement Activities

Ref: ADRP 3-0, Operations (Nov '16), pp. 1-7 to 1-8.

Security cooperation is all Department of Defense interactions with foreign defense establishments to build defense relationships that promote specific United States security interests, develop allied and friendly military capabilities for self-defense and multinational operations, and provide United States forces with peacetime and contingency access to a host nation (JP 3-22). Security cooperation provides the means to build partner capacity. The interactions of security cooperation encourage and enable international partners to work with the United States to achieve strategic objectives. These objectives include—

- Building defensive and security relationships that promote specific U.S. security interests, including all international armaments cooperation activities and security assistance activities.
- Developing capabilities for self-defense and multinational operations.
- Providing U.S. forces with peacetime and contingency access to host nations in order to increase situational understanding of the operational environment.

Supported by appropriate policy, legal frameworks, and authorities, Army forces support the objectives of the combatant commander's campaign plan. The plan supports those objectives by leading security cooperation interactions, specifically those involving security force assistance and foreign internal defense for partner units, institutions, and security sector functions. Security force assistance is the Department of Defense activities that contribute to unified action by the United States Government to support the development of the capacity and capability of foreign security forces and their supporting institutions (JP 3-22). Foreign internal defense is participation by civilian and military agencies of a government in any of the action programs taken by another government or other designated organization to free and protect its society from subversion, lawlessness, insurgency, terrorism, and other threats to its security (JP 3-22).

Through security force assistance and foreign internal defense, operating forces and the institutional Army contribute to security sector programs. These programs professionalize and develop secure partner capacity to enable synchronized and sustained operations. Army security cooperation interactions enable other interorganizational efforts to build partner capacity. Army forces—including special operations forces—advise, assist, train, and equip partner units to develop unit and individual proficiency in security operations. The institutional Army advises and trains partner Army activities to build institutional capacity for professional education, force generation, and force sustainment.

Refer to FM 3-22 for more information on Army support to security cooperation.

Refer to TAA2: Military Engagement, Security Cooperation & Stability SMARTbook (Foreign Train, Advise, & Assist) for further discussion. Topics include the Range of Military Operations (JP 3-0), Security Cooperation & Security Assistance (Train, Advise, & Assist), Stability Operations (ADRP 3-07), Peace Operations (JP 3-07.3), Counterinsurgency Operations (JP & FM 3-24), Civil-Military Operations (JP 3-57), Multinational Operations (JP 3-16), Interorganizational Coordination (JP 3-08), and more.

A contractor is a person or business operating under a legal agreement who provides products or services for pay. A contractor furnishes supplies and services or performs work at a certain price or rate based on contracted terms. Contracted support includes traditional goods and services support, but it may also include interpreter communications, infrastructure, and other related support. Contractor employees include contractors authorized to accompany the force as a formal part of the force and local national employees who normally have no special legal status.

Refer to ATP 4-10 for more information on contractors.

Civilian organizations—such as other government agencies, intergovernmental organizations, and nongovernmental organizations—bring resources and capabilities that can help establish host-nation civil authority and capabilities. Most civilian organizations are not under military control, nor does the American ambassador or a United Nations commissioner control them. Civilian organizations have different organizational cultures and norms. Some may be willing to work with Army forces; others may not. Also, civilian organizations may arrive well after military operations have begun. Thus, personal contact and team building are essential. Command emphasis on immediate and continuous coordination encourages effective cooperation. Commanders should establish liaison with civilian organizations to integrate their efforts as much as possible with Army and joint operations. Civil affairs units typically establish this liaison.

Refer to FM 3-57 for more information on civil affairs units.

B. Joint Operations

Single Services may accomplish tasks and missions in support of Department of Defense objectives. However, the Department of Defense primarily employs two or more Services (from two military departments) in a single operation, particularly in combat, through joint operations. Joint operations is a general term to describe military actions conducted by joint forces and those Service forces employed in specified command relationships with each other, which of themselves, do not establish joint forces (JP 3-0). A joint force is a general term applied to a force composed of significant elements, assigned or attached, of two or more Military Departments operating under a single joint force commander (JP 3-0). Joint operations exploit the advantages of interdependent Service capabilities through unified action, and joint planning integrates military power with other instruments of national power (diplomatic, informational, and economic) to achieve a desired military end state. The end state is the set of required conditions that defines achievement of the commander's objectives (JP 3-0). Joint planning connects the strategic end state to the joint force commander's operational campaign design and ultimately to tactical missions. Joint force commanders use campaigns and major operations to translate their operational-level actions into strategic results. A campaign is a series of related major operations aimed at achieving strategic and operational objectives within a given time and space (JP 5-0). A major operation is a series of tactical actions (battles, engagements, strikes) conducted by combat forces of a single or several Services, coordinated in time and place, to achieve strategic or operational objectives in an operational area (JP 3-0). Planning for a campaign is appropriate when the contemplated military operations exceed the scope of a single major operation. Campaigns are always joint operations. Army forces do not conduct campaigns unless they are designated as a joint task force. However, Army forces contribute to campaigns through the conduct of land operations.

See following pages 1-14 to 1-15 for a discussion of joint operations from JP 3-0.

C. Multinational Operations

Ref: ADRP 3-0, Operations (Nov '16), pp. 1-7 to 1-8.

Multinational operations is a collective term to describe military actions conducted by forces of two or more nations, usually undertaken within the structure of a coalition or alliance (JP 3-16). While each nation has its own interests and often participates within limitations of national caveats, all nations bring value to an operation. Each nation's force has unique capabilities, and each usually contributes to the operation's legitimacy in terms of international or local acceptability. Army forces should anticipate participating in multinational operations and plan accordingly.

Refer to FM 3-16 for more information on multinational operations.

Alliance

An alliance is the relationship that results from a formal agreement between two or more nations for broad, long-term objectives that further the common interests of the members (JP 3-0). Military alliances, such as the North Atlantic Treaty Organization (commonly known as NATO), allow partners to establish formal, standard agreements.

Coalition

A coalition is an arrangement between two or more nations for common action (JP 5-0). Nations usually form coalitions for focused, short-term purposes. A coalition action is an action outside the bounds of established alliances, usually for single occasions or longer cooperation in a narrow sector of common interest. Army forces may conduct coalition actions under the authority of a United Nations resolution.

Soldiers assigned to a multinational force face many demands. These include dealing with cultural issues, different languages, interoperability challenges, national caveats on the use of respective forces and sharing of information and intelligence, rules of engagement, and underdeveloped methods and systems for commanding and controlling. Commanders analyze the mission's particular requirements to exploit the multinational force's advantages and compensate for its limitations. Establishing effective liaison with multinational partners is an important means for increasing the commander's understanding.

Multinational sustainment requires detailed planning and coordination. Normally, each nation provides a national support element to sustain its deployed forces. However, integrated multinational sustainment may improve efficiency and effectiveness. When authorized and directed, an Army theater sustainment command can provide logistics and other support to multinational forces. Integrating support requirements of several nations' forces, often spread over considerable distances and across international boundaries, is challenging. Commanders consider multinational force capabilities, such as mine clearance, that may exceed U.S. forces' capabilities.

Refer to TAA2: Military Engagement, Security Cooperation & Stability SMARTbook (Foreign Train, Advise, & Assist) for further discussion. Topics include the Range of Military Operations (JP 3-0), Security Cooperation & Security Assistance (Train, Advise, & Assist), Stability Operations (ADRP 3-07), Peace Operations (JP 3-07.3), Counterinsurgency Operations (JP & FM 3-24), Civil-Military Operations (JP 3-57), Multinational Operations (JP 3-16), Interorganizational Coordination (JP 3-08), and more.

Joint Operations, Unified Action, & the Range of Military Operations (ROMO)

Ref: JP 3-0, Joint Operations (Aug '11), chap. 1.

Services may accomplish tasks and missions in support of Department of Defense (DOD) objectives. However, the DOD primarily employs two or more services in a single operation, particularly in combat, through joint operations. The general term, joint operations, describes military actions conducted by joint forces or by Service forces employed under command relationships. A joint force is one composed of significant elements, assigned or attached, of two or more military departments operating under a single joint force commander. Joint operations exploit the advantages of interdependent Service capabilities through unified action, and joint planning integrates military power with other instruments of national power to achieve a desired military end state.

Unified Action

Whereas the term joint operations focuses on the integrated actions of the Armed Forces of the United States in a unified effort, the term unified action has a broader connotation. JFCs are challenged to achieve and maintain operational coherence given the requirement to operate in conjunction with interorganizational partners. CCDRs play a pivotal role in unifying joint force actions, since all of the elements and actions that comprise unified action normally are present at the CCDR's level. However, subordinate JFCs also integrate and synchronize their operations directly with the operations of other military forces and the activities of nonmilitary organizations in the operational area to promote unified action.

Unified action is a comprehensive approach that synchronizes, coordinates, and when appropriate, integrates military operations with the activities of other governmental and nongovernmental organizations to achieve unity of effort.

When conducting operations for a joint force commander, Army forces achieve unified action by synchronizing actions with the activities of components of the joint force and unified action partners.

The Range of Military Operations (ROMO)

The range of military operations is a fundamental construct that provides context. Military operations vary in scope, purpose, and conflict intensity across a range that extends from military engagement, security cooperation, and deterrence activities to crisis response and limited contingency operations and, if necessary, to major operations and campaigns. Use of joint capabilities in military engagement, security cooperation, and deterrence activities helps shape the operational environment and keep the day-to-day tensions between nations or groups below the threshold of armed conflict while maintaining US global influence.

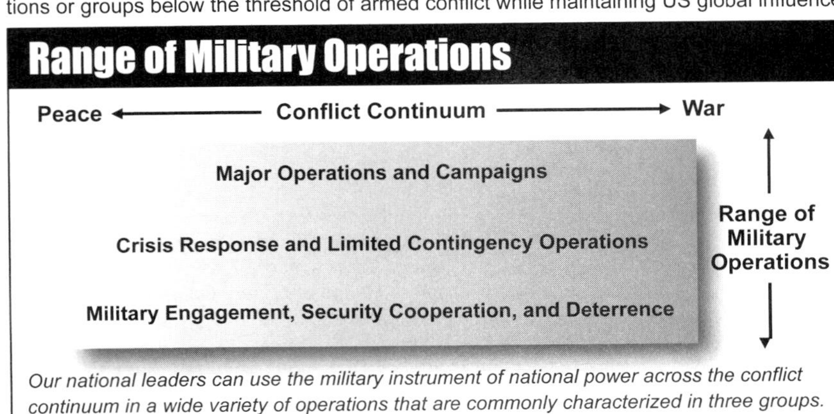

Our national leaders can use the military instrument of national power across the conflict continuum in a wide variety of operations that are commonly characterized in three groups.

A. Military Engagement, Security Cooperation, and Deterrence

These ongoing activities establish, shape, maintain, and refine relations with other nations and domestic civil authorities (e.g., state governors or local law enforcement). The general strategic and operational objective is to protect US interests at home and abroad.

B. Crisis Response & Limited Contingency Operations

A crisis response or limited contingency operation can be a single small-scale, limited-duration operation or a significant part of a major operation of extended duration involving combat. The associated general strategic and operational objectives are to protect US interests and/or prevent surprise attack or further conflict.

C. Major Operations and Campaigns

When required to achieve national strategic objectives or protect national interests, the US national leadership may decide to conduct a major operation or campaign normally involving large-scale combat. During major operations, joint force actions are conducted simultaneously or sequentially in accordance with a common plan and are controlled by a single commander. A campaign is a series of related major operations aimed at achieving strategic and operational objectives within a given time and space.

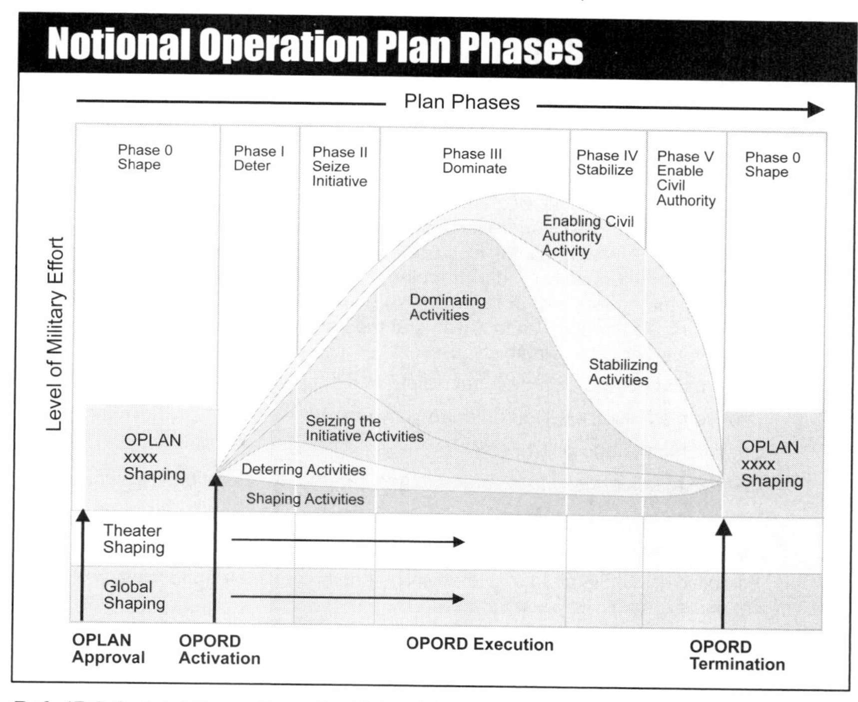

Ref: JP 3-0, Joint Operations, fig. V-3, p V-6.

Refer to JFODS4:The Joint Forces Operations & Doctrine SMARTbook (Guide to Joint, Multinational & Interagency Operations) for complete discussion of joint operations and unified action, the range of military operations and operation plan phases. Additional topics include joint doctrine fundamentals, joint operation planning, joint logistics, joint task forces, information operations, multinational operations, & IGO/NGO coordination.

IV. Land Operations

An operation is a sequence of tactical actions with a common purpose or unifying theme (JP 1). The Army's primary mission is to organize, train, and equip forces to conduct prompt and sustained land combat operations and perform such other duties, not otherwise assigned by law, as may be prescribed by the President or the Secretary of Defense (as described in Title 10, United States Code). The Army does this through its operational concept of unified land operations. Army doctrine aligns with joint doctrine and takes into account the nature of land operations. The command and control of operations on land fundamentally differs from other types of military operations.

Army forces, with unified action partners, conduct land operations to protect the homeland and engage regionally to prevent conflict, shape security environments, and create multiple options for responding to and resolving crises. Army forces defeat enemy organizations, control terrain, secure populations, consolidate gains, and preserve joint force freedom of movement and action through multi-domain battle in the land, air, maritime, space, and cyberspace domains.

The dynamic relationships among friendly forces, enemy forces, adversaries, and the environment make land operations exceedingly complex. Understanding each of these elements separately is necessary, but not sufficient, to understand the relationships among them. Friendly forces strive to achieve dominance against the enemy to attain operational advantages in both the physical and information environments. Advantages in the physical environment allow Army forces to close with and destroy the enemy with minimal losses. These advantages allow Army forces to decisively defeat enemy forces.

Joint doctrine discusses traditional war as a confrontation between nation-states or coalitions of nation-states. This confrontation typically involves small-scale to large-scale, force-on-force military operations in which enemies use various conventional and unconventional military capabilities against each other. Landpower normally solidifies the outcome, even when it is not the definitive instrument. Landpower is the ability—by threat, force, or occupation—to gain, sustain, and exploit control over land, resources, and people. Landpower is at the very heart of unified land operations. Landpower includes the ability to—

- Protect and defend U.S. national assets and interests.
- Impose the Nation's will on an enemy, by force if necessary.
- Sustain high tempo operations.
- Engage to influence, shape, prevent, and deter in an operational environment.
- Defeat enemy organizations and control terrain.
- Secure populations and consolidate gains.
- Establish and maintain a stable environment that sets the conditions for political and economic development.
- Address the consequences of catastrophic events—both natural and man-made—to restore infrastructure and reestablish basic civil services.

A. Characteristics of Land Operations

Ref: ADRP 3-0, Operations (Nov '16), pp. 1-9 to 1-10.

Land operations may involve destroying or dislocating enemy forces on land, or seizing or securing key land objectives that reduce the enemy's ability to conduct operations. Five characteristics distinguish land operations: scope, duration, terrain, permanence, and civilian presence.

The characteristics of land operations increase the uncertainty of the environment in which Army forces conduct operations. Commanders organize, train, and equip their forces to persevere through casualties and setbacks. Decentralized execution—based on a shared understanding of the commander's intent, mission orders, and sharing available information—allows lower level commanders to cope with uncertainty by exercising disciplined initiative.

Scope

Land operations can occur across the entire expanse of the land domain and across the range of military operations. Land combat may involve close combat—warfare carried out on land in a direct-fire fight, supported by direct and indirect fires and other assets. Units involved in close combat employ direct fire weapons, supported by indirect fire, air-delivered fires, and nonlethal engagement means. Units in close combat defeat or destroy enemy forces, or seize and retain ground. Close combat at lower echelons contains many more interactions between friendly and enemy forces than any other form of combat.

Duration

Land operations are repetitive and continuous. With few exceptions (such as ambushes or raids), Army forces do not execute an operation and return to a base; they remain in contact with enemy forces almost continuously. Doing this allows them to destroy enemies or render them incapable or unwilling to conduct further action. The duration of land combat operations contributes to the large number of interactions between friendly and enemy forces, as well as between friendly forces and the civilian population.

Terrain

Land operations take place in the densest of all media—the ground environment. The complex variety of natural and manmade features of the ground environment contrasts significantly with the relative transparency of air, sea, space, and cyberspace environments. In addition to considering the visibility limits resulting from clutter and other terrain features, effective plans for land combat also account for the effects of weather and climate.

Permanence

Land operations frequently require seizing or securing terrain. With control of terrain comes control of the local population and its productive capabilities. Thus, Army forces in land operations make permanent the often temporary effects of other operations.

Civilian Presence

Land operations affect civilians by disrupting routine life patterns and potentially placing civilians in harm's way. Additionally, land combat often impacts civilian access to necessary items such as food, water, and medical supplies. Army forces must plan to conduct minimum-essential stability tasks (providing security, food, water, shelter, and medical treatment) as an integral part of land combat.

B. Army Forces—Expeditionary Capability and Campaign Quality

Future conflicts will place a premium on promptly deploying landpower and constantly adapting to each campaign's unique circumstances as they occur and change. But swift campaigns, however desirable, are the exception. Whenever objectives involve controlling populations or dominating terrain, campaign success usually requires employing landpower for protracted periods. Therefore, the Army combines expeditionary capability and campaign quality to contribute crucial, sustained landpower to unified action. Army forces provide the joint force commander the capability to conduct prompt and sustained land combat operations.

Expeditionary capability is the ability to promptly deploy combined arms forces on short notice to any location in the world, capable of conducting operations immediately upon arrival. Expeditionary operations require the ability, with joint air and maritime support, to deploy quickly with little notice, rapidly shape conditions in an area of operations, and operate immediately on arrival, exploiting success and consolidating gains while sustaining operational reach. Operational reach is the distance and duration across which a joint force can successfully employ military capabilities (JP 3-0). Adequate operational reach is a necessity in order to conduct decisive action. Extending operational reach is a paramount concern for commanders. To achieve the desired end state, forces must possess the necessary operational reach to establish and maintain conditions that define success. Commanders and staffs increase operational reach through deliberate, focused planning, well in advance of operations if possible, and the appropriate sustainment to facilitate endurance.

See pp. 1-33 to 1-37 for discussion of decisive action.

Expeditionary capabilities are more than physical attributes; they begin with a mindset that pervades the force. Expeditionary capabilities assure friends, multinational partners, enemies, and adversaries that the Nation is able and willing to deploy the right combination of Army forces to the right place at the right time. Forward deployed units, forward positioned capabilities, and force projection—from anywhere in the world—all contribute to the Army's expeditionary capabilities. Providing joint force commanders with expeditionary capabilities requires forces organized and equipped to be modular, versatile, and rapidly deployable as well as able to conduct operations with institutions capable of supporting them.

Campaign quality is the Army's ability to sustain operations as long as necessary and to conclude operations successfully. Army forces are organized, trained, and equipped for endurance, and they are foundational and essential to the joint force to conduct campaigns. The Army's campaign quality extends its expeditionary capability well beyond deploying combined arms forces that are effective upon arrival to include theater-enabling capabilities. Campaign quality is an ability to conduct sustained operations for as long as necessary, adapting to unpredictable and often profound changes in an operational environment as the campaign unfolds. The Army's Sustainable Readiness Model (formerly the Army Force Generation Model) provides force generation policies and processes that extend expeditionary capabilities and campaign quality to precombat and postcombat campaign periods. Campaigning requires a mindset and vision that complements expeditionary requirements. Soldiers understand that no matter how long they are deployed, the Army will take care of them and their families. They are confident that the loyalty they pledge to their units will be returned to them, no matter what happens on the battlefield or in what condition they return home. Army leaders understand the effects of protracted land operations on Soldiers and adjust the tempo of operations whenever circumstances allow. Senior joint commanders plan effective campaigns and large-scale combat operations in conjunction with senior Army leaders.

C. Close Combat

Ref: ADRP 3-0, Operations (Nov '16), p. 1-11.

Close combat is indispensable and unique to land operations. Only on land do combatants routinely and in large numbers come face-to-face with one another. When other means fail to drive enemy forces from their positions, Army forces close with and destroy or capture them. The outcome of battles and engagements depends on Army forces' ability to prevail in close combat.

The complexity of urban terrain and density of noncombatants reduce the effectiveness of advanced sensors and long-range and air-delivered weapons. Thus, a weaker enemy often attempts to negate Army advantages by engaging Army forces in urban environments. Operations in large, densely populated areas require special considerations. From a planning perspective, commanders view cities as both topographic features and a dynamic system of varying operational entities containing hostile forces, local populations, and infrastructure.

Regardless of the importance of technological capabilities, success in operations requires Soldiers to accomplish the mission. Today's operational environment requires professional Soldiers and leaders whose character, commitment, and competence represent the foundation of a values-based, trained, and ready Army. Today's Soldiers and leaders adapt and learn while training to perform tasks both individually and collectively. Soldiers and leaders develop the ability to exercise judgment and disciplined initiative under stress. Army leaders and their subordinates must remain—

- Honorable servants of the Nation.
- Competent and committed professionals.
- Dedicated to living by and upholding the Army Ethic.
- Able to articulate mission orders to operate within their commander's intent.
- Committed to developing their subordinates and creating shared understanding while building mutual trust and cohesion.
- Courageous enough to accept prudent risk and exercise disciplined initiative while seeking to exploit opportunities in a dynamic and complex operational environment.
- Trained to operate across the range of military operations.
- Able to operate in combined arms teams within unified action and leverage other capabilities in achieving their objectives.
- Apply cultural understanding to make the right decisions and take the right actions.
- Opportunistic and offensively minded.

Effective close combat relies on lethality with a high degree of situational understanding. The capacity for physical destruction is a foundation of all other military capabilities, and it is the most basic building block of military operations. Army leaders organize, equip, train, and employ their formations for unmatched lethality over a wide range of conditions. Lethality is a persistent requirement for Army organizations, even in conditions where only the implicit threat of violence suffices to accomplish the mission through nonlethal engagements and activities. An inherent, complementary relationship exists between using lethal force and applying military capabilities for nonlethal purposes.

V. Readiness Through Training

Ref: ADRP 3-0, Operations (Nov '16), p. 1-12.

Effective training is the cornerstone of operational success. As General Mark A. Milley, Chief of Staff of the Army, wrote in his initial message to the Army, "Readiness for ground combat is—and will remain—the U.S. Army's #1 priority. We will always be ready to fight today, and we will always prepare to fight tomorrow." Through training and leader development, Soldiers, leaders, and units achieve the tactical and technical competence that builds confidence and allows them to conduct successful operations across the continuum of conflict. Achieving this competence requires specific, dedicated training on offensive, defensive, and stability or defense support of civil authorities tasks. Training continues in deployed units to sustain skills and to adapt to changes in an operational environment.

Army training includes a system of techniques and standards that allow Soldiers and units to determine, acquire, and practice necessary skills. Candid assessments, after action reviews, and applied lessons learned and best practices produce quality Soldiers and versatile units, ready for all aspects of a situation. Through training and experiential practice and learning, the Army prepares Soldiers to win in land combat. Training builds teamwork and cohesion within units. It recognizes that Soldiers ultimately fight for one another and their units. Training instills discipline. It conditions Soldiers to operate within the law of war and rules of engagement.

Army training produces formations that fight and win with overwhelming combat power against any enemy. However, the complexity of integrating all unified action partners' demands that Army forces maintain a high degree of preparedness at all times, as it is difficult to achieve proficiency quickly. Leaders at all levels seek and require training opportunities between the Regular Army and Reserve Components, and their unified action partners at home station, at combat training centers, and when deployed.

The Army as a whole must be flexible enough to operate successfully across the range of military operations. Units must be agile enough to adapt quickly and be able to shift with little effort from a focus on one portion of the continuum of conflict to focus on another portion. Change and adaptation that once required years to implement must now be recognized, communicated, and enacted far more quickly. Technology, having played an increasingly important role in increasing the lethality of the industrial age battlefield, will assume more importance and require greater and more rapid innovation in tomorrow's conflicts. No longer can responses to hostile asymmetric approaches be measured in months; solutions must be anticipated and rapidly fielded across the force—and then be adapted frequently and innovatively as the enemy adapts to counter new-found advantages.

U.S. responsibilities are global; therefore, Army forces prepare to operate in any environment. Army training develops confident, competent, and agile leaders and units. Commanders focus their training time and other resources on tasks linked to their mission. Because Army forces face diverse threats and mission requirements, commanders adjust their training priorities based on a likely operational environment. As units prepare for deployment, commanders adapt training priorities to address tasks required by actual or anticipated operations.

Refer to TLS5: The Leader's SMARTbook, 5th Ed. for complete discussion of Military Leadership (ADP/ADRP 6-22); Leader Development (FM 6-22); Counsel, Coach, Mentor (ATP 6-22.1); Army Team Building (ATP 6-22.6); Military Training (ADP/ADRP 7-0); Train to Win in a Complex World (FM 7-0); Unit Training Plans, Meetings, Schedules, and Briefs; Conducting Training Events and Exercises; Training Assessments, After Action Reviews (AARs); and more!

Chap 1 II. Operational Art

Ref: ADRP 3-0, Operations (Nov '16), chap. 2.

I. The Application Of Operational Art

Prior to conducting land operations, Army commanders seek to thoroughly analyze an operational environment and determine the most effective and efficient methods for applying decisive action in various locations across multiple echelons and multiple domains. They use operational art and the principles of joint operations to envision how to establish conditions that accomplish their missions and achieve assigned objectives. Actions and interactions across the levels of warfare influence these conditions.

A. Operational Art

Operational art is the cognitive approach by commanders and staffs—supported by their skill, knowledge, experience, creativity, and judgment—to develop strategies, campaigns, and operations to organize and employ military forces by integrating ends, ways, and means (JP 3-0). For Army forces, operational art is the pursuit of strategic objectives, in whole or in part, through the arrangement of tactical actions in time, space, and purpose. This approach enables commanders and staffs to use skill, knowledge, experience, and judgment to overcome the ambiguity and intricacies of a complex, ever-changing, and uncertain operational environment. Operational art applies to all aspects of operations and integrates ends, ways, and means, while accounting for risk. Operational art applies to all levels of warfare, strategic, operational, and tactical. Army commanders focus on planning and executing operations and activities to achieve military objectives in support of the joint force commander's campaign plan.

B. Principles of Joint Operations

The twelve principles of joint operations represent important factors that affect the conduct of operations across the levels of warfare. (See table 2-1 on page 2-2.) Rather than a checklist, the principles are considerations. While commanders consider the principles in all operations, they do not apply in the same way to every situation. Nor do all principles apply to all situations. Rather, these principles summarize characteristics of successful operations. Their greatest value lies in educating military professionals. Applied to the study of past operations, these principles are powerful tools that can assist commanders in analyzing pending operations. While considering the principles of joint operations, commanders synchronize efforts and determine if or when to deviate from the principles based on the current situation.

See following page (p. 1-23) for a detailed discussion on the principles of joint operations from JP 3-0.

When applying operational art, commanders and staff must create a shared understanding of purpose. This begins with open, continuous collaboration and dialogue between commanders at various echelons of command. Such collaboration and dialogue enables commanders to share an understanding of the problems and conditions of an operational environment. Effective collaboration facilitates assessment, fosters critical analysis, and anticipates opportunities and risk.

Operational art spans a continuum—from comprehensive strategic direction to tactical actions. Bridging this continuum requires creative vision coupled with broad experience and knowledge. Through operational art, commanders translate their operational approach into a concept of operations—a verbal or graphic statement that clearly and concisely expresses what the joint force commander intends to accomplish and how it will be done using available resources (JP 5-0)—and ultimately into tactical tasks. Commanders then array forces and maneuver them to achieve a desired end state.

C. Army Design Methodology

Army design methodology assists commanders in developing their operational approach. Applying operational art requires a shared understanding of an operational environment with the problem analyzed through the Army design methodology. This understanding enables commanders to develop an operational approach to guide the force in establishing those conditions to win and accomplish the mission. The operational approach is a description of the broad actions the force must take to transform current conditions into those desired at end state (JP 5-0). Commanders use a common doctrinal language to visualize and describe their operational approach. The operational approach provides a framework that relates tactical tasks to the desired end state. It provides a unifying purpose and focus to all operations.

Refer to ATP 5-0.1 for more information on Army design methodology.

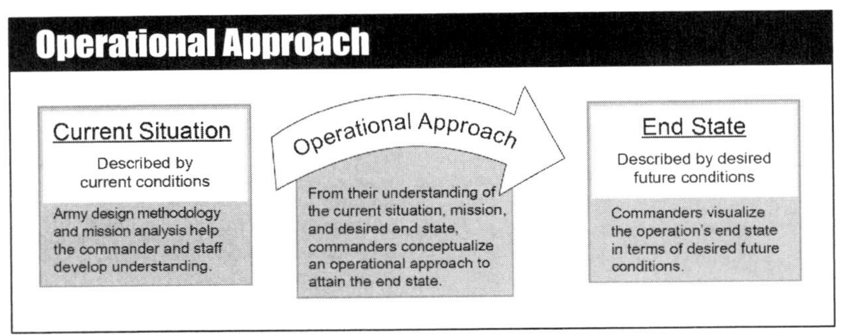

Ref: ADRP 3-0 (2016), fig. 2-1. Operational approach.

II. Defeat and Stability Mechanisms

When developing an operational approach, commanders consider how to employ a combination of defeat mechanisms and stability mechanisms. Defeat mechanisms are dominated by offensive and defensive tasks, while stability mechanisms are dominant in stability tasks that establish and maintain security and facilitate consolidating gains in an area of operations.

A. Defeat Mechanism

A defeat mechanism is a method through which friendly forces accomplish their mission against enemy opposition. Army forces at all echelons use combinations of four defeat mechanisms: destroy, dislocate, disintegrate, and isolate. Applying focused combinations produces complementary and reinforcing effects not attainable with a single mechanism. Used individually, a defeat mechanism achieves results proportional to the effort expended. Used in combination, the effects are likely to be both synergistic and lasting.

When commanders destroy, they apply lethal combat power on an enemy capability so that it can no longer perform any function. Destroy is a tactical mission task that physically renders an enemy force combat-ineffective until it is reconstituted. Alternatively, to destroy a combat system is to damage it so badly that it cannot perform any function or be restored to a usable condition without being entirely rebuilt. The enemy cannot restore a destroyed force to a usable condition without entirely rebuilding it.

Principles of Joint Operations

Ref: ADRP 3-0, Operations (Nov '16), table 2-1 and JP 3-0, Joint Operations (Aug '11), app. A.

Objective
Direct every military operation toward a clearly defined, decisive, and achievable goal.

Offensive
Seize, retain, and exploit the initiative.

Mass
Concentrate the effects of combat power at the most advantageous place and time to produce decisive results.

Maneuver
Place the enemy in a position of disadvantage through the flexible application of combat power.

Economy of Force
Expend minimum essential combat power on secondary efforts in order to allocate the maximum possible combat power on primary efforts.

Unity of Command
Ensure unity of effort under one responsible commander for every objective.

Security
Prevent the enemy from acquiring unexpected advantage.

Surprise
Strike at a time or place or in a manner for which the enemy is unprepared.

Simplicity
Increase the probability that plans and operations will be executed as intended by preparing clear, uncomplicated plans and concise orders.

Restraint
Limit collateral damage and prevent the unnecessary use of force.

Perseverance
Ensure the commitment necessary to attain the national strategic end state.

Legitimacy
Maintain legal and moral authority in the conduct of operations.

Refer to JFODS4: The Joint Forces Operations & Doctrine SMARTbook (Guide to Joint, Multinational & Interorganizational Operations) for further discussion. Topics and chapters include joint doctrine fundamentals, joint operations, joint operation planning, joint logistics, joint task forces, information operations, multinational operations, and interorganizational coordination.

Dislocate

Dislocate is to employ forces to obtain significant positional advantage, rendering the enemy's dispositions less valuable, perhaps even irrelevant. Commanders often achieve dislocation by placing forces in locations where the enemy does not expect them.

Disintegrate

Disintegrate means to disrupt the enemy's command and control system, degrading its ability to conduct operations while leading to a rapid collapse of the enemy's capabilities or will to fight. Commanders often achieve disintegration by specifically targeting the enemy's command structure and communications systems.

Isolate

Isolate is a tactical mission task that requires a unit to seal off—both physically and psychologically—an enemy from sources of support, deny the enemy freedom of movement, and prevent the isolated enemy force from having contact with other enemy forces (FM 3-90-1). When commanders isolate, they deny an enemy or adversary access to capabilities that enable an enemy unit to maneuver in time and space at will.

Commanders describe a defeat mechanism as the physical, temporal, or psychological effects it produces. Operational art formulates the most effective, efficient way to defeat enemy aims. Physically defeating the enemy deprives enemy forces of the ability to achieve those aims. Temporally defeating the enemy anticipates enemy reactions and counters them before they can become effective. Psychologically defeating the enemy deprives the enemy of the will to continue the conflict.

In addition to defeating an enemy, Army forces often seek to stabilize an area of operations by performing stability tasks. There are six primary stability tasks:

- Establish civil security.
- Establish civil control.
- Restore essential services.
- Support governance.
- Support economic and infrastructure development.
- Conduct security cooperation.

The combination of stability tasks conducted during operations depends on the situation. In some operations, the host nation can meet most or all of the population's requirements. In those cases, Army forces work with and through host-nation authorities. Commanders use civil affairs operations to mitigate how the military presence affects the population and vice versa. Conversely, Army forces operating in a failed state may need to support the well-being of the local population. That situation requires Army forces to work with civilian organizations to restore basic capabilities. Civil affairs operations prove essential in establishing the trust between Army forces and civilian organizations required for effective, working relationships.

B. Stability Mechanism

A stability mechanism is the primary method through which friendly forces affect civilians in order to attain conditions that support establishing a lasting, stable peace. As with defeat mechanisms, combinations of stability mechanisms produce complementary and reinforcing effects that accomplish the mission more effectively and efficiently than single mechanisms do alone.

The four stability mechanisms are compel, control, influence, and support. Compel means to use, or threaten to use, lethal force to establish control and dominance, effect behavioral change, or enforce compliance with mandates, agreements, or civil authority. Control involves imposing civil order. Influence means to alter the opinions, attitudes, and ultimately the behavior of foreign friendly, neutral, adversary, and enemy targets and audiences through messages, presence, and actions. Support is to establish, reinforce, or set the conditions necessary for the instruments of national power to function effectively.

III. The Elements of Operational Art

In applying operational art, Army commanders and their staffs use intellectual tools to help them understand an operational environment as well as visualize and describe their approach for conducting an operation. Collectively, this set of tools is known as the elements of operational art. These tools help commanders understand, visualize, and describe the integration and synchronization of the elements of combat power as well as their commander's intent and guidance. Commanders selectively use these tools in any operation. However, the tools' broadest application applies to long-term operations.

Not all elements of operational art apply at all levels of warfare. For example, a company commander may be concerned about the tempo of an upcoming operation but is probably not concerned with an enemies' center of gravity. On the other hand, a corps commander may consider all elements of operational art in developing a plan in support of the joint force commander. As such, the elements of operational art are flexible enough to be applicable when pertinent.

As some elements of operational design apply only to joint force commanders, the Army modifies the elements of operational design into elements of operational art, adding Army-specific elements. During the planning and execution of Army operations, Army commanders and staffs consider the elements of operational art as they assess the situation. They adjust current and future operations and plans as the operation unfolds, and reframe as necessary.

Elements of Operational Art

A. End State and Conditions
B. Center Of Gravity (COG)
C. Decisive Points
D. Lines of Operations and Lines of Effort
E. Basing
F. Tempo
G. Phasing and Transitions
H. Culmination
I. Operational Reach
J. Risk

Ref: ADRP 3-0 (2016), table 2-2. Elements of operational art.

A. End State and Conditions

The end state is a set of desired future conditions the commander wants to exist when an operation ends. Commanders include the end state in their planning guidance. A clearly defined end state promotes unity of effort; facilitates integration, synchronization, and disciplined initiative; and helps mitigate risk.

Army operations typically focus on achieving the military end state that may include contributions to establishing nonmilitary conditions. Commanders explicitly describe the end state and its conditions for every operation. Otherwise, missions become vague, and operations lose focus. Successful commanders direct every operation toward a clearly defined, conclusive, and attainable end state (the objective).

The end state may evolve as an operation progresses. Commanders continuously monitor operations and evaluate their progress. Commanders use formal and informal assessment methods to assess their progress in achieving the end state and determine if they need to reframe. The end state should anticipate future operations and set conditions for transitions. The end state should help commanders think through the conduct of operations to best facilitate transitions.

B. Center Of Gravity (COG)

A center of gravity is the source of power that provides moral or physical strength, freedom of action, or will to act (JP 5-0). The loss of a center of gravity can ultimately result in defeat. The center of gravity is a vital analytical tool for planning operations. It provides a focal point and identifies sources of strength and weakness. However, the concept of center of gravity is only meaningful when considered in relation to the objectives of the mission.

Centers of gravity are not limited to military forces and can be either physical or moral. They are part of a dynamic perspective of an operational environment. Physical centers of gravity, such as a capital city or military force, are tangible and typically easier to identify, assess, and target than moral centers of gravity. Physical centers of gravity can often be influenced solely by military means. In contrast, moral centers of gravity are intangible and more difficult to influence. They can include a charismatic leader, powerful ruling elite, or strong-willed population. Military means alone usually prove ineffective when targeting moral centers of gravity. Affecting them requires collective, integrated efforts of all instruments of national power.

A center of gravity often has subcomponents, such as command and control or logistics that may be targetable by information collection. This targeting may lead to identifying critical vulnerabilities, such as communications or enemy morale, on which commanders can apply lethal or nonlethal capabilities.

Commanders analyze a center of gravity thoroughly and in detail. Faulty conclusions drawn from hasty or abbreviated analyses can adversely affect operations, waste critical resources, and incur undue risk. Thoroughly understanding an operational environment helps commanders identify and target enemy centers of gravity. This understanding encompasses how enemies organize, fight, and make decisions. It also includes their physical and moral strengths and weaknesses. In addition, commanders should understand how military forces interact with other government and civilian organizations. This understanding helps planners identify centers of gravity, their associated decisive points, and the best approach for achieving the desired end state.

C. Decisive Points

A decisive point is a geographic place, specific key event, critical factor, or function that, when acted upon, allows commanders to gain a marked advantage over an adversary or contribute materially to achieving success (JP 5-0). Decisive points help commanders select clear, conclusive, attainable objectives that directly contribute to achieving the end state. Geographic decisive points can include port facilities, distribution networks and nodes, and bases of operation. Specific events and elements of an enemy force may also be decisive points. Examples of such events include commitment of the enemy operational reserve and reopening a major oil refinery.

A common characteristic of decisive points is their importance to a center of gravity. A decisive point's importance requires the enemy to commit significant resources to defend it. The loss of a decisive point weakens a center of gravity and may expose more decisive points, eventually leading to an attack on the center of gravity itself. Decisive points are not centers of gravity; they are key to attacking or protecting centers of gravity. Commanders identify the decisive points that offer the greatest physical, temporal, or psychological advantage against centers of gravity.

Decisive points apply to both the operational and tactical levels when shaping the concept of operations. Decisive points enable commanders to seize, retain, or exploit the initiative. Controlling them is essential to mission accomplishment. Enemy control of a decisive point may stall friendly momentum, force early culmination, or allow an enemy counterattack.

D. Lines of Operations and Lines of Effort

Ref: ADRP 3-0, Operations (Nov '16), pp. 2-5 to 2-6.

Lines of operations and lines of effort link objectives to the end state. Commanders may describe an operation along lines of operations, lines of effort, or a combination of both. The combination of them may change based on the conditions within an area of operations. Commanders synchronize and sequence actions, deliberately creating complementary and reinforcing effects. The lines then converge on the well-defined, commonly understood end state outlined in the commander's intent.

Commanders at all levels may use lines of operations and lines of effort to develop tasks and allocate resources. Commanders may designate one line as the decisive operation and others as shaping operations. Commanders synchronize and sequence related actions along multiple lines. Seeing these relationships helps commanders assess progress toward achieving the end state as forces perform tasks and accomplish missions.

Lines of Operations

A line of operations is a line that defines the directional orientation of a force in time and space in relation to the enemy and links the force with its base of operations and objectives. Lines of operations connect a series of decisive points that lead to control of a geographic or force-oriented objective. Operations designed using lines of operations generally consist of a series of actions executed according to a well-defined sequence. A force operates on interior and exterior lines. Interior lines are lines on which a force operates when its operations diverge from a central point. Interior lines allow commanders to move quickly against operations converge on the enemy. Exterior lines allow commanders to concentrate forces against multiple positions on the ground, thus presenting multiple dilemmas to the enemy. Lines of operations tie offensive and defensive tasks to the geographic and positional references in the area of operations.

Lines of Effort

A line of effort is a line that links multiple tasks using the logic of purpose rather than geographical reference to focus efforts toward establishing a desired end state. Lines of effort are essential to long-term planning when positional references to an enemy or adversary have little relevance. In operations involving many nonmilitary factors, lines of effort may be the only way to link tasks to the end state. Lines of effort are often essential to helping commanders visualize how military capabilities can support the other instruments of national power.

Commanders use lines of effort to describe how they envision their operations creating the intangible end state conditions. These lines of effort show how individual actions relate to each other and to achieving the end state. Commanders often use stability and defense support of civil authorities tasks along lines of effort. These tasks link military actions with the broader interagency or interorganizational effort across the levels of warfare. As operations progress, commanders may modify the lines of effort after assessing conditions. Commanders use measures of performance and measures of effectiveness when continually assessing operations. A measure of performance is a criterion used to assess friendly actions that is tied to measuring task accomplishment (JP 3-0). A measure of effectiveness is a criterion used to assess changes in system behavior, capability, or operational environment that is tied to measuring the attainment of an end state, achievement of an objective, or creation of an effect (JP 3-0).

Combining Lines of Operations and Lines of Effort

Commanders use lines of operations and lines of effort to connect objectives to a central, unifying purpose. The difference between lines of operations and lines of effort is that lines of operations are oriented on physical linkages, while lines of effort are oriented on logical linkages. Combining lines of operations and lines of effort allows a commander to include stability or defense support of civil authorities tasks in the long-term plan.

E. Basing

Army basing overseas typically falls into two general categories: permanent (bases or installations) and nonpermanent (base camps). A base is a locality from which operations are projected or supported (JP 4-0). Generally, bases are in host nations where the United States has a long-term lease agreement and a status-of-forces agreement. A base camp is an evolving military facility that supports the military operations of a deployed unit and provides the necessary support and services for sustained operations. Basing locations are nonpermanent by design and designated as a base when the intention is to make them permanent. Bases or base camps may have a specific purpose (such as serving as an intermediate staging base, a logistics base, or a base camp) or they may be multifunctional. The longer base camps exist, the more they exhibit many of the same characteristics as bases in terms of the support and services provided and types of facilities developed. A base or base camp has a defined perimeter, has established access controls, and takes advantage of natural and manmade features.

Basing may be joint or single Service and will routinely support both U.S. and multinational forces, as well as interagency partners, operating anywhere along the range of military operations. Commanders often designate a specific area as a base or base camp and assign responsibility to a single commander for protection and terrain management within the base. Units located within the base or base camp are under the tactical control of the base or base camp commander for base security and defense. Within large echelon support areas or joint security areas, controlling commanders may designate base clusters for mutual protection and mission command.

See p. 6-29 for further discussion of basing.

Support Areas

When a base camp expands to include clusters of sustainment, headquarters, and other supporting units, commanders may designate a support area. Echelon commanders designate support areas. These specific areas of operations facilitate the positioning, employment, and protection of resources required to sustain, enable, and control tactical operations. Army forces typically rely on a mix of bases and base camps to serve as intermediate staging bases, lodgments (subsequently developed into base camps or potentially bases), and forward operating bases. These bases and base camps deploy and employ landpower simultaneously to operational depth. They establish and maintain strategic reach for deploying forces and ensure sufficient operational reach to extend operations in time and space.

Intermediate Staging Base (ISB)

An intermediate staging base is a tailorable, temporary location used for staging forces, sustainment and/or extraction into and out of an operational area (JP 3-35). At the intermediate staging base, units are unloaded from intertheater lift, reassembled and integrated with their equipment, and then moved by intratheater lift into the area of operations. The theater army commander provides extensive support to Army forces transiting the base. The combatant commander may designate the theater army commander to command the base or provide a headquarters suitable for the task. Intermediate staging bases are established near, but normally not in, the joint operations area. They often are located in the supported combatant commander's area of responsibility. For land forces, intermediate staging bases may be located in the area of operations. However, if possible, they are established outside the range of direct and most indirect enemy fire systems and beyond the enemy's political sphere of influence.

Base Camp

A base camp that expands to include an airfield may become a forward operating base. A forward operating base is an airfield used to support tactical operations without establishing full support facilities (JP 3-09.3). Forward operating bases may be used for an extended time and are often critical to security. During protracted operations, they may be further expanded and improved to establish a more permanent presence. The scale and complexity of a forward operating base, however, directly relate to the size of the force required to maintain it. A large forward operating base with extensive facilities requires a much larger security force than a smaller, austere base. Commanders weigh whether to expand and improve a forward operating base against the type and number of forces available to secure it, the expected length of the forward deployment, the force's sustainment requirements, and the enemy threat.

Lodgement

A lodgment is a designated area in a hostile or potentially hostile operational area that, when seized and held, makes the continuous landing of troops and materiel possible and provides maneuver space for subsequent operations (JP 3-18). Identifying and preparing the initial lodgment significantly influences the conduct of an operation. Lodgments should expand to allow easy access to strategic sealift and airlift, offer adequate space for storage, facilitate transshipment of supplies and equipment, and be accessible to multiple lines of communications. Typically, deploying forces establish lodgments near key points of entry in the operational area that offer central access to air, land, and sea transportation hubs.

Refer to SMFLS4: Sustainment & Multifunctional Logistics SMARTbook (Guide to Logistics, Personnel Services, & Health Services Support). Includes ATP 4-94 Theater Sustainment Command (Jun '13), ATP 4-93 Sustainment Brigade (Aug '13), ATP 4-90 Brigade Support Battalion (Aug '14), Sustainment Planning, JP 4-0 Joint Logistics (Oct '13), ATP 3-35 Army Deployment and Redeployment (Mar '15), and more than a dozen new/updated Army sustainment references.

F. Tempo

Tempo is the relative speed and rhythm of military operations over time with respect to the enemy. It reflects the rate of military action. Controlling tempo helps commanders keep the initiative during combat operations or rapidly establish a sense of normalcy during humanitarian crises. During certain operations, commanders normally seek to maintain a higher tempo than the enemy does; a rapid tempo can overwhelm an enemy's ability to counter friendly actions. During other operations, commanders act quickly to control events and deny the enemy positions of advantage. By acting faster than the situation deteriorates, commanders can change the dynamics of a crisis and restore stability.

Commanders control tempo throughout the conduct of operations. First, they formulate operations that stress the complementary and reinforcing effects of simultaneous and sequential operations. They synchronize those operations in time and space to degrade enemy capabilities throughout the area of operations. Second, commanders avoid unnecessary engagements. This practice includes bypassing resistance at times and avoiding places commanders do not consider decisive. Third, through mission command, commanders enable subordinates to exercise initiative and act independently. Controlling tempo requires both audacity and patience. Audacity initiates the actions needed to develop a situation; patience allows a situation to develop until the force can strike at the most crucial time and place. Ultimately, the goal is maintaining a tempo appropriate to retaining and exploiting the initiative and achieving the end state.

G. Phasing and Transitions

Ref: ADRP 3-0, Operations (Nov '16), pp. 2-58 to 2-9.

Phasing

A phase is a planning and execution tool used to divide an operation in duration or activity. A change in phase usually involves a change of mission, task organization, or rules of engagement. Phasing helps in planning and controlling, and it may be indicated by time, distance, terrain, or an event. The ability of Army forces to extend operations in time and space, coupled with a desire to dictate tempo, often presents commanders with more objectives and decisive points than the force can engage simultaneously. This may require commanders and staffs to consider sequencing operations.

Phasing is critical to arranging all tasks of an operation that cannot be conducted simultaneously. It describes how the commander envisions the overall operation unfolding. It is the logical expression of the commander's visualization in time. Within a phase, a large portion of the force executes similar or mutually supporting activities. Achieving a specified condition or set of conditions typically marks the end of a phase.

Simultaneity, depth, and tempo are vital to all operations. However, they cannot always be attained to the degree desired. In such cases, commanders limit the number of objectives and decisive points engaged simultaneously. They deliberately sequence certain actions to maintain tempo while focusing combat power at a decisive point in time and space. Commanders combine the simultaneous and sequential tasks of an operation to establish end state conditions.

Phasing can extend operational reach. Only when the force lacks the capability to accomplish the mission in a single action do commanders phase the operation. Each phase should strive to focus effort, concentrate combat power in time and space at a decisive point, and achieve its objectives deliberately and logically.

Transitions

Transitions mark a change of focus between phases or between the ongoing operation and execution of a branch or sequel. Shifting priorities between offensive, defensive, stability, and defense support of civil authorities tasks also involve a transition. Transitions require planning and preparation well before their execution, so the force can maintain the momentum and tempo of operations. The force is vulnerable during transitions, and commanders establish clear conditions for their execution.

A transition occurs for several reasons. Transitions occur with the delivery of essential services, retention of infrastructure needed for reconstruction, or when consolidating gains. An unexpected change in conditions may require commanders to direct an abrupt transition between phases. In such cases, the overall composition of the force remains unchanged despite sudden changes in mission, task organization, and rules of engagement. Typically, task organization evolves to meet changing conditions; however, transition planning must also account for changes in mission. Commanders continuously assess the situation, and they task-organize and cycle their forces to retain the initiative. Commanders strive to achieve changes in emphasis without incurring an operational pause.

Commanders identify potential transitions during planning and account for them throughout execution. Considerations for identifying potential transitions should include—

- Forecasting in advance when and how to transition.
- Arranging tasks to facilitate transitions.
- Creating a task organization that anticipates transitions.
- Rehearsing certain transitions such as from defense to counterattack or from offense to consolidating gains.
- Ensuring the force understands different rules of engagement during transitions.

Army forces expend more energy and resources when operating at a high tempo. Commanders assess their force's capacity to operate at a higher tempo based on its performance and available resources. An effective operational design varies tempo throughout an operation to increase endurance while maintaining appropriate speed and momentum. There is more to tempo than speed. While speed can be important, commanders mitigate speed to achieve endurance and optimize operational reach.

H. Culmination

The culminating point is a point at which a force no longer has the capability to continue its form of operations, offense or defense (JP 5-0). Culmination represents a crucial shift in relative combat power. It is relevant to both attackers and defenders at each level of warfare. While conducting offensive tasks, the culminating point occurs when the force cannot continue the attack and must assume a defensive posture or execute an operational pause. While conducting defensive tasks, it occurs when the force can no longer defend itself and must withdraw or risk destruction. The culminating point is more difficult to identify when Army forces conduct stability tasks. Two conditions can result in culmination: units being too dispersed to achieve security and units lacking required resources to achieve the end state. While conducting defense support of civil authorities tasks, culmination may occur if forces must respond to more catastrophic events than they can manage simultaneously. That situation results in culmination due to exhaustion.

A culmination may be a planned event. In such cases, the concept of operations predicts which part of the force will culminate, and the task organization includes additional forces to assume the mission after culmination. Typically, culmination is caused by direct combat actions or higher echelon resourcing decisions. Culmination relates to the force's ability to generate and apply combat power, and it is not a lasting condition. To continue operations after culminating, commanders may reinforce or reconstitute tactical units.

I. Operational Reach

Applicable to Army forces as part of the joint force, operational reach reflects the ability to achieve success through a well-conceived operational approach. Operational reach is a tether; it is a function of intelligence, protection, sustainment, endurance, and relative combat power. The limit of a unit's operational reach is its culminating point. It balances the natural tension among endurance, momentum, and protection.

Endurance refers to the ability to employ combat power anywhere for protracted periods. It stems from the ability to create, protect, and sustain a force, regardless of the distance from its base and the austerity of the environment. Endurance involves anticipating requirements and making the most effective, efficient use of available resources. Their endurance gives Army forces their campaign quality. Endurance contributes to Army forces' ability to make permanent the transitory effects of other capabilities.

Momentum comes from retaining the initiative and executing high-tempo operations that overwhelm enemy resistance. Commanders control momentum by maintaining focus and pressure. They set a tempo that prevents exhaustion and maintains sustainment. A sustainable tempo extends operational reach. Commanders maintain momentum by anticipating and transitioning rapidly between any combination of offensive, defensive, stability, or defense support of civil authorities tasks. Sometimes commanders push the force to its culminating point to take maximum advantage of an opportunity. For example, exploitations and pursuits often involve pushing all available forces to the limit of their endurance to capitalize on momentum and retain the initiative.

Protection is an important contributor to operational reach. Commanders anticipate how enemy actions and environmental factors might disrupt operations and then determine the protection capabilities required to maintain sufficient reach. Protection closely relates to endurance and momentum. It also contributes to the commander's ability to extend operations in time and space. The protection warfighting function helps commanders maintain the force's integrity and combat power.

Commanders and staffs consider operational reach to ensure Army forces accomplish their missions before culminating. Commanders continually strive to extend operational reach. They assess friendly and enemy force status and civil considerations, anticipate culmination, consolidate gains, and plan operational pauses if necessary. Commanders have studied and reflected on the challenge of conducting and sustaining operations over long distances and times. History contains many examples of operations hampered by inadequate operational reach. Achieving the desired end state requires forces with the operational reach to establish and maintain security, so they can successfully transition to the end state conditions.

J. Risk

Risk is the probability and severity of loss linked to hazards (JP 5-0). Risk, uncertainty, and chance are inherent in all military operations. When commanders accept risk, they create opportunities to seize, retain, and exploit the initiative and achieve decisive results. The willingness to incur risk is often the key to exposing enemy weaknesses that the enemy considers beyond friendly reach. Understanding risk requires assessments coupled with boldness and imagination. Successful commanders assess and mitigate risk continuously throughout the operations process.

Inadequate planning and preparation risks forces, and it is equally rash to delay action while waiting for perfect intelligence and synchronization. Reasonably estimating and intentionally accepting risk is fundamental to conducting operations and essential to mission command. Experienced commanders balance audacity and imagination with risk and uncertainty to strike at a time and place and in a manner wholly unexpected by enemy forces. This is the essence of surprise. It results from carefully considering and accepting risk.

Commanders accept risks and seek opportunities to create and maintain the conditions necessary to seize, retain, and exploit the initiative and achieve decisive results. During execution, opportunities are fleeting. The surest means to create opportunity is to accept risk while minimizing hazards to friendly forces. A good operational approach considers risk and uncertainty equally with friction and chance. The final plans and orders then provide the flexibility commanders need to facilitate subordinate initiative and take advantage of opportunity in a highly competitive and dynamic environment throughout the conduct of unified land operations.

Refer to ATP 5-19 for detailed discussion on risk management.

Refer to BSS5: The Battle Staff SMARTbook, 5th Ed. for further discussion. BSS5 covers the operations process (ADRP 5-0); commander's activities (Understand, Visualize, Describe, Direct, Lead, Assess); the military decisionmaking process and troop leading procedures (MDMP & TLP); integrating processes and continuing activities (IPB, targeting, risk management); plans and orders (WARNOs/FRAGOs/OPORDs); mission command, command posts, liaison; rehearsals & after action reviews; and operational terms & symbols.

III. The Army's Operational Concept

Ref: ADRP 3-0, Operations (Nov '16), chap. 3.

I. The Goal of Unified Land Operations

Unified land operations is the Army's operational concept and the Army's contribution to unified action. Unified land operations are simultaneous offensive, defensive, and stability or defense support of civil authorities tasks to seize, retain, and exploit the initiative and consolidate gains to prevent conflict, shape the operational environment, and win our Nation's wars as part of unified action. The goal of unified land operations is to apply landpower as part of unified action to defeat the enemy on land and establish conditions that achieve the joint force commander's end state. Unified land operations is how the Army applies combat power through 1) simultaneous offensive, defensive, and stability, or defense support of civil authorities tasks, to 2) seize, retain, and exploit the initiative, and 3) consolidate gains. Where possible, military forces working with unified action partners seek to prevent or deter threats. However, if necessary, military forces possess the capability in unified land operations to prevail over aggression.

II. Decisive Action

Decisive action is the continuous, simultaneous combinations of offensive, defensive, and stability or defense support of civil authorities tasks. In unified land operations, commanders seek to seize, retain, and exploit the initiative while synchronizing their actions to achieve the best effects possible. Operations conducted outside the United States and its territories simultaneously combine three elements—offense, defense, and stability. Within the United States and its territories, decisive action combines the elements of defense support of civil authorities and, as required, offense and defense to support homeland defense.

Decisive action begins with the commander's intent and concept of operations. As a single, unifying idea, decisive action provides direction for an entire operation. Based on a specific idea of how to accomplish the mission, commanders and staffs refine the concept of operations during planning and determine the proper allocation of resources and tasks. They adjust the allocation of resources and tasks to specific units throughout the operation, as subordinates develop the situation or conditions change.

The simultaneity of the offensive, defensive, and stability or defense support of civil authorities tasks is not absolute. The higher the echelon, the greater the possibility of simultaneous offensive, defensive, and stability tasks. At lower echelons, an assigned task may require all of the echelons' combat power to execute a specific task. For example, a higher echelon, such as a division, always performs offensive, defensive, and stability tasks simultaneously in some form. Subordinate brigades perform some combination of offensive, defensive, and stability tasks, but they may not perform all three simultaneously.

For any organization assigned an area of operations, there will always be implied or even specified minimum-essential stability tasks of security, food, water, shelter, and medical treatment. If the organization cannot perform these tasks on its own, it must either request additional resources from higher headquarters or request relief from those tasks.

Unified land operations addresses more than combat between armed opponents. Army forces conduct operations amid populations. This requires Army forces to defeat the enemy and simultaneously shape civil conditions. Offensive and defensive tasks defeat enemy forces, whereas stability tasks shape civil conditions. Winning battles and engagements is important, but that alone may not be the most significant task. Shaping civil conditions (in concert with civilian organizations, civil authorities, and multinational forces) often proves just as important to campaign success. In many joint operations, stability or defense support of civil authorities tasks often prove more important than offensive and defensive tasks.

The emphasis on different tasks of decisive action changes with echelon, time, and location. In an operation dominated by stability, part of the force might conduct simultaneous offensive and defensive tasks in support of establishing stability. Within the United States, defense support of civil authorities may be the only activity actually conducted. Simultaneous combinations of the tasks, which commanders constantly adapt to conditions, are the key to successful land operations in achieving the end state.

Operations require versatile, adaptive units and flexible leaders who exhibit sound judgment. These qualities develop primarily from training that prepares individuals and units for challenging operational environments. Managing training for unified land operations challenges leaders at all echelons. Training for decisive action tasks develops discipline, endurance, unit cohesion, tolerance for uncertainty, and mutual support. It prepares Soldiers and units to address ambiguities inherent in stability and defense support of civil authorities tasks as well.

However, operational experience demonstrates that forces trained exclusively for offensive and defensive tasks are not as proficient at stability tasks as those trained specifically for stability tasks. For maximum effectiveness, tasks for stability and defense support of civil authorities require dedicated training, similar to training for offensive and defensive tasks. Likewise, forces involved in protracted stability or defense support of civil authorities tasks require intensive training to regain proficiency in offensive or defensive tasks before engaging in large-scale combat operations. Effective training reflects a balance among the tasks of decisive action that produce and sustain Soldier, leader, and unit proficiency in individual and collective tasks.

A. The Tasks of Decisive Action

Decisive action requires simultaneous combinations of offense, defense, and stability or defense support of civil authorities tasks.

See following pages (pp. 1-36 to 1-37) for discussion of the tasks of decisive action.

B. The Purpose of Simultaneity

Simultaneously conducting offensive, defensive, and stability or defense support of civil authorities tasks requires the synchronized application of combat power. Simultaneity means doing multiple things at the same time. It requires the ability to conduct operations in depth and to integrate them so that their timing multiplies their effectiveness throughout an area of operations and across the multiple domains.

See p. 1-38 for further discussion.

C. Transitioning in Decisive Action

Conducting decisive action involves more than simultaneous execution of all its tasks. It requires commanders and staffs to consider their units' capabilities and capacities relative to each task. Commanders consider their missions; decide which tactics, techniques, and procedures to use; and balance the tasks of decisive action while preparing their commander's intent and concept of operations. They determine which tasks the force can accomplish simultaneously, if phasing is required, what additional resources it may need, and how to transition from one task to another.

See p. 1-38 for further discussion.

D. Homeland Defense and Decisive Action

Ref: ADRP 3-0, Operations (Nov '16), p. 3-5.

Homeland defense is the protection of United States sovereignty, territory, domestic population, and critical infrastructure against external threats and aggression or other threats as directed by the President (JP 3-27). The Department of Defense has lead responsibility for homeland defense. The strategy for homeland defense (and defense support of civil authorities) calls for defending the U.S. territory against attack by state and nonstate actors through an active, layered defense—a global defense that aims to deter and defeat aggression abroad and simultaneously protect the homeland. The Army supports this strategy with capabilities in the forward regions of the world, in the geographic approaches to U.S. territory, and within the U.S. homeland.

Homeland defense operations conducted in the land domain could be the result of extraordinary circumstances and decisions by the President. In homeland defense, Department of Defense and Army forces work closely with federal, state, territorial, tribal, local, and private agencies. Land domain homeland defense could consist of offensive and defensive tasks as part of decisive action. Homeland defense is a defense-in-depth that relies on collection, analysis, and sharing of information and intelligence; strategic and regional deterrence; military presence in forward regions; and the ability to rapidly generate and project warfighting capabilities to defend the United States, its allies, and its interests. These means may include support to civil law enforcement; antiterrorism and force protection; counterdrug; air and missile defense; chemical, biological, radiological, nuclear, and high-yield explosives; and defensive cyberspace operations; as well as security cooperation with other partners to build an integrated, mutually supportive concept of protection.

Refer to The Homeland Defense & DSCA SMARTbook (Protecting the Homeland / Defense Support to Civil Authority) for complete discussion. Topics and references include homeland defense (JP 3-28), defense support of civil authorities (JP 3-28), Army support of civil authorities (ADRP 3-28), multi-service DSCA TTPs (ATP 3-28.1/MCWP 3-36.2), DSCA liaison officer toolkit (GTA 90-01-020), key legal and policy documents, and specific hazard and planning guidance.

Refer to CTS1: The Counterterrorism, WMD & Hybrid Threat SMARTbook for further discussion. CTS1 topics and chapters include: the terrorist threat (characteristics, goals & objectives, organization, state-sponsored, international, and domestic), hybrid and future threats, forms of terrorism (tactics, techniques, & procedures), counterterrorism, critical infrastructure, protection planning and preparation, countering WMD, and consequence management (all hazards response).

Refer to CYBER: The Cyberspace Operations SMARTbook (in development). U.S. armed forces operate in an increasingly network-based world. The proliferation of information technologies is changing the way humans interact with each other and their environment, including interactions during military operations. This broad and rapidly changing operational environment requires that today's armed forces must operate in cyberspace and leverage an electromagnetic spectrum that is increasingly competitive, congested, and contested.

Tasks of Decisive Action

Ref: ADRP 3-0, Operations (Nov '16), pp. 3-4 to 3-5 and table 3-1, p. 3-2.

Decisive action requires simultaneous combinations of offense, defense, and stability or defense support of civil authorities tasks.

1. Offensive Tasks

An offensive task is a task conducted to defeat and destroy enemy forces and seize terrain, resources, and population centers. Offensive tasks impose the commander's will on the enemy. Against a capable, adaptive enemy, the offense is the most direct and sure means of seizing, retaining, and exploiting the initiative to gain physical and psychological advantages and achieve definitive results. In the offense, the decisive operation is a sudden, shattering action against an enemy weakness that capitalizes on speed, surprise, and shock. If that operation does not destroy the enemy, operations continue until enemy forces disintegrate or retreat to where they no longer pose a threat. Executing offensive tasks compels the enemy to react, creating or revealing additional weaknesses that the attacking force can exploit.

See pp. 3-2 to 3-3 and SUTS2: The Small Unit Tactics SMARTbook, 2nd Ed (ADRP 3-90)

Offensive Tasks

Primary Tasks

- Movement to contact
- Attack
- Exploitation
- Pursuit

Purposes

- Dislocate, isolate, disrupt and destroy enemy forces
- Seize key terrain
- Deprive the enemy of resources
- Refine intelligence
- Deceive and divert the enemy
- Provide a secure environment for stability operations

Defensive Tasks

Primary Tasks

- Mobile defense
- Area defense
- Retrograde

Purposes

- Deter or defeat enemy offensive operations
- Gain time
- Achieve economy of force
- Retain key terrain
- Protect the populace, critical assets and infrastructure
- Refine intelligence

2. Defensive Tasks

A defensive task is a task conducted to defeat an enemy attack, gain time, economize forces, and develop conditions favorable for offensive or stability tasks. Normally the defense alone cannot achieve a decisive victory. However, it can set conditions for a counteroffensive or counterattack that enables Army forces to regain the initiative. Defensive tasks are a counter to the enemy offense. They defeat attacks, destroying as much of the attacking enemy as possible. They also preserve and maintain control over land, resources, and populations. The purpose of defensive tasks is to retain key terrain, guard populations, protect lines of communications, and protect critical capabilities against enemy attacks and counterattacks. Commanders can conduct defensive tasks to gain time and economize forces so offensive tasks can be executed elsewhere.

See pp. 3-2 to 3-3 and SUTS2: The Small Unit Tactics SMARTbook, 2nd Ed. (ADRP 3-90)

3. Stability Tasks

Stability tasks are tasks conducted as part of operations outside the United States in coordination with other instruments of national power to maintain or reestablish a safe and secure environment and provide essential governmental services, emergency infrastructure reconstruction, and humanitarian relief (ADP 3-07). These tasks support governance, whether it is imposed by a host nation, an interim government, or military government. Stability tasks involve both coercive and constructive actions. They help to establish or maintain a safe and secure environment and facilitate reconciliation among local or regional adversaries. Stability tasks assist in building relationships among unified action partners, and promote specific U.S. security interests. Stability tasks can also help establish political, legal, social, and economic institutions while supporting the transition to legitimate host-nation governance. Stability tasks cannot succeed if they only react to enemy initiatives. Stability tasks must maintain the initiative by pursuing objectives that resolve the causes of instability.

See pp. 3-4 to 3-5 and TAA2: The Military Engagement, Security Cooperation & Stability SMARTbook, 2nd Ed. (ADRP 3-07)

Stability Tasks

Primary Tasks

- Establish civil security
- Establish civil control
- Restore essential services
- Support to governance
- Support to economic and infrastructure development
- Conduct security cooperation

Purposes

- Provide a secure environment
- Secure land areas
- Meet the critical needs of the populace
- Gain support for host-nation government
- Shape the environment for interagency and host-nation success
- Promote security, build partner capacity, and provide access
- Refine intelligence

4. Defense Support of Civil Authority Tasks

Defense support of civil authorities is support provided by United States Federal military forces, Department of Defense civilians, Department of Defense contract personnel, Department of Defense component assets, and National Guard forces (when the Secretary of Defense, in coordination with the Governors of the affected States, elects and requests to use those forces in Title 32, United States Code, status) in response to requests for assistance from civil authorities for domestic emergencies, law enforcement support, and other domestic activities, or from qualifying entities for special events (DODD 3025.18). For Army forces, defense support of civil authorities is a task that takes place only in the homeland and U.S. territories. Defense support of civil authorities is conducted in support of another primary or lead federal agency, or in some cases, local authorities.

See pp. 3-6 to 3-7 and HDS1: Homeland Defense & DSCA SMARTbook. (JP 3-28)

Defense Support of Civil Authorities Tasks

Primary Tasks

- Provide support for domestic disasters
- Provide support for domestic CBRN incidents
- Provide support for domestic civilian law enforcement agencies
- Provide other designated support

Purposes

- Save lives
- Restore essential services
- Maintain or restore law and order
- Protect infrastructure and property
- Support maintenance or restoration of local government
- Shape the environment for interagency success

Simultaneity & Transitioning

Ref: ADRP 3-0, Operations (Nov '16), pp. 3-3 to 3-4 and 3-5 to 3-6.

The Purpose of Simultaneity

Simultaneously conducting offensive, defensive, and stability or defense support of civil authorities tasks requires the synchronized application of combat power. Simultaneity means doing multiple things at the same time. It requires the ability to conduct operations in depth and to integrate them so that their timing multiplies their effectiveness throughout an area of operations and across the multiple domains. Commanders consider their entire area of operations, the enemy, and the information collection activities necessary to shape an operational environment and civil conditions. Then they mount simultaneous operations that immobilize, suppress, or surprise the enemy. Such actions nullify the enemy's ability to conduct synchronized, mutually supporting reactions. Simultaneity presents the enemy with multiple dilemmas. Then, the enemy cannot focus on a single problem, but must address multiple dilemmas, presenting the enemy with more than it can deal with effectively.

See p. 1-50 for discussion of simultaneity as a tenet of unified land operations.

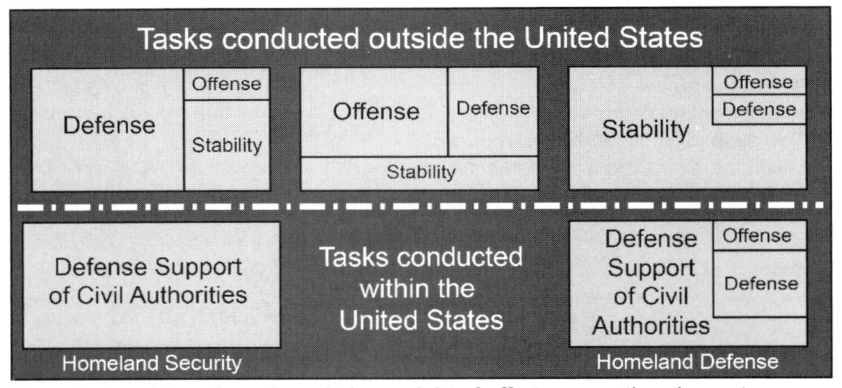

The mission determines the relative weight of effort among the elements.

Ref: ADRP 3-0 (2016), fig. 3-1. Decisive action.

Army forces increase the **depth** of their operations in time and space through combined arms, economy of force, continuous reconnaissance, advanced information systems, and joint capabilities. Because Army forces conduct operations across large areas, the enemy faces many potential friendly actions. Executing operations in depth is equally important in security; commanders act to keep threats from operating outside the reach of friendly forces. In defense support of civil authorities and some stability tasks, depth includes conducting operations that reach all citizens in the area of operations, bringing relief as well as hope.

See p. 1-51 for discussion of depth as a tenet of unified land operations.

Transitioning in Decisive Action

Conducting decisive action involves more than simultaneous execution of all its tasks. It requires commanders and staffs to consider their units' capabilities and capacities relative to each task. Commanders consider their missions; decide which tactics, techniques, and procedures to use; and balance the tasks of decisive action while preparing their commander's intent and concept of operations. They determine which tasks the force can accomplish simultaneously, if phasing is required, what additional resources it may need, and how to transition from one task to another.

Examples of Combining Elements

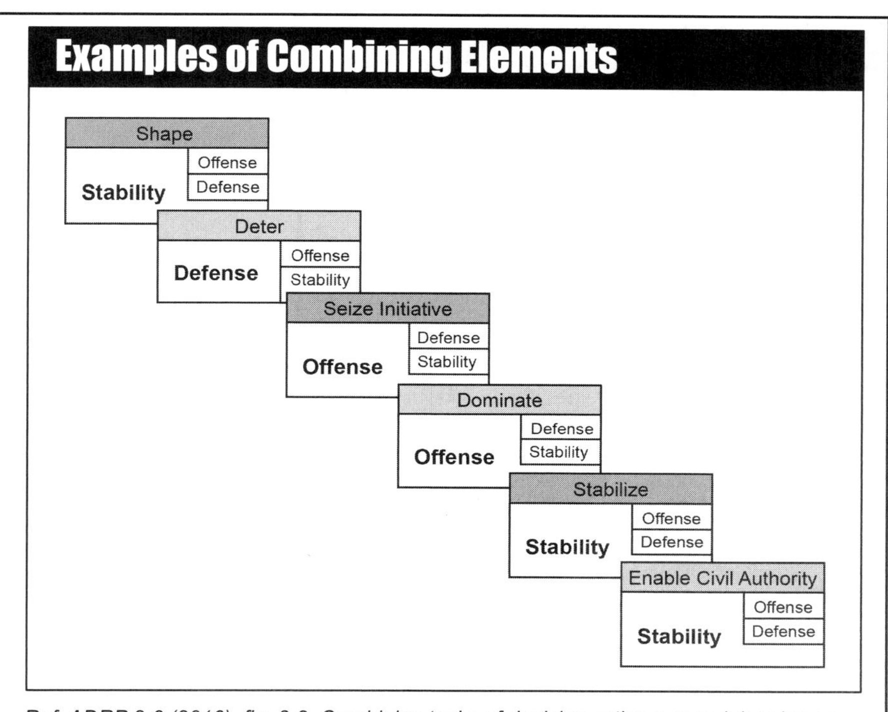

Ref: ADRP 3-0 (2016), fig. 3-2. Combining tasks of decisive action across joint phases.

The transitions between tasks of decisive action require careful assessment, prior planning, and unit preparation as commanders shift their combinations of offensive, defensive, stability, or defense support of civil authorities tasks. Commanders first assess the situation to determine applicable tasks and the priority for each. When conditions change, commanders adjust the combination of tasks of decisive action in the concept of operations. When an operation is phased, the plan includes these changes. The relative weight given to each element varies with the actual or anticipated conditions. It is reflected in tasks assigned to subordinates, resource allocation, and task organization.

Decisive action is not a phasing method. Commanders consider the concurrent conduct of each task—offensive, defensive, and stability or defense support of civil authorities—in every phase of an operation. Figure 3-2 illustrates combinations and weighting of the tasks of decisive action across the phases of a joint campaign or operation. The phases shown are examples. An actual campaign may name and array phases differently. Also, operations do not necessarily move linearly through the phases. For example, a unit may move from stability to seizing the initiative with little or no time for deterrence.

Unanticipated changes in or an improved understanding of an operational environment may result in commanders reframing the problem and modifying operations to adapt to the changing situation. Unforeseen success in an offense resulting in collapse of enemy opposition illustrates one unanticipated change. Another example is degradation in peace operations resulting in a requirement to transition to defensive tasks, or even offensive tasks, to reestablish stability. Commanders need to adjust the task organization to meet changing requirements. In some instances, they incorporate additional forces provided by higher headquarters to assist in the conduct of operations. When transitioning in operations, subordinate commanders must clearly understand the higher commander's intent, concept of operations, and desired end state. Successful commanders understand which transitions involve risks, how much risk to accept, and where risk is accepted.

III. Seize, Retain, and Exploit the Initiative

Army forces seize, retain, and exploit the initiative by forcing the enemy to respond to friendly action. By presenting the enemy multiple dilemmas, commanders force the enemy to react continuously until the enemy is finally driven into untenable positions. Seizing the initiative pressures enemy commanders into abandoning their preferred options and making costly mistakes. As enemy mistakes occur, friendly forces seize opportunities and create new avenues for exploitation. Throughout operations, commanders focus combat power to protect populations, friendly forces, and infrastructure; to deny the enemy positions of advantage; and to consolidate gains to retain the initiative.

Commanders create conditions for seizing the initiative by acting. Without action, seizing the initiative is impossible. Faced with an uncertain situation, there is a natural tendency to hesitate and gather more information to reduce uncertainty. However, waiting and gathering information might reduce uncertainty but not eliminate it. Waiting may even increase uncertainty by providing the enemy with time to seize the initiative. It is far better to manage uncertainty by acting and developing the situation.

Seizing the initiative means setting and dictating the terms of action throughout the operation. Commanders plan to seize the initiative as early as possible. Effective planning determines where, when, and how to do so. However, enemies will actively try to prevent this and disrupt friendly plans. Seizing the initiative requires effective plans to counter enemy efforts. During preparation, commanders set conditions that lead to seizing the initiative and assessing the effectiveness of actions. During execution, commanders and staffs recognize and exploit projected opportunities to attack the command and control elements of enemy forces in order to prevent their synchronization of combat power, including use of deception to achieve surprise. Seizing the initiative often requires accepting risk. Commanders and staffs evaluate enemy and friendly actions to determine who has the initiative. They determine what friendly actions will enable friendly forces to retain and exploit the initiative if they have it and seize the initiative if they do not. The following are general indicators that friendly forces have the initiative:

- Friendly forces are no longer decisively engaged or threatened with decisive engagement.
- Subordinate commanders are able to mass combat power or concentrate forces at times and places of their choosing.
- Enemy forces are not offering effective resistance and do not appear capable of reestablishing resistance.
- Friendly forces encounter lighter-than-anticipated enemy resistance or large numbers of prisoners.
- Friendly rates of advance suddenly accelerate or casualty rates suddenly drop.

Retaining the initiative involves applying unrelenting pressure on the enemy. Commanders do this by synchronizing the warfighting functions to present enemy commanders with continuously changing combinations of combat power at a tempo they cannot effectively counter. Commanders and staffs use information collection assets to identify enemy attempts to regain the initiative. Effective information management processes this information quickly enough to keep commanders inside the enemy's decision-making cycle. Combined with effective planning, information management helps commanders to anticipate key events and likely enemy actions hours or days beforehand and to develop branches, sequels, or adjustments to the plan. During execution, commanders create a seamless, uninterrupted series of actions that force enemies to react immediately and do not allow them to regain synchronization. Ideally, these actions present enemies with multiple critical problems that require more resources to solve than they have.

Exploiting the initiative means following through on initial successes to realize long-term decisive success. Once friendly forces seize the initiative, they immediately plan to exploit it by conducting continuous operations to accelerate the enemy's complete defeat. Internal to the organization, commanders identify any disorganization among friendly forces and direct reorganization or reconstitution to restore those forces to combat readiness and to develop options to exploit the initiative.

IV. Consolidate Gains

Consolidate gains is the activities to make permanent any temporary operational success and set the conditions for a sustainable stable environment allowing for a transition of control to legitimate civil authorities. Army forces provide the joint force commander the ability to capitalize on operational success by consolidating gains. Consolidate gains is an integral part of winning armed conflict and achieving success across the range of military operations; it is essential to retaining the initiative over determined enemies and adversaries. To consolidate gains, Army forces reinforce and integrate the efforts of all unified action partners.

Consolidate gains is not a mission, but rather a capability that Army forces provide to the joint force commander. Consolidate gains is demonstrated by the execution of tasks when emphasis shifts from actions addressing the immediate threats in an operational environment to those measures that address the long term needs of the host nation and its population. It is a transition from the occupation of territory and control of populations by Army forces, gained as a result of military operations, to the transfer of control to legitimate authorities building the capability to govern and secure the host nation. Activities to consolidate gains occur across the range of military operations and are often continuous throughout all phases of an operation.

Army forces consolidate gains in support of the host nation and its civilian population. These gains may include the relocation of displaced civilians, reestablishment of law and order, performance of humanitarian assistance, and restoration of key infrastructure. Concurrently, Army forces must be able to accomplish such activities while sustaining, repositioning, and reorganizing forces to continue operations.

Army forces must analyze the host nation's capability and capacity to provide services, or determine the ability of other agencies of government (both host nation and United States), international agencies, nongovernmental organizations, and contractors to provide services that can enable, complement, or conduct as needed. The goal is to address drivers of conflict, foster host-nation resiliencies, and create conditions that enable sustainable peace and security by transitioning capabilities to local authorities and returning to normal peacetime engagement characterized by security cooperation.

Army forces must deliberately plan and prepare for consolidating gains to capitalize on operational success prior to an operation. Planning considerations include the changes to task organization and the additional assets required. These assets may include engineers, military police, civil affairs, and medical support, especially those assets required for the potential increase in stability tasks. In some instances, Army forces will be in charge of integrating and synchronizing activities. In other situations, the Army will be in support. The Army's campaign quality gives it the capability and capacity to conduct decisive action associated with consolidating gains for a sustained period over large land areas.

To consolidate gains, Army forces take specific actions. These actions include—

- **Consolidation**: Forces organize and strengthen their newly occupied positions so that they can be used for subsequent operations.
- **Area security**: Forces conduct security tasks to protect friendly forces, installation routes, critical infrastructure, populations, and actions within an assigned area of operations.

- **Stability tasks**: Forces first conduct minimum-essential stability tasks, then maintain or reestablish a safe and secure environment and provide essential governmental services, emergency infrastructure reconstruction, and humanitarian relief.
- **Influence over local and regional audiences (when authorized)**: Commanders ensure that supporting and credible narratives are developed and communicated to the intended population to assist them in understanding the overall goal of military actions and the benefits of those actions for the population.
- **Security from external threats**: Commanders use forward presence in an area of operations to allow a transition in operations to occur without disruption from nascent threats.

Army forces routinely conduct consolidation upon occupying a position on the battlefield or achieving success. Consolidation is the organizing and strengthening in newly captured position so that it can be used against the enemy (FM 3-90-1). Normally, an attacking unit tries to exploit success, but in some situations, the unit may have to consolidate before exploiting its gains. Consolidation activities include—

- Conducting reconnaissance.
- Establishing security.
- Eliminating enemy pockets of resistance.
- Positioning forces to enable them to conduct a hasty defense by blocking possible enemy counterattacks.
- Adjusting fire planning.
- Preparing for potential additional missions.

Refer to FM 3-90-1 for an additional discussion of consolidation.

However, when consolidating gains, commanders ensure security is both established and can be sustained throughout transition. Army forces conduct continuous reconnaissance and, if necessary, gain or maintain contact with the enemy to defeat or preempt enemy actions and retain the initiative. Consolidating gains may include actions required to eliminate or neutralize isolated or bypassed threat forces (including the processing of enemy prisoners and civilian detainees) to increase area security and protect lines of communications.

The Multi-Domain Battle

Army forces conduct multi-domain battle, as part of a joint force, to seize, retain, and exploit control over enemy forces. Army forces deter adversaries, restrict enemy freedom of action, and ensure freedom of maneuver and action in multiple domains for the joint force commander. For example, Army forces use aviation and unmanned aircraft systems in the air domain and protect vital communications networks in cyberspace, while retaining dominance in the land domain. Army forces operate dispersed over wide areas while retaining the ability to concentrate rapidly, presenting multiple dilemmas to enemy forces. Key considerations for operating in multiple domains include—

- Effective—
  - Communications.
  - Long-range systems.
  - Systems with reduced sustainment demands.
  - Leaders well versed in the principles of mission command.
- Mobility on land and air to maneuver rapidly over larger distances.
- Combined arms and cross-domain capabilities pushed down to the lowest tactical levels to enable maneuver and survivability. Commanders understanding multi-domain opportunities employ land capabilities to influence or enable operations by unified action partners in other domains.

Stability Tasks

Ref: ADRP 3-0, Operations (Nov '16), pp. 3-8 to 3-9.

During the consolidation of gains, Army forces are responsible for accomplishing both the minimum-essential stability tasks and the Army primary stability tasks. Commanders must quickly ensure the provision of minimum-essential stability tasks of security, food, water, shelter, and medical treatment. Once conditions allow, these tasks are a legal responsibility of Army forces. However, commanders may not need to have Army forces conduct all essential tasks if a military unit or appropriate civilian organization exists that can adequately conduct those tasks. For example, there may be sufficient civilian or military governance in place to ensure that the population has adequate food and medical care. However, Army forces will continue consolidating gains by conducting the Army's primary stability tasks: establish civil security, establish civil control, restore essential services, support to governance, support to economic and infrastructure development, and conduct security cooperation. The tasks associated with the primary stability tasks will evolve over time. The military will retain the lead and execution to establish civil security through the conduct of security force assistance. Eventually, the lead for the other four tasks will transfer to another military or civilian organization, although the Army may retain a supporting role.

It is important for commanders to understand that activities to begin and sustain consolidating gains may occur over a significant time. Gradually, emphasis will shift from actions to ensure the defeat of remaining threat forces to those measures that address the needs of the urban population, manage their perceptions, and allow responsibility to shift from Army forces to organizations such as local governing groups, interorganizational groups, or interagency partners. Additionally, there may be interim transitions during consolidating gains when Army forces may transition various tasks to another military force that is task organized to sustain consolidating gains or to a civilian agency to conduct various tasks. If a host-nation government is nonexistent, overall control may transfer to an interim civilian government or to a reconstituted host-nation government.

Ultimately, commanders conduct decisive action when consolidating gains to make permanent any temporary successes while continuing pressure on enemy forces. Consolidating gains provides security and protection for both friendly forces and the population, facilitates reorganization, and allows forces to prepare for handover of the area of operations and population control to legitimate civil authorities.

V. Principles Of Unified Land Operations

A principle is a comprehensive and fundamental rule or an assumption of central importance that guides how an organization or function approaches and thinks about the conduct of operations (ADP 1-01). By integrating the six principles of unified land operations—mission command, develop the situation through action, combined arms, adherence to the law of war, establish and maintain security, and create multiple dilemmas for the enemy—Army commanders can achieve operational and strategic success. Success requires fully integrating U.S. military operations with the efforts of unified action partners. Success also requires commanders to exercise disciplined initiative. Initiative is used to gain a position of relative advantage that degrades and defeats the enemy throughout the depth of an organization.

A. Mission Command

Mission command is the exercise of authority and direction by the commander using mission orders to enable disciplined initiative within the commander's intent to empower agile and adaptive leaders in the conduct of unified land operations (ADP 6-0). Mission command is exercised by Army commanders. It blends the art of command and the science of control while integrating the warfighting functions to conduct the tasks of decisive action. Mission command has six fundamental principles:

- Build cohesive teams through mutual trust.
- Create shared understanding.
- Provide a clear commander's intent.
- Exercise disciplined initiative.
- Use mission orders.
- Accept prudent risk.

See chap. 2, Mission Command Warfighting Function, for further discussion.

Through mission command, commanders integrate and synchronize operations. Commanders understand that they do not operate independently but as part of a larger force. They integrate and synchronize their actions with the rest of the force to achieve the overall objective of the operation. Commanders create and sustain shared understanding and purpose through collaboration and dialogue within their organization and with unified action partners to facilitate unity of effort. They provide a clear commander's intent and use mission orders to assign tasks, allocate resources, and issue broad guidance.

Guided by the commander's intent and the purpose of the mission, subordinates use disciplined initiative and take actions that will best accomplish the mission. They take appropriate actions and perform the necessary coordination without needing new orders. Often, subordinates acting on the commander's intent develop the situation in ways that exploit unforeseen opportunities. Commander's intent is a clear and concise expression of the purpose of the operation and the desired military end state that supports mission command, provides focus to the staff, and helps subordinate and supporting commanders act to achieve the commander's desired results without further orders, even when the operation does not unfold as planned (JP 3-0).

Mission command requires commanders to convey a clear commander's intent and concept of operations. These become essential in operations where multiple operational and mission variables interact with the lethal application of ground combat power. Such dynamic interaction often compels subordinate commanders to make difficult decisions in unforeseen circumstances.

Mission command emphasizes the critical contributions of leaders at every echelon. It establishes a mindset among Army leaders that the best understanding comes from a synthesis of information and an understanding from all echelons and unified action partners—bottom-up input is as important as top-down guidance. Mission command emphasizes the importance of creating shared understanding and purpose. It highlights how commanders—through disciplined initiative within the commander's intent—transition among offensive, defensive, and stability or defense support of civil authorities tasks, look for fleeting opportunities to exploit, and vary the level of control to account for changes in an operational environment.

Successful mission command fosters adaptability and a greater understanding of an operational environment. Adaptability reflects a quality that Army leaders and forces exhibit through critical thinking, their comfort with ambiguity and uncertainty, their willingness to accept prudent risk, and their ability to rapidly adjust while continuously assessing the situation. A greater understanding enables commanders to make better decisions and develop courses of action that more quickly accomplish missions and achieve the overall end state.

B. Develop the Situation Through Action

During operations, commanders develop the situation through action. Commanders fight for information to develop the situation while in contact with the enemy and gain information through close association with the population. Developing the situation through action is the responsibility of each and every Soldier. Commanders assign information collection tasks (reconnaissance, surveillance, security operations, and intelligence operations) to collect information requirements, thus enhancing situational awareness and understanding. A thorough understanding of the operational environment is imperative to identifying conflicting interests and information collection requirements that are developed through reconnaissance, security, and intelligence collection tasks.

When units encounter an enemy force or an obstacle, commanders must quickly determine the threat they face. For an enemy force, commanders must determine the enemy's composition, dispositions, activities, and movements and then assess the implications of that information. For an obstacle, commanders must determine the type and extent of the obstacle and if it is covered by fire. Obstacles can provide the attacker with information concerning the location of enemy forces, weapon capabilities, and organization of fires. Often this information can only be provided by close combat that forces the enemy to reveal locations, troops, and intentions.

In planning, commanders actively seek answers to information gaps for developing information requirements that are satisfied through information collection tasks within a given area. Through information collection and analysis, staffs develop options for the commander to further inform the population, influence various actors, seize opportunities, and maintain initiative.

Commanders take enemy plans, capabilities, and reaction times into account when making decisions. They ensure that plans delegate decision-making authority to the lowest echelon possible to obtain faster and more suitable decisions in battle. Subordinates can then use their initiative to make decisions that further their higher commander's intent. Empowered with trust, authority, and a shared understanding, they can develop the situation through action, adapt, and act decisively.

In execution, commanders make decisions quickly—even with incomplete information. Commanders who can make and implement decisions faster than the enemy, even to a small degree, gain an accruing advantage that becomes significant over time. Commanders should not delay a decision in hopes of finding a perfect solution to a battlefield problem.

Timely decisions and actions are essential for effective command. Commanders who consistently decide and act more quickly than the enemy have a significant advantage. By the time the slower commander decides and acts, the faster one has already changed the situation, rendering the slower one's actions inappropriate. With such an advantage, the commander can maintain the initiative and dictate the tempo.

To make timely decisions, commanders must understand the effects of their decisions on a complex operational environment. To help them understand, staffs work together to develop the environment input to the common operational picture. They must understand the terrain and weather and its impact on operations. They must also understand the population and its needs. Understanding an operational environment includes civil considerations—such as the population (with demographics and culture), the government, economics, nongovernmental organizations, and history—among other factors. Commanders make decisions that start and govern actions by subordinate forces throughout the operations process.

C. Combined Arms

Combined arms is the synchronized and simultaneous application of all elements of combat power that together achieve an effect greater than if each element was used separately or sequentially. Combined arms integrates leadership, information, and each of the warfighting functions and their supporting systems, as well as joint weapon systems. Used destructively, combined arms integrates different capabilities so that counteracting one makes the enemy vulnerable to another. Used constructively, combined arms uses all assets available to the commander to multiply the effectiveness and efficiency of Army capabilities used in stability or defense support of civil authorities tasks.

Operations against elusive and capable enemies demand an extension of the concept of combined arms from two or more arms or elements of one Service to include the application of unified action partner capabilities. Combined arms uses the capabilities of all Army and joint weapons systems, including cyberspace operations and multinational assets in complementary and reinforcing capabilities. Complementary capabilities protect the weaknesses of one system or organization with the capabilities of a different warfighting function. For example, commanders use artillery (fires) to suppress an enemy bunker complex, pinning down an enemy infantry unit. Protected by integrated air defense systems, the commander's infantry unit then closes with and destroys the enemy (movement and maneuver). In this example, the fires warfighting function complements the movement and maneuver warfighting function. Also, ground maneuver makes enemy forces vulnerable to joint weapon systems. Electronic warfare assets prevent the enemy from communicating or relaying information about friendly maneuver. Finally, information obtained from nongovernmental organizations can facilitate effective distribution of supplies during humanitarian assistance and disaster relief operations.

Reinforcing capabilities combine similar systems or capabilities within the same warfighting function to increase the function's overall capabilities. In urban operations, for example, infantry, aviation, and armor units (movement and maneuver) often operate close to each other. This combination reinforces the protection, maneuver, and direct fire capabilities of each. The infantry protects tanks from enemy infantry and antitank systems; tanks provide protection and firepower for the infantry. Attack helicopters maneuver above buildings to fire from positions of advantage, while other aircraft help sustain, extract, or air assault ground forces. Army space-enabled capabilities and services such as communications and Global Positioning System enable communication, navigation, situational awareness, protection, and sustainment of land forces. Army operations are supported by close air support, air interdiction, air defense, and, in some cases, naval surface fire support. Finally, unified action partners bring skills, knowledge, and capabilities that enhance the impact of combined arms on enemy forces.

Combined arms operations create multiple dilemmas for the enemy. Combined arms operations allow Army forces to gain a position of relative advantage while denying the enemy a relative advantage. Army forces achieve surprise by maneuvering across operational and strategic distances and arriving at unexpected locations. Army forces are reliant on other Services to accomplish such maneuvers. Army forces have the mobility, protection, and firepower necessary to strike the enemy from unexpected directions. In anti-access and area denial environments, dispersion allows Army forces to evade enemy attacks, deceive the enemy, and achieve surprise.

D. Adherence to Law of War

The law of war is that part of international law that regulates the conduct of armed hostilities (JP 1-04). The law of war's evolution was largely humanitarian and designed to reduce the evils of war. The main purposes of the law of war are to—

- Protect combatants, noncombatants, and civilians from unnecessary suffering.
- Provide certain fundamental protections for persons who fall into the hands of the enemy, particularly prisoners of war, civilians, and military wounded, sick, and shipwrecked.
- Facilitate the restoration of peace.
- Assist military commanders in ensuring the disciplined and efficient use of military force.
- Preserve the professionalism and humanity of combatants.

Soldiers consider five important principles that govern the law of war when planning and executing operations: military necessity, humanity, distinction, proportionality, and honor. Three interdependent principles—military necessity, humanity, and honor—provide the foundation for other law of war principles—such as proportionality and distinction—and most of the treaty and customary rules of the law of war. Law of war principles work as interdependent and reinforcing parts of a coherent system. Military necessity justifies certain actions necessary to defeat the enemy as quickly and efficiently as possible. Conversely, humanity forbids actions unnecessary to achieve that object. Proportionality requires that even when actions may be justified by military necessity, such actions not be unreasonable or excessive. Distinction underpins the parties' responsibility to comport their behavior with military necessity, humanity, and proportionality by requiring parties to a conflict to apply certain legal categories, principally the distinction between the armed forces and the civilian population. Lastly, honor supports the entire system and gives parties confidence in it.

Rules of engagement are directives issued by competent military authority that delineate the circumstances and limitations under which United States forces will initiate and/or continue combat engagement with other forces encountered (JP 1-04). Rules of engagement always recognize the inherent right of self-defense. These rules vary between operations, may vary between types of units in the same area of operations, and may change during an operation. Adherence to them ensures Soldiers act consistently with international law, national policy, and military regulations.

Soldiers deployed to a combat zone overseas follow rules of engagement established by the Secretary of Defense and adjusted for theater conditions by the joint force commander. Within the United States and its territories, Soldiers adhere to rules for the use of force. Rules for the use of force consist of directives issued to guide U.S. forces on the use of force during various operations. These directives may take the form of execute orders, deployment orders, memoranda of agreement, or plans. There are many similarities among these directives, for example in the inherent right of self-defense, but they differ in intent. Rules of engagement are by nature permissive measures intended to allow the maximum use of destructive combat power appropriate for the mission. Rules for the use of force are restrictive measures intended to allow only the minimum force necessary to accomplish the mission. The underlying principle is a "continuum of force," a carefully graduated level of response determined by the behavior of possible threats.

Refer to HDS1: The Homeland Defense & DSCA SMARTbook for discussion on rules for the use of force.

Soldiers use discipline when applying lethal and nonlethal actions, and successful operations require disciplined Soldiers. Today's threats challenge the morals and ethics of Soldiers. Often an enemy does not respect international laws or conventions and commits atrocities simply to provoke retaliation in kind. Enemy and adversary forces, as well as neutral and friendly forces, will take any loss of discipline on the part of Soldiers, distort and exploit it in propaganda, and magnify it through the media. It is therefore crucial that all personnel operate at all times within applicable U.S., international, and in some cases host-nation laws and regulations. The challenge of ensuring Soldiers remain within legal, moral, and ethical boundaries at all times is a leadership concern and priority. This challenge rests heavily on small-unit and company-grade leaders charged with maintaining good order and discipline within their respective units. The Soldier's Rules in AR 350-1 distill the essence of the law of war. They outline the ethical and lawful conduct required of Soldiers in operations, and all Soldiers should follow them. Table 3-2 lists the Soldier's Rules.

The Soldier's Rules

- Soldiers fight only enemy combatants.
- Soldiers do not harm enemies who surrender. They disarm them and turn them over to their superior.
- Soldiers do not kill or torture any personnel in their custody.
- Soldiers collect and care for the wounded, whether friend or foe.
- Soldiers do not attack medical personnel, facilities, or equipment.
- Soldiers destroy no more than the mission requires.
- Soldiers treat civilians humanely.
- Soldiers do not steal. Soldiers respect private property and possessions.
- Soldiers should do their best to prevent violations of the law of war.
- Soldiers report all violations of the law of war to their superior.

Ref: ADRP 3-0 (2016), table 3-2. The Soldier's Rules.

E. Establish and Maintain Security

Army forces conduct area security to ensure freedom of movement and action and deny the enemy the ability to disrupt operations. Commanders combine reconnaissance; raids; and offensive, defensive, and stability tasks to protect populations, friendly forces, installations, borders, extended infrastructure, and activities critical to mission accomplishment. Army forces integrate with partner military, law enforcement, and civil capabilities to establish and maintain security. Army forces conduct area security to deny the enemy use of terrain, protect populations, and enable the joint force to project power from land into the air, maritime, space, and cyberspace domains. The Army's ability to establish control on land prevents the enemy from disrupting activities and efforts critical to consolidating gains in the wake of successful military operations.

Security operations during preparation prevent surprise and reduce uncertainty. They provide early and accurate warning of enemy operations to provide the force being protected with time and maneuver space within which to react to the enemy and to develop the situation to allow the commander to effectively use the protected force. Security operations are designed to prevent enemies from discovering the friendly force's plan and to protect the force from unforeseen enemy actions. Security elements direct their main effort toward preventing the enemy from gathering essential elements of friendly information. As with reconnaissance, security is a dynamic effort that anticipates and thwarts enemy collection efforts. When successful, security operations provide the force enough time and maneuver space to react to enemy attacks. To accomplish this, staffs coordinate security operations among the units that conduct them and concurrently synchronize them with local unit security.

F. Create Multiple Dilemmas for the Enemy

Army forces present the enemy with multiple dilemmas because they possess the simultaneity to overwhelm the enemy physically and psychologically, the depth to prevent enemy forces from recovering, and the endurance to sustain operations. Simultaneous operations in depth and across multiple domains, supported by military deception, present the enemy with multiple dilemmas, degrade enemy freedom of action, reduce enemy flexibility and endurance, and upset enemy plans and coordination. At the same time, these operations place critical enemy functions at risk and deny the enemy the ability to synchronize or generate combat power. The simultaneous application of combat power throughout the area of operations is preferable to the attritional nature of sequential operations. Army forces use joint and multinational capabilities in a complementary and reinforcing fashion to create multiple dilemmas.

The capability to project power across operational distances allows forces to present the enemy with multiple dilemmas as forces with mobility, protection, and lethality arrive at unexpected locations, bypassing enemy anti-access and aerial denial systems and strong points. Forcible entry operations can create multiple dilemmas by creating threats that exceed the enemy's capability to respond.

To create multiple dilemmas, commanders must know the positioning and dispersion of enemy forces. Commanders commit forces to conduct reconnaissance as part of a focused effort to collect information on enemy activities and resources; geographical, hydrological, and meteorological characteristics; and civilian considerations. The information gained is used to inform situational understanding, decision making, intelligence preparation of the battlefield, course of action development, and target development and refinement.

Reconnaissance efforts, by nature, are not conducted with the expressed purpose to delay, disrupt, divert, or destroy enemy forces. However, reconnaissance efforts develop the situation through actions, such as reconnaissance in force, to test the enemy's strength, dispositions, and reactions or to obtain other information. A reconnaissance in force is a deliberate combat operation designed to discover or test the enemy's strength, dispositions, and reactions or to obtain other information (ADRP 3-90). This operation is an aggressive reconnaissance that is conducted as an offensive operation. A commander assigns a reconnaissance in force mission when the enemy is known to be operating within an area, and the commander cannot obtain adequate intelligence by any other means. Because of the lack of information about the enemy, a commander normally conducts a reconnaissance in force across a broad front as a movement to contact, deliberate attack, or raid.

With knowledge of how the enemy is arrayed, Army forces achieve surprise through maneuver across vast distances and arrival at unexpected locations. Army forces have the mobility, protection, and firepower necessary to strike the enemy from unexpected directions. Army forces operate dispersed while maintaining mutual support. Dispersion allows Army forces to evade enemy attacks, deceive the enemy, and achieve surprise. Even when operating dispersed, combined arms teams are able to concentrate rapidly to isolate the enemy, attack critical enemy assets, and seize upon fleeting opportunities.

Army forces conduct continuous reconnaissance and security operations to seize, retain, and exploit the initiative over the enemy while protecting the force against dangers. As part of the joint force, Army forces maneuver and project power to ensure joint force freedom of action and deny the enemy the ability to operate freely. Army leaders synchronize the efforts of unified action partners to ensure unity of effort.

VI. Tenets of Unified Land Operations

Ref: ADRP 3-0, Operations (Nov '16), pp. 3-14 to 3-16.

Tenets of operations are desirable attributes that should be built into all plans and operations and are directly related to the Army's operational concept (ADP 1-01, see following page). The tenets of unified land operations describe the Army's approach to generating and applying combat power across the range of military operations through the four tasks of decisive action. For Army forces, an operation is a military action, consisting of two or more related tactical actions, designed to achieve a strategic objective, in whole or in part. A tactical action is a battle or engagement employing lethal and nonlethal actions designed for a specific purpose relative to the enemy, the terrain, friendly forces, or other entities. Tactical actions include widely varied activities. They can include an attack to seize a piece of terrain or destroy an enemy unit, the defense of a population, and the training of other militaries to assist security forces as part of building partner capacity. In the homeland, Army forces apply the tenets of operations when in support of civil authorities in order to save lives, alleviate suffering, and protect property. Army forces may provide assistance to civil authorities in situations such as natural disasters, chemical or biological incidents, or major public events.

Army operations are characterized by four tenets:

Tenets of Unified Land Operations

- **A** Simultaneity
- **B** Depth
- **C** Synchronization
- **D** Flexibility

A. Simultaneity

Simultaneity is the execution of related and mutually supporting tasks at the same time across multiple locations and domains. Operating simultaneously across the land, air, maritime, space, and cyberspace domains allows Army forces to deliver multiple blows to the enemy while reassuring allies and influencing neutrals. The simultaneous application of joint and combined arms capabilities across the range of military operations aims to overwhelm the enemy physically and psychologically. Combined arms operations create multiple dilemmas for the enemy. Army forces achieve surprise by maneuvering across strategic distances and arriving at unexpected locations. Simultaneity extends efforts beyond physical battlefields into other contested spaces such as public perception, political subversion, illicit financing, and criminality.

Interdependence gained by the right mix of complementary conventional and special operations forces, at the appropriate echelon, enhances success throughout the range of military operations and all phases of joint operations. Simultaneity requires creating shared understanding and purpose through collaboration with all elements of the friendly force.

B. Depth

Depth is the extension of operations in time, space, or purpose to achieve definitive results. Army leaders engage enemy forces throughout their depth, preventing the effective employment of reserves, command and control nodes, logistics, and other capabilities not in direct contact with friendly forces. Operations in depth can disrupt the enemy's decision cycle. These operations contribute to protecting the force by destroying enemy capabilities before the enemy can use them. Commanders balance their forces' tempo and momentum to produce simultaneous results throughout their areas of operations. Commanders provide depth within the commander's intent (the purpose), which empowers subordinates to act on initiative, resulting in increased tempo. To achieve simultaneity, commanders establish a high tempo to target enemy capabilities located at the limit of a force's operational reach.

Unified land operations achieve the best results when the enemy must cope with U.S. actions throughout the enemy's entire physical, temporal, and organizational depth. Army forces use combined arms, advanced information systems, and joint capabilities to increase the depth of friendly operations.

C. Synchronization

Synchronization is the arrangement of military actions in time, space, and purpose to produce maximum relative combat power at a decisive place and time. (JP 2-0). Synchronization is not the same as simultaneity; it is the ability to execute multiple related and mutually supporting tasks in different locations at the same time, producing greater effects than executing each in isolation. For example, synchronization of information collection, obstacles, direct fires, and indirect fires results in the destruction of an enemy formation. When conducting offensive tasks, synchronizing forces along multiple lines of operations temporarily disrupts the enemy organization and allows for exploitation.

Information networks greatly enhance the potential for synchronization. They do this by allowing commanders to quickly understand an operational environment and communicate their commander's intent. Subordinate and adjacent units use that common understanding to synchronize their actions with those of other units without direct control from the higher headquarters. Information networks do not guarantee synchronization; however, they provide a powerful tool for leaders to use in synchronizing their efforts.

Commanders determine the degree of control necessary to synchronize their operations. They balance synchronization with agility and initiative, never surrendering the initiative for the sake of synchronization. Rather, they synchronize activities to best facilitate mission accomplishment. Excessive synchronization can lead to too much control, which limits the initiative of subordinates and undermines mission command.

D. Flexibility

Flexibility is the employment of a versatile mix of capabilities, formations, and equipment for conducting operations. To achieve tactical, operational, and strategic success, commanders seek to demonstrate flexibility. Flexibility is an important trait of effective leaders. Commanders enable adaptive forces through flexibility, which facilitates collaborative planning and decentralized execution. They exercise mission command to achieve maximum flexibility and foster individual initiative. To adapt, leaders constantly learn from experience (their own and that of others) and apply new knowledge to each situation. Flexible plans help units adapt quickly to changing circumstances in operations. Commanders build opportunities for initiative by anticipating events that allow them to operate inside of the enemy's decision cycle or react promptly to deteriorating situations.

Flexibility and innovation are at a premium, as are creative and adaptive leaders. As knowledge increases, Army forces continuously adapt to changes in an operational environment. Such adaptation enhances flexibility across the range of military operations. Army forces require flexibility in thought, plans, and operations to be successful in unified land operations.

VII. Success Through Unified Land Operations

Ultimately, the operational concept of unified land operations aims to accomplish the mission. Execution of unified land operations through decisive action requires the following:

- A clear commander's intent and concept of operations that establishes the role of each element and its contribution to accomplishing the mission.
- A flexible mission command system.
- A shared understanding of an operational environment and the purpose of the operation.
- Aggressive information collection and intelligence analysis.
- Aggressive planning for, and when authorized, execution of cyberspace operations.
- Aggressive security operations.
- Units that can quickly change their task organization.
- Operational and disciplined initiative.
- An ability to respond quickly.
- Planned and responsive sustainment.
- Combat power applied through combined arms.
- Well-trained, cohesive teams and bold, adaptive, and imaginative leaders.
- The acceptance of prudent risk.
- An ability to liaise and coordinate operations with unified action partners.
- An ability to consolidate gains to make success in operations permanent.

Chap 1 IV. Operations Structure

Ref: ADRP 3-0, Operations (Nov '16), chap. 4.

The operations structure—the operations process, warfighting functions, and the operational framework—is the Army's common construct for unified land operations. It allows Army leaders to rapidly and effectively organize efforts in a manner commonly understood across the Army. The operations process provides a broadly defined approach to developing and executing operations. The warfighting functions provide a common organization for critical functions. The operational framework provides Army leaders with basic conceptual options for arraying forces and visualizing and describing operations.

I. The Operations Process

The operations process consists of the major mission command activities performed during operations: planning, preparing, executing, and continuously assessing the operation. The operations process is a commander-led activity, informed by mission command. These activities may be sequential or simultaneous. In fact, they are rarely discrete and often involve a great deal of overlap. Commanders use the operations process to drive the planning necessary to understand, visualize, and describe their unique operational environments; make and articulate decisions; and direct, lead, and assess military operations.

See following page (p. 1-55) for an overview of the operations process from ADP 5-0.

Planning is the art and science of understanding a situation, envisioning a desired future, and laying out effective ways of bringing that future about (ADP 5-0). Planning consists of two separate but interrelated components: a conceptual component and a detailed component. Successful planning requires the integration of both these components. Army leaders employ three methodologies for planning: the Army design methodology, the military decisionmaking process, and troop leading procedures. Commanders determine how much of each methodology to use based on the scope of the problem, their familiarity with the methodology, the echelon, and the time available to the staffs.

See p. 1-57 for an overview of the Army planning methodologies.

Preparation consists of activities that units perform to improve their ability to execute an operation. Preparation creates conditions that improve friendly forces' opportunities for success. It requires commander, staff, unit, and Soldier actions to ensure the force is trained, equipped, and ready to execute operations. Preparation activities help commanders, staffs, and Soldiers understand a situation and their roles in upcoming operations, as well as setting conditions for successful execution.

Execution puts a plan into action by applying combat power to accomplish the mission and by using situational understanding to assess progress and make execution and adjustment decisions. In execution, commanders, staffs, and subordinate commanders focus their efforts on translating decisions into actions. They apply combat power to seize, retain, and exploit the initiative to gain and maintain a position of relative advantage. This is the essence of unified land operations.

Assessment is determination of the progress toward accomplishing a task, creating a condition, or achieving an objective (JP 3-0). Assessment precedes and then occurs during the other activities of the operations process. Assessment involves deliberately comparing forecasted outcomes with actual events to determine the overall effectiveness of force employment. More specifically, assessment helps the commander determine progress toward achieving the desired end state, attaining objectives, and performing tasks.

While units execute numerous tasks throughout the operations process, commanders and staffs have responsibility for special tasks. Commanders and staffs always plan for and coordinate the following continuing activities:

Liaison

Liaison is that contact or intercommunication maintained between elements of military forces or other agencies to ensure mutual understanding and unity of purpose and action (JP 3-08). Most commonly used for establishing and maintaining close communication, liaison continuously enables direct, physical communication between commands. Commanders use liaison during operations to help facilitate communication between organizations, preserve freedom of action, and maintain flexibility.

Information Collection

Information collection is an activity that synchronizes and integrates the planning and employment of sensors and assets as well as the processing, exploitation, and dissemination systems in direct support of current and future operations (FM 3-55). Information collection is the acquisition of information and the provision of this information to processing elements. It integrates the functions of the intelligence and operations staffs focused on answering the commander's critical information requirements. Joint operations refer to this as intelligence, surveillance, and reconnaissance.

Security Operations

Security operations are those operations undertaken by a commander to provide early and accurate warning of enemy operations, to provide the force being protected with time and maneuver space within which to react to the enemy, and to develop the situation to allow the commander to effectively use the protected force (ADRP 3-90). The five forms of security operations are screen, guard, cover, area security, and local security. (See FM 3-90-2 for a detailed discussion of security operations.) Local security for units in the deep area (such as reconnaissance forces) and units supporting a deep operation (for example, units establishing a forward refueling and rearming point or field artillery units in forward positioning areas) is an important consideration. Planners may augment these units by attaching additional security elements (for example, infantry or military police units) or assign tasks to subordinate brigades to provide local security.

Protection

Protection is preservation of the effectiveness and survivability of mission-related military and nonmilitary personnel, equipment, facilities, information, and infrastructure deployed or located within or outside the boundaries of a given operational area (JP 3-0). Protection preserves capability, momentum, and tempo. Commanders and staffs synchronize, integrate, and organize capabilities and resources throughout the operations process to preserve combat power and mitigate the effects of threats and hazards. Protection is both a warfighting function and a continuing activity of the operations process. Commanders ensure the various tasks of protection are integrated into all aspects of operations to safeguard the force, personnel (both combatants and noncombatants), systems, and physical assets.

Terrain Management

Terrain management is the process of allocating terrain by establishing areas of operations, designating assembly areas, and specifying locations for units and activities to deconflict activities that might interfere with each other (ADRP 5-0). Throughout the operations process, commanders manage terrain within the boundaries of their assigned area of operations. Through terrain management, commanders identify and locate units in the area. The operations officer, with support from others in the staff, deconflicts operations, controls movements, and deters fratricide as units execute their missions.

Activities of the Operations Process

Ref: ADP 5-0, The Operations Process (Mar '12), pp. 2 to 6.

The Army's framework for exercising mission command is the operations process -- the major mission command activities performed during operations: planning, preparing, executing, and continuously assessing the operation.

The Operations Process (Underlying Logic)

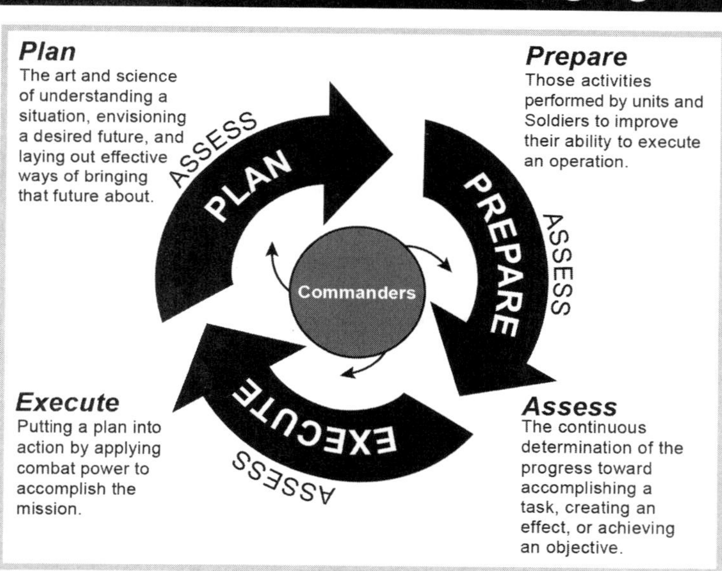

Central idea...

Commanders, supported by their staffs, use the **operations process** to drive the conceptual and detailed planning necessary to understand, visualize, and describe their operational environment; make and articulate decisions; and direct, lead, and assess military operations.

guided by...

Principles

- Commanders drive the operations process
- Apply critical and creative thinking
- Build and maintain situational understanding
- Encourage collaboration and dialogue

Ref: ADP 5-0, The Operations Process, fig. 1, p. iv.

Refer to BSS5: The Battle Staff SMARTbook (Guide to Designing, Planning & Conducting Military Operations) for discussion of the operations process. Commanders, supported by their staffs, use the operations process to drive the conceptual and detailed planning necessary to understand, visualize, and describe their operational environment; make and articulate decisions; and direct, lead, and assess military operations.

Airspace Control

Airspace control is capabilities and procedures used to increase operational effectiveness by promoting the safe, efficient, and flexible use of airspace (JP 3-52). Airspace elements participate in various working groups and provide expertise on methods to maximize airspace use for information collection, targeting, and protection purposes. Throughout the operations process, commanders and staffs must integrate and synchronize forces and warfighting functions within an area of operations (ground and air). Through airspace control, commanders and staffs establish both positive and procedural controls to maximize the use of airspace to facilitate the simultaneity of air-ground operations and joint fires.

II. The Warfighting Functions

To execute operations, commanders conceptualize capabilities in terms of combat power. Combat power has eight elements: leadership, information, mission command, movement and maneuver, intelligence, fires, sustainment, and protection. The Army collectively describes the last six elements as the warfighting functions. Commanders apply combat power through the warfighting functions using leadership and information.

See p. 1-60 for an overview of the Army Warfighting Functions (and chapters five through seven), and pp. 1-61 to 1-64 for discussion of combat power.

III. Army Operational Framework

Army leaders are responsible for clearly articulating their concept of operations in time, space, purpose, and resources. They do this through an operational framework and associated vocabulary. An operational framework is a cognitive tool used to assist commanders and staffs in clearly visualizing and describing the application of combat power in time, space, purpose, and resources in the concept of operations (ADP 1-01). An operational framework establishes an area of geographic and operational responsibility for the commander and provides a way to visualize how the commander will employ forces against the enemy. To understand this framework is to understand the relationship between the area of operations and operations in depth. Proper relationships allow for simultaneous operations and massing of effects against the enemy.

The operational framework has four components. First, commanders are assigned an area of operations for the conduct of operations. Second, a commander can designate a deep, close, and support areas to describe the physical arrangement of forces in time and space. Third, within this area, commanders conduct decisive, shaping, and sustaining operations to articulate the operation in terms of purpose. Finally, commanders designate the main and supporting efforts to designate the shifting prioritization of resources.

See following pages (pp. 1-58 to 1-59) for an overview and further discussion.

Army Planning Methodologies

Ref: ADRP 3-0, Operations (Nov '16), p. 4-3.

Planning is the art and science of understanding a situation, envisioning a desired future, and laying out effective ways of bringing that future about (ADP 5-0). Planning consists of two separate but interrelated components: a conceptual component and a detailed component. Successful planning requires the integration of both these components. Army leaders employ three methodologies for planning: the Army design methodology, the military decisionmaking process, and troop leading procedures.

A. The Army Design Methodology (ADM)

Army design methodology is a methodology for applying critical and creative thinking to understand, visualize, and describe unfamiliar problems and approaches to solving them (ADP 5-0). The Army design methodology is particularly useful as an aid to conceptual thinking about unfamiliar problems. To produce executable plans, commanders integrate the Army design methodology with the detailed planning typically associated with the military decisionmaking process. Commanders who use the Army design methodology may gain a greater understanding of an operational environment and its problems, and then they can visualize an appropriate operational approach. With this greater understanding, commanders can provide a clear commander's intent and concept of operations. Such clarity enables subordinate units and commanders to take initiative.

Army design methodology is iterative and collaborative. As the operations process unfolds, the commander, staff, subordinates, and other partners continue to learn and collaborate to improve their shared understanding. An improved understanding may lead to modifications to the commander's operational approach or an entirely new approach altogether.

B. The Military Decision-making Process (MDMP)

The military decisionmaking process is also an iterative planning methodology. It integrates activities of the commander, staff, subordinate headquarters, and other partners. This integration enables them to understand the situation and mission; develop, analyze, and compare courses of action; decide on a course of action that best accomplishes the mission; and produce an operation order for execution. The military decisionmaking process applies both conceptual and detailed approaches to thinking, but it is most closely associated with detailed planning.

For unfamiliar problems, executable solutions typically require integrating the Army design methodology with the military decisionmaking process. The military decisionmaking process helps leaders apply thoroughness, clarity, sound judgment, logic, and professional knowledge so they understand situations, develop options to solve problems, and reach decisions.

C. Troop Leading Procedures (TLP)

Troop leading procedures are a dynamic process used by small-unit leaders to analyze a mission, develop a plan, and prepare for an operation. Heavily weighted in favor of familiar problems and short time frames, organizations with staffs typically do not employ troop leading procedures. More often, leaders use troop leading procedures to solve tactical problems when working alone or with a small group. For example, a company commander may use the executive officer, first sergeant, fire support officer, supply sergeant, and communications sergeant to assist during troop leading procedures.

Refer to BSS5: The Battle Staff SMARTbook (Guide to Designing, Planning & Conducting Military Operations) for complete discussion of the three Army planning methodologies. Additional related topics include the operations process, integrating processes and continuing activities, plans and orders, mission command, rehearsals and after action reviews, and operational terms and military symbols.

The Army Operational Framework

Ref: ADRP 3-0, Operations (Nov '16), pp. 4-4 to 4-8.

Army leaders are responsible for clearly articulating their concept of operations in time, space, purpose, and resources. They do this through an operational framework and associated vocabulary. An operational framework is a cognitive tool used to assist commanders and staffs in clearly visualizing and describing the application of combat power in time, space, purpose, and resources in the concept of operations (ADP 1-01). An operational framework establishes an area of geographic and operational responsibility for the commander and provides a way to visualize how the commander will employ forces against the enemy.

The operational framework has four components. First, commanders are assigned an area of operations for the conduct of operations. Second, a commander can designate a deep, close, and support areas to describe the physical arrangement of forces in time and space. Third, within this area, commanders conduct decisive, shaping, and sustaining operations to articulate the operation in terms of purpose. Finally, commanders designate the main and supporting efforts to designate the shifting prioritization of resources.

Area of Operations

An area of operations is an operational area defined by the joint force commander for land and maritime forces that should be large enough to accomplish their missions and protect their forces (JP 3-0). For land operations, an area of operations includes subordinate areas of operations assigned by Army commanders to their subordinate echelons as well. In operations, commanders use control measures to assign responsibilities, coordinate fires and maneuver, and control combat operations. A control measure is a means of regulating forces or warfighting functions (ADRP 6-0). One of the most important control measures is the area of operations. The Army commander or joint force land component commander is the supported commander within an area of operations designated by the joint force commander for land operations. Within their areas of operations, commanders integrate and synchronize combat power. To facilitate this integration and synchronization, commanders designate targeting priorities, effects, and timing within their areas of operations.

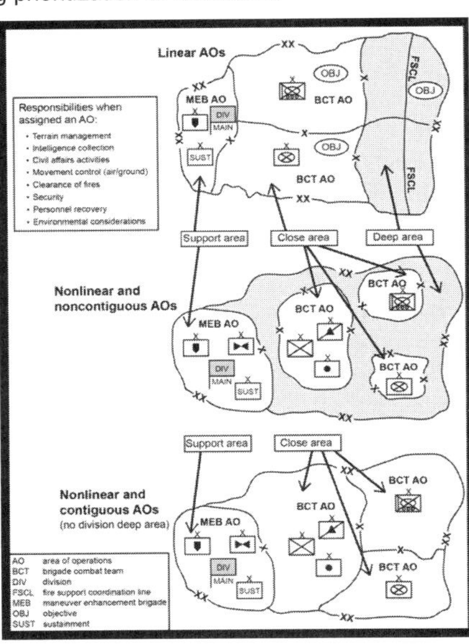

Area of Influence

Commanders consider a unit's area of influence when assigning it an area of operations. An area of influence is a geographical area wherein a commander is directly capable of influencing operations by maneuver or fire support systems normally under the commander's command or control (JP 3-0).

Understanding the area of influence helps the commander and staff plan branches to the current operation in which the force uses capabilities outside the area of operations.

Area of Interest

An area of interest is that area of concern to the commander, including the area of influence, areas adjacent thereto, and extending into enemy territory. This area also includes areas occupied by enemy forces who could jeopardize the accomplishment of the mission (JP 3-0). An area of interest for stability or DSCA tasks may be much larger than that area associated with the offense and defense.

Deep, Close and Support Areas

- A **deep area** is the portion of the commander's area of operations that is not assigned to subordinate units. Operations in the deep area involve efforts to prevent uncommitted enemy forces from being committed in a coherent manner. A commander's deep area generally extends beyond subordinate unit boundaries out to the limits of the commander's designated area of operations. The purpose of operations in the deep area is frequently tied to other events distant in time, space, or both time and space.
- The **close area** is the portion of a commander's area of operations assigned to subordinate maneuver forces. Operations in the close area are operations that are within a subordinate commander's area of operations. Commanders plan to conduct decisive operations using maneuver and fires in the close area, and they position most of the maneuver force within it. Within the close area, depending on the echelon, one unit may conduct the decisive operation while others conduct shaping operations. A close operation requires speed and mobility to rapidly concentrate overwhelming combat power at the critical time and place and to exploit success.
- In operations, a commander may refer to a **support area**. The support area is the portion of the commander's area of operations that is designated to facilitate the positioning, employment, and protection of base sustainment assets required to sustain, enable, and control operations.

Decisive–Shaping–Sustaining Operations

Decisive, shaping, and sustaining operations lend themselves to a broad conceptual orientation.

- The **decisive operation** is the operation that directly accomplishes the mission. It determines the outcome of a major operation, battle, or engagement. The decisive operation is the focal point around which commanders design an entire operation. Multiple subordinate units may be engaged in the same decisive operation. Decisive operations lead directly to the accomplishment of a commander's intent. Commanders typically identify a single decisive operation, but more than one subordinate unit may play a role in a decisive operation.
- A **shaping operation** is an operation that establishes conditions for the decisive operation through effects on the enemy, other actors, and the terrain. Information operations, for example, may integrate Soldier and leader engagement tasks into the operation to reduce tensions between Army units and different ethnic groups through direct contact between Army leaders and local leaders. In combat, synchronizing the effects of aircraft, artillery fires, and obscurants to delay or disrupt repositioning forces illustrates shaping operations. Shaping operations may occur throughout the area of operations and involve any combination of forces and capabilities. Shaping operations set conditions for the success of the decisive operation. Commanders may designate more than one shaping operation.
- A **sustaining operation** is an operation at any echelon that enables the decisive operation or shaping operations by generating and maintaining combat power. Sustaining operations differ from decisive and shaping operations in that they focus internally (on friendly forces) rather than externally (on the enemy or environment).

Main and Supporting Efforts

Commanders designate main and supporting efforts to establish clear priorities of support and resources among subordinate units.

- The **main effort** is a designated subordinate unit whose mission at a given point in time is most critical to overall mission success. It is usually weighted with the preponderance of combat power. Typically, commanders shift the main effort one or more times during execution. Designating a main effort temporarily prioritizes resource allocation. When commanders designate a unit as the main effort, it receives priority of support and resources in order to maximize combat power. Commanders may designate a unit conducting a shaping operation as the main effort until the decisive operation commences. However, the unit with primary responsibility for the decisive operation then becomes the main effort upon the execution of the decisive operation.
- A **supporting effort** is a designated subordinate unit with a mission that supports the success of the main effort. Commanders resource supporting efforts with the minimum assets necessary to accomplish the mission. Forces often realize success of the main effort through success of supporting efforts.

The Six Warfighting Functions

1. Mission Command

The mission command warfighting function is the related tasks and systems that develop and integrate those activities enabling a commander to balance the art of command and the science of control in order to integrate the other warfighting functions. Commanders, assisted by their staffs, integrate numerous processes and activities within the headquarters and across the force as they exercise mission command.

See chap. 2 for further discussion.

2. Movement and Maneuver

The movement and maneuver warfighting function is the related tasks and systems that move and employ forces to achieve a position of relative advantage over the enemy and other threats. Direct fire and close combat are inherent in maneuver. The movement and maneuver warfighting function includes tasks associated with force projection related to gaining a position of advantage over the enemy. Movement is necessary to disperse and displace the force as a whole or in part when maneuvering. Maneuver is the employment of forces in the operational area.

See chap. 3 for further discussion.

3. Intelligence

The intelligence warfighting function is the related tasks and systems that facilitate understanding the enemy, terrain, and civil considerations. This warfighting function includes understanding threats, adversaries, and weather. It synchronizes information collection with the primary tactical tasks of reconnaissance, surveillance, security, and intelligence operations. Intelligence is driven by commanders and is more than just collection. Developing intelligence is a continuous process that involves analyzing information from all sources and conducting operations to develop the situation.

See chap. 4 for further discussion.

4. Fires

The fires warfighting function is the related tasks and systems that provide collective and coordinated use of Army indirect fires, air and missile defense, and joint fires through the targeting process. Army fires systems deliver fires in support of offensive and defensive tasks to create specific lethal and nonlethal effects on a target.

See chap. 5 for further discussion.

5. Sustainment

The sustainment warfighting function is the related tasks and systems that provide support and services to ensure freedom of action, extend operational reach, and prolong endurance. The endurance of Army forces is primarily a function of their sustainment. Sustainment determines the depth and duration of Army operations. It is essential to retaining and exploiting the initiative. Sustainment provides the support necessary to maintain operations until mission accomplishment.

See chap. 6 for further discussion.

6. Protection

The protection warfighting function is the related tasks and systems that preserve the force so the commander can apply maximum combat power to accomplish the mission. Preserving the force includes protecting personnel (combatants and noncombatants) and physical assets of the United States and multinational military and civilian partners, to include the host nation. The protection warfighting function enables the commander to maintain the force's integrity and combat power. Protection determines the degree to which potential threats can disrupt operations and then counters or mitigates those threats.

See chap. 7 for further discussion.

V. Combat Power (and Warfighting Functions)

Ref: ADRP 3-0, Operations (Nov '16), chap. 5.

I. The Elements of Combat Power

Operations executed through simultaneous offensive, defensive, stability, or defense support of civil authorities tasks require continuously generating and applying combat power, often for extended periods. Combat power is the total means of destructive, constructive, and information capabilities that a military unit or formation can apply at a given time. To an Army commander, Army forces generate combat power by converting potential into effective action. Combat power includes all capabilities provided by unified action partners that are integrated and synchronized with the commander's objectives to achieve unity of effort in sustained operations.

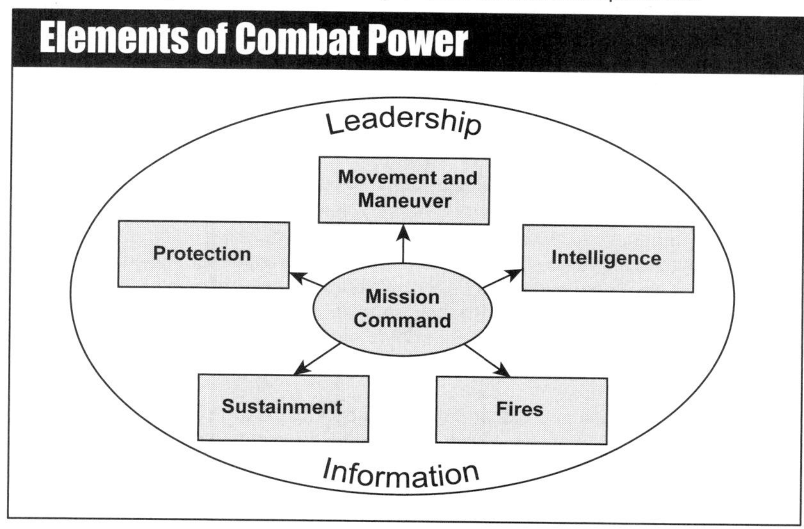

Ref: ADRP 3-0 (2016), fig. 5-1. The elements of combat power.

To execute combined arms operations, commanders conceptualize capabilities in terms of combat power. Combat power has eight elements: leadership, information, mission command, movement and maneuver, intelligence, fires, sustainment, and protection. These elements facilitate Army forces accessing joint and multinational fires and assets. The Army collectively describes the last six elements as the warfighting functions. Commanders apply combat power through the warfighting functions using leadership and information.

Generating and maintaining combat power throughout an operation is essential to success. Factors contributing to generating combat power include employing reserves, rotating committed forces, operating in cyberspace, and focusing joint support. Also, training forces on the conduct of operations, both when deployed and when not deployed, helps commanders to maintain and sustain combat power. Commanders balance the ability to mass lethal and nonlethal effects with the need to deploy and sustain the units that produce those effects. They balance the ability of accomplishing the mission with the ability to project and sustain the force.

Commanders apply leadership through mission command. Leadership is the multiplying and unifying element of combat power. The Army defines leadership as the process of influencing people by providing purpose, direction, and motivation to accomplish the mission and improve the organization (ADP 6-22). An Army commander, by virtue of assumed role or assigned responsibility, inspires and influences people to accomplish organizational goals. Information enables commanders at all levels to make informed decisions on how best to apply combat power. Ultimately, this creates opportunities to achieve definitive results. Knowledge management enables commanders to make informed, timely decisions despite the uncertainty of operations. Information management helps commanders make and disseminate effective decisions faster than the enemy can. Information management uses procedures and information systems to facilitate the collection, processing, storing, displaying, disseminating, and protecting of knowledge and information. Every operation requires complementary tasks of information operations that affect the commander's intent and concept of operations.

Commanders and their units must coordinate what they do, say, and portray. Fundamental to that process is the development of information themes and messages in support of an operation and military action. An information theme is a unifying or dominant idea or image that expresses the purpose for military action. Information themes are tied to objectives, lines of effort, and end state conditions. Information themes are overarching and apply to the capabilities of public affairs, military information support operations, and Soldier and leader engagements. A message is a verbal, written, or electronic communication that supports an information theme focused on a specific actor and in support of a specific action. Commanders employ themes and messages as part of planned activities designed to influence specific foreign audiences for various purposes that support current or planned operations.

Every operation also requires cyberspace electromagnetic activities. Cyberspace electromagnetic activities is the process of planning, integrating, and synchronizing cyberspace and electronic warfare operations in support of unified land operations. (This is also known as CEMA.) Cyberspace operations is the employment of cyberspace capabilities where the primary purpose is to achieve objectives in or through cyberspace (JP 3-0). Electronic warfare is military action involving the use of electromagnetic and directed energy to control the electromagnetic spectrum or to attack the enemy (JP 3-13.1).

Army cyberspace and electronic warfare operations are conducted to seize, retain, and exploit an advantage in cyberspace and the electromagnetic spectrum. These operations support decisive action through the conduct of six core missions: offensive cyberspace operations, defensive cyberspace operations, Department of Defense information network operations, electronic attack, electronic protection, and electronic warfare support. Commanders and their staffs conduct cyberspace electromagnetic activities to project power in and through cyberspace and the electromagnetic spectrum; secure and defend friendly force networks; and protect personnel, facilities, and equipment. Spectrum management operations are a critical enabler of the integration of cyberspace operations and electronic warfare.

II. The Six Warfighting Functions

A warfighting function is a group of tasks and systems united by a common purpose that commanders use to accomplish missions and training objectives. The warfighting functions are the physical means that tactical commanders use to execute operations and accomplish missions assigned by superior tactical- and operational-level commanders. The purpose of warfighting functions is to provide an intellectual organization for common critical capabilities available to commanders and staffs at all echelons and levels of war. Commanders integrate and synchronize these capabilities with other warfighting functions to achieve objectives and accomplish missions.

Each of the six warfighting function has a corresponding chapter within The Army Operations & Doctrine SMARTbook.

The Six Warfighting Functions

1. Mission Command

The mission command warfighting function is the related tasks and systems that develop and integrate those activities enabling a commander to balance the art of command and the science of control in order to integrate the other warfighting functions. Commanders, assisted by their staffs, integrate numerous processes and activities within the headquarters and across the force as they exercise mission command.

See chap. 2 for further discussion.

2. Movement and Maneuver

The movement and maneuver warfighting function is the related tasks and systems that move and employ forces to achieve a position of relative advantage over the enemy and other threats. Direct fire and close combat are inherent in maneuver. The movement and maneuver warfighting function includes tasks associated with force projection related to gaining a position of advantage over the enemy. Movement is necessary to disperse and displace the force as a whole or in part when maneuvering. Maneuver is the employment of forces in the operational area.

See chap. 3 for further discussion.

3. Intelligence

The intelligence warfighting function is the related tasks and systems that facilitate understanding the enemy, terrain, and civil considerations. This warfighting function includes understanding threats, adversaries, and weather. It synchronizes information collection with the primary tactical tasks of reconnaissance, surveillance, security, and intelligence operations. Intelligence is driven by commanders and is more than just collection. Developing intelligence is a continuous process that involves analyzing information from all sources and conducting operations to develop the situation.

See chap. 4 for further discussion.

4. Fires

The fires warfighting function is the related tasks and systems that provide collective and coordinated use of Army indirect fires, air and missile defense, and joint fires through the targeting process. Army fires systems deliver fires in support of offensive and defensive tasks to create specific lethal and nonlethal effects on a target.

See chap. 5 for further discussion.

5. Sustainment

The sustainment warfighting function is the related tasks and systems that provide support and services to ensure freedom of action, extend operational reach, and prolong endurance. The endurance of Army forces is primarily a function of their sustainment. Sustainment determines the depth and duration of Army operations. It is essential to retaining and exploiting the initiative. Sustainment provides the support necessary to maintain operations until mission accomplishment.

See chap. 6 for further discussion.

6. Protection

The protection warfighting function is the related tasks and systems that preserve the force so the commander can apply maximum combat power to accomplish the mission. Preserving the force includes protecting personnel (combatants and noncombatants) and physical assets of the United States and multinational military and civilian partners, to include the host nation. The protection warfighting function enables the commander to maintain the force's integrity and combat power. Protection determines the degree to which potential threats can disrupt operations and then counters or mitigates those threats.

See chap. 7 for further discussion.

III. Organizing Combat Power

Ref: ADRP 3-0, Operations (Nov '16), pp. 5-7 to 5-9.

Commanders employ three means to organize combat power: force tailoring, task-organizing, and mutual support.

A. Force Tailoring

Force tailoring is the process of determining the right mix of forces and the sequence of their deployment in support of a joint force commander. It involves selecting the right force structure for a joint operation from available units within a combatant command or from the Army force pool. Commanders then sequence selected forces into the area of operations as part of force projection. Joint force commanders request and receive forces for each campaign phase, adjusting the quantity of Service component forces to match the weight of effort required. Army Service component commanders tailor Army forces to meet land force requirements determined by joint force commanders. Army Service component commanders also recommend forces and a deployment sequence to meet those requirements. Force tailoring is continuous. As new forces rotate into the area of operations, forces with excess capabilities return to the supporting combatant and Army Service component commands. Force tailoring is continuous. As new forces rotate into the area of operations, forces with excess capabilities return to the supporting combatant and Army Service component commands.

B. Task Organization

Task-organizing is the act of designing a force, support staff, or sustainment package of specific size and composition to meet a unique task or mission. Characteristics to examine when task-organizing the force include, but are not limited to, training, experience, equipment, sustainability, operational environment, enemy threat, and mobility. Task-organizing includes allocating assets to subordinate commanders and establishing their command and support relationships. It occurs within a tailored force package as commanders organize subordinate units for specific missions employing doctrinal command and support relationships. As task-organizing continues, commanders reorganize units for subsequent missions. The ability of Army forces to task-organize gives them extraordinary agility. It lets commanders configure their units to best use available resources. It also allows Army forces to match unit capabilities to the priority assigned to offensive, defensive, and stability or defense support of civil authorities tasks.

C. Mutual Support

Commanders consider mutual support when task-organizing forces, assigning areas of operations, and positioning units. Mutual support is that support which units render each other against an enemy, because of their assigned tasks, their position relative to each other and to the enemy, and their inherent capabilities (JP 3-31). Understanding mutual support and accepting risk during operations are fundamental to the art of tactics. In Army doctrine, mutual support is a planning consideration related to force disposition, not a command relationship. Mutual support has two aspects—supporting range and supporting distance. When friendly forces are static, supporting range equals supporting distance.

Supporting range is the distance one unit may be geographically separated from a second unit yet remain within the maximum range of the second unit's weapons systems. It depends on available weapons systems and is normally the maximum range of the supporting unit's indirect fire weapons. For small units (such as squads, sections, and platoons), it is the distance between two units that their direct fires can cover effectively. Visibility may limit the supporting range. If one unit cannot effectively or safely fire in support

Supporting distance is the distance between two units that can be traveled in time for one to come to the aid of the other and prevent its defeat by an enemy or ensure it regains control of a civil situation. The following factors affect supporting distance: terrain and mobility, distance, enemy capabilities, friendly capabilities, and reaction time.

Mission Command Warfighting Function

Ref: ADP 6-0, Mission Command (May '12) and ADRP 3-0, Operations (Nov '16), p. 5-3.

The mission command warfighting function is the related tasks and systems that develop and integrate those activities enabling a commander to balance the art of command and the science of control in order to integrate the other warfighting functions. Commanders, assisted by their staffs, integrate numerous processes and activities within the headquarters and across the force as they exercise mission command.

Mission command encourages the greatest possible freedom of action from subordinates. While the commander remains the central figure, mission command enables subordinates to develop the situation. Through exercising disciplined initiative in dynamic conditions within the commander's intent, subordinates adapt and act decisively. Mission command creates a shared understanding of an operational environment and the commander's intent to establish the appropriate degree of control.

The art of command is the creative and skillful exercise of authority through decision making and leadership. As commanders exercise the art of command, they—

- Drive the operations process through their activities of understanding, visualizing, describing, directing, leading, and assessing operations.
- Develop teams, both within their own organizations and with unified action partners.
- Inform and influence audiences, both inside and outside their organizations.

The commander leads the staff's tasks under the science of control. The science of control consists of systems and procedures to improve the commander's understanding and to support accomplishing missions. The four primary staff tasks are—

- Conduct the operations process: plan, prepare, execute, and assess.
- Conduct knowledge management, information management, and foreign disclosure.
- Synchronize information-related capabilities.
- Conduct cyberspace electromagnetic activities.

In addition to mission command warfighting function tasks, six additional tasks reside within the mission command warfighting function. These tasks are—

- Conduct civil affairs operations.
- Conduct military deception.
- Install, operate, and maintain the Department of Defense information network.
- Conduct airspace control.
- Conduct information protection.
- Plan and conduct space activities.

Commanders and staffs work with unified action partners to perform mission command warfighting function tasks that contribute to mission accomplishment. In addition to the principles of mission command in ADRP 6-0, commanders consider the following when performing mission command warfighting function tasks:

- Clear and established command and support relationships that are understood by commanders, staffs, and subordinate units will help in the exercise of mission command.
- The commanders' presence is vital to understanding intent and purpose.
- Effective collaboration enhances mission command by sharing knowledge and aiding the creation of shared understanding. This is especially true when sharing information with multinational partners through the foreign disclosure process.

I. The Exercise of Mission Command

Ref: ADRP 6-0, Mission Command (May '12), pp. 1-2 to 1-5.

To function effectively and have the greatest chance for mission accomplishment, commanders, supported by their staffs, exercise mission command throughout the conduct of operations. In this discussion, the "exercise of mission command" refers to an overarching idea that unifies the mission command philosophy of command and the mission command war fighting function. The exercise of mission command encompasses how Army commanders and staffs apply the foundational mission command philosophy together with the mission command war fighting function, guided by the principles of mission command.

An effective approach to mission command must be comprehensive, without being rigid. Military operations are affected by human interactions and as a whole defy orderly, efficient, and precise control. People are the basis of all military organizations. Commanders understand that some decisions must be made quickly and are better made at the point of action. Mission command concentrates on the objectives of an operation, not how to achieve it. Commanders provide subordinates with their intent, the purpose of the operation, the key tasks, the desired end state, and resources. Subordinates then exercise disciplined initiative to respond to unanticipated problems. Mission command is based on mutual trust and shared understanding and purpose. It demands every Soldier be prepared to assume responsibility, maintain unity of effort, take prudent action, and act resourcefully within the commander's intent.

Mission Command as a Philosophy

As the Army's philosophy of command, mission command emphasizes that command is essentially a human endeavor. Successful commanders understand that their leadership directs the development of teams and helps to establish mutual trust and shared understanding throughout the force. Commanders provide a clear intent to their forces that guides subordinates' actions while promoting freedom of action and initiative. Subordinates, by understanding the commander's intent and the overall common objective, are then able to adapt to rapidly changing situations and exploit fleeting opportunities. They are given the latitude to accomplish assigned tasks in a manner that best fits the situation. Subordinates understand that they have an obligation to act and synchronize their actions with the rest of the force. Likewise, commanders influence the situation and provide direction and guidance while synchronizing their own operations. They encourage subordinates to take action, and they accept prudent risks to create opportunity and to seize the initiative. Commanders at all levels need education, rigorous training, and experience to apply these principles effectively. Mission command operates more on self-discipline than imposed discipline.

Mission Command as a Warfighting Function

Mission command—as a warfighting function—assists commanders in balancing the art of command with the science of control, while emphasizing the human aspects of mission command. A warfighting function is a group of tasks and systems (people, organizations, information, and processes) united by a common purpose that commanders use to accomplish missions (ADRP 3-0). The mission command war fighting function consists of the mission command war fighting function tasks and the mission command system.

Mission Command System

Commanders need support to exercise mission command effectively. At every echelon of command, each commander establishes a mission command system—the arrangement of personnel, networks, information systems, processes and procedures, and facilities and equipment that enable commanders to conduct operations (ADP 6-0). Commanders organize the five components of their mission command system to support decision making and facilitate communication. The most important of these components is personnel.

The Exercise of Mission Command

Unified Land Operations

How the Army seizes, retains, and exploits the initiative to gain and maintain a position of relative advantage in sustained land operations through simultaneous offensive, defensive, and stability operations in order to prevent or deter conflict, prevail in war, and create the conditions for favorable conflict resolution.

One of the foundations is...

Nature of Operations

Military operations are human endeavors.

They are contests of wills characterized by continuous and mutual adaptation by all participants.

Army forces conduct operations in complex, ever-changing, and uncertain operational environments.

To account for this, the Army exercises....

Mission Command Philosophy

Exercise of authority and direction by the commander using mission orders to enable disciplined initiative within the commander's intent to empower agile and adaptive leaders in the conduct of unified land operations.

Guided by the principles of...

- Build cohesive teams through mutual trust
- Create shared understanding
- Provide a clear commander's intent
- Exercise disciplined initiative
- Use mission orders
- Accept prudent risk

The principles of mission command assist commanders and staff in balancing the ***art of command*** *with the* ***science of control****.*

Executed through the...

Mission Command Warfighting Function

The related tasks and systems that develop and integrate those activities enabling a commander to balance the art of command and the science of control in order to integrate the other warfighting functions.

A series of mutually supported tasks...

Commander Tasks:

- Drive the operations process through the activities of understand, visualize, describe, direct, lead, and assess
- Develop teams, both within their own organizations and with unified action partners
- Inform and influence audiences, inside and outside their organizations

Leads →

← **Supports**

Staff Tasks:

- Conduct the operations process (plan, prepare, execute, and assess)
- Conduct knowledge management and information management
- Conduct inform and influence activities
- Conduct cyber electromagnetic activities

Additional Tasks:

- Conduct military deception
- Conduct civil affairs operations
- Conduct airspace control
- Install, operate, and maintain the network
- Conduct information protection

Enabled by a system...

Mission Command System:

- Personnel
- Networks
- Information systems
- Processes and procedures
- Facilities and equipment

Together, the mission command philosophy and warfighting function guide, integrate, and synchronize Army forces throughout the conduct of unified land operations.

Ref: ADRP 6-0, Mission Command, fig. 1-1, p. 1-3.

ADRP 6-0, Mission Command (May '12)

Ref: ADRP 6-0, Mission Command (May '12), introduction.

Historically, military commanders have employed variations of two basic concepts of command: mission command and detailed command. While some have favored detailed command, the nature of operations and the patterns of military history point to the advantages of mission command. Mission command has been the Army's preferred style for exercising command since the 1980s. The concept traces its roots back to the German concept of Auftragstaktik, which translates roughly to mission-type tactics. Auftragstaktik held all German commissioned and noncommissioned officers duty bound to do whatever the situation required, as they personally saw it. Understanding and achieving the broader purpose of a task was the central idea behind this style of command. Commanders expected subordinates to act when opportunities arose.

Army Doctrine Reference Publication (ADRP) 6-0 develops the concept of mission command to help Army forces function effectively and accomplish missions. This publication expands on the principles of mission command found in ADP 6-0. ADRP 6-0 updates mission command doctrine to incorporate the Army's operational concept of unified land operations, found in ADP 3-0. ADRP 6-0 remains generally consistent with the doctrine in the 2011 edition of Field Manual (FM) 6-0, Mission Command, on key topics, while adopting updated terminology and concepts as necessary. These topics include mission command as a foundation of unified land operations and updated mission command warfighting function tasks.

The significant change from FM 6-0, 2011, is the restructuring of doctrinal information. The principles of the mission command philosophy of command and the mission command warfighting function are now found in ADP 6-0 and ADRP 6-0. Under the restructuring plan, several new FMs will address the specific tactics and procedures associated with mission command.

Chapter 1

Chapter 1 discusses the exercise of mission command. First, it describes the general nature of military operations, including the complex challenges for which mission command doctrine must provide solutions. Then it discusses mission command as a foundation of the Army's operational concept, unified land operations. Next, it explains the Army's approach to the exercise of mission command, including an introduction to mission command as a philosophy of command and as a warfighting function.

Chapter 2

Chapter 2 addresses the mission command philosophy of command in greater depth. First, it discusses the principles of mission command that guide commanders and staffs. Next, it elaborates on the art of command, including authority, decision making, and leadership. Then it explains the science of control, including information, communication, structure, and degree of control. It concludes with a short discussion of how commanders apply the philosophy of mission command to balance the art of command with the science of control.

Chapter 3

Chapter 3 addresses the mission command warfighting function in greater depth. First, it defines the mission command warfighting function and describes its purpose. Next, it discusses the tasks of the mission command warfighting function, including commander tasks, staff tasks, and additional tasks. The chapter concludes with a discussion of the commander's mission command system, including personnel, networks, information systems, processes and procedures, and facilities and equipment.

Significant Changes

This mission command doctrine makes some significant changes from FM 6-0. Changes include revising the mission command warfighting function tasks. The commander tasks are:

- Drive the operations process through their activities of understanding, visualizing, describing, directing, leading, and assessing operations
- Develop teams, both within their own organizations and with joint, interagency, and multinational partners
- Inform and influence audiences, inside and outside their organizations

The staff tasks are:

- Conduct the operations process: plan, prepare, execute, and assess
- Conduct knowledge management and information management
- Conduct inform and influence activities
- Conduct cyber electromagnetic activities

The additional tasks are:

- Conduct military deception
- Conduct civil affairs operations
- Install, operate, and maintain the network
- Conduct airspace control
- Conduct information protection

ADRP 6-0 provides a starting point for the exercise of mission command. It establishes how commanders, supported by their staffs, apply the foundational mission command philosophy with the mission command warfighting function to lead forces toward mission accomplishment. The doctrine in this publication is a guide for action rather than a set of fixed rules. In operations, effective leaders recognize when and where doctrine, training, or even their experience, no longer fits the situation, and they adapt accordingly.

New, Modified and Rescinded Terms

Based on current doctrinal changes, certain terms for which ADP 6-0 and ADRP 6-0 are proponent have been modified. The glossary contains defined terms. See introductory table-1 below for specific term changes.

Introductory Table-1. Modified Army terms

Term	Remarks
art of command	Modifies the definition.
ASCOPE	Retained as an acronym; no longer formally defined.
commander's intent	Adopted the joint definition; Army definition no longer used.
exceptional information	No longer formally defined.
information requirement	Modifies the definition. No longer plural.
knowledge management	Modifies the definition.
METT-TC	Retained as an acronym; no longer a formally defined term.
mission command	Modifies the definition.
OAKOC	Retained as an acronym; no longer a formally defined term.
PMESII-PT	Retained as an acronym; no longer a formally defined term.
science of control	Modifies the definition.

II. Mission Command Warfighting Function Tasks

The mission command warfighting function tasks define what commanders and staffs do to integrate the other warfighting functions. They include mutually supporting commander, staff, and additional tasks. The commander leads the staff tasks, and the staff tasks fully support the commander in executing the commander tasks. Commanders, assisted by their staffs, integrate numerous processes and activities within the headquarters and across the force as they exercise mission command.

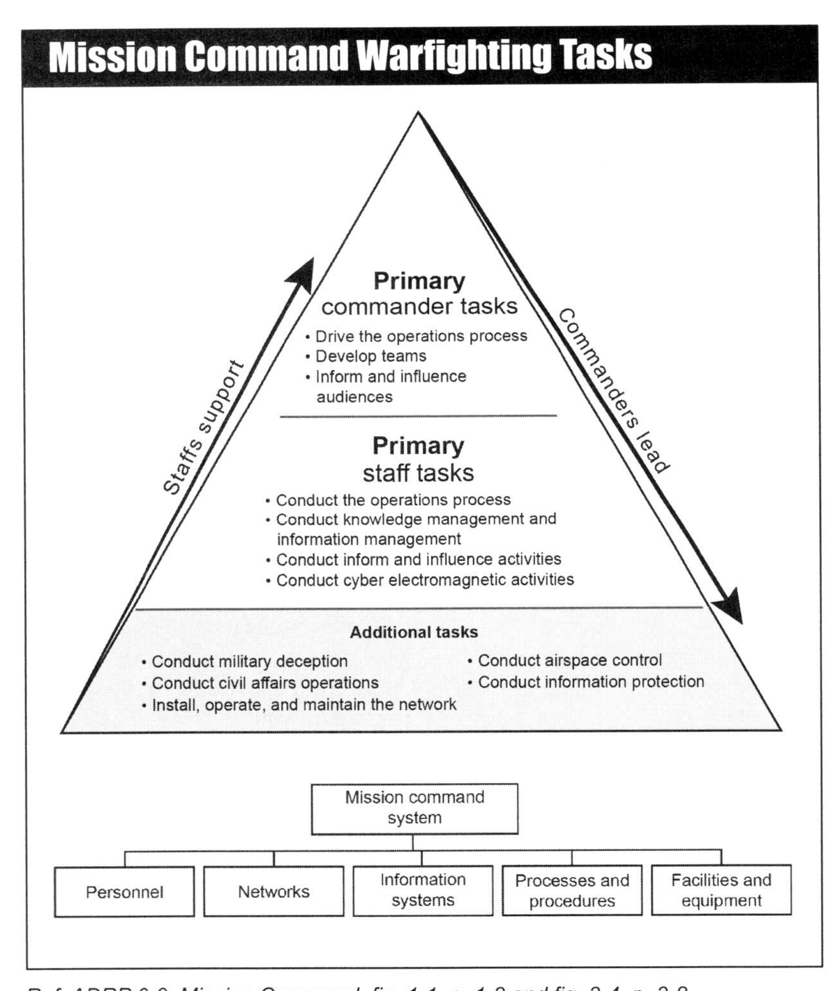

Ref: ADRP 6-0, Mission Command, fig. 1-1, p. 1-3 and fig. 3-4, p. 3-8.

III. Mission Command System

Ref: ADP 6-0, Mission Command (May '12), pp. 11 to 12.

At every echelon of command, each commander establishes a mission command system—the arrangement of personnel, networks, information systems, processes and procedures, and facilities and equipment that enable commanders to conduct operations. Commanders organize their mission command system to support decision making and facilitate communication.

1. Personnel

A commander's mission command system begins with people. Therefore, commanders base their mission command system on human characteristics and abilities more than on equipment and procedures. An effective mission command system requires trained personnel; commanders must not underestimate the importance of providing training. Key personnel dedicated to mission command include seconds in command, command sergeants major, and staff.

2. Networks

Social and technical networks enable commanders to communicate information and control forces, leading to successful operations. Generally, a network is a grouping of people or things interconnected for a purpose. Commanders develop and leverage various social networks—individuals and organizations interconnected by a common interest—to exchange information and ideas, build teams, and promote unity of effort. Technical networks also connect people and allow sharing of resources and information. For example, LandWarNet (the Army's portion of the Global Information Grid) is a technical network. It encompasses all Army information management systems and information systems that collect, process, store, display, disseminate, and protect information worldwide.

3. Information Systems

Commanders determine their information requirements and focus their staffs and organizations on using information systems to meet these requirements. An information system consists of equipment that collects, processes, stores, displays, and disseminates information. This includes computers—hardware and software—and communications, as well as policies and procedures for their use. Staffs use information systems to process, store, and disseminate information according to the commander's priorities. These capabilities relieve the staff of handling routine data. Information systems—especially when merged into a single, integrated network—enable extensive information sharing.

4. Processes and Procedures

Processes and procedures help commanders organize the activities within the headquarters and throughout the force. Processes and procedures govern actions within a mission command system to make it more effective and efficient. A process is a series of actions directed to an end state. One example is the military decision making process. Procedures are standard, detailed steps, often used by staffs, which describe how to perform specific tasks to achieve the desired end state. One example is a standard operating procedure. Adhering to processes and procedures minimizes confusion, misunderstanding, and hesitation as commanders make frequent, rapid decisions to meet operational requirements.

5. Facilities and Equipment

Facilities and equipment include command posts, signal nodes, and all mission command support equipment, excluding information systems. A facility is a structure or location that provides a work environment and shelter for the personnel within the mission command system. Facilities range from a command post composed of vehicles and tentage to hardened buildings. Examples of equipment needed to sustain a mission command system include vehicles, generators, and lighting.

A. Commander Tasks

Ref: ADRP 6-0, Mission Command (May '12), pp. 3-2 to 3-5.

Commanders are the central figures in mission command. Throughout operations, commanders balance their time between leading their staffs through the operations process and providing purpose, direction, and motivation to subordinate commanders and Soldiers. Commanders encourage disciplined initiative through a clear commander's intent while providing enough direction to integrate and synchronize the actions of the force at the decisive place and time. Commanders create positive command climates that foster mutual trust and shared understanding within their command and with unified action partners. The commander tasks are:

- Drive the operations process through their activities of understanding, visualizing, describing, directing, leading, and assessing operations
- Develop teams, both within their own organizations and with joint, interagency, and multinational partners
- Inform and influence audiences, inside and outside their organizations

1. Drive the Operations Process through Understanding, Visualizing, Describing, Directing, Leading and Assessing

The Army's overarching framework for exercising mission command is the operations process—the major mission command activities performed during operations: planning, preparing, executing, and continuously assessing the operation (ADP 5-0). Commanders, assisted by their staffs, integrate activities within the headquarters and across the force, as they exercise mission command. Commanders drive the operations process through the activities of understanding, visualizing, describing, directing, leading, and assessing operations. Throughout the operations process, commanders apply leadership to translate decisions into action. They do this by synchronizing forces and war fighting functions in time, space, and purpose, to accomplish the mission.

Commanders understand, visualize, describe, direct, lead, and assess throughout operations. Commanders continuously develop, test, and update their understanding throughout the conduct of operations. They actively collaborate with other commanders, the staff, and unified action partners, to create a shared understanding. As commanders begin to develop an understanding of the operational environment, they start visualizing the operation's end state and potential solutions to solve problems. After commanders visualize an operation, they describe it to their staffs and subordinates. This description facilitates shared understanding of the situation, mission, and intent. Based on this understanding, commanders make decisions and direct action throughout the operations process. Commanders use the operations process to lead Soldiers and forces by providing direction and guidance. Commanders assess operations continuously to better understand current conditions and determine how operations are progressing. Commanders incorporate the assessments of the staff, subordinate commanders, and unified action partners into their personal assessment of the situation. Based on their assessment, commanders modify plans and orders to better accomplish the mission. If their assessment reveals a significant variance from their original commander's visualization, commanders reframe the problem and develop a new operational approach.

The commander's focus on understanding, visualizing, describing, directing, leading, or assessing throughout operations varies during different operations process activities. For example, during planning commanders focus more on understanding, visualizing, and describing while directing, leading, and assessing. During execution, commanders often focus more on directing, leading, and assessing—while improving their understanding and modifying their visualization as needed. (See ADRP 5-0 for a detailed discussion of the operations process.)

2. Develop Teams within Their Own Organizations and with Joint, Interagency, and Multinational Partners

Successful mission command relies on teams and teamwork. A team is a group of individuals or organizations that work together toward a common goal. Teams range from informal groups of peers to structured, hierarchical groups. Teams may form in advance or gradually as the situation develops.

Commanders cannot always rely on habitual or pre-established relationships, and they must be able to build teams. In some cases, commanders must overcome biases that inhibit trust and cooperation. Commanders use their team-building skills to form effective teams and foster unity of effort. Successful team builders establish mutual trust, shared understanding, and cohesion. They instill a supportive attitude and a sense of responsibility among the team, and they appropriately distribute authority. Additionally, commanders expect to join pre-existing teams as host-nation and civilian organizations often are present before military forces arrive and remain long after forces leave. Overall, team building is a worthwhile investment because good teams complete missions on time with given resources and a minimum of wasted effort.

Effective teams synchronize individual efforts toward a common goal. They promote the exchange of ideas, creativity, and the development of collective solutions. They collaborate across the team to develop and improve processes. The variety of knowledge, talent, expertise, and resources in a team can produce better understanding and alternative options faster than one individual can achieve alone. Successful mission command fosters a greater understanding of the operational environment and solution development through teamwork. This results in teams that:

- Are adaptive and anticipate transitions
- Accept risks to create opportunities
- Influence friendly, neutrals, adversaries, enemies, and unified action partners

The ultimate team outcome is successful mission accomplishment.

3. Inform and Influence Audiences, Inside and Outside Their Organizations

Commanders use inform and influence activities to ensure actions, themes, and messages compliment and reinforce each other to accomplish objectives. Inform and influence activities are the integration of designated information-related capabilities in order to synchronize themes, messages, and actions with operations to inform United States and global audiences, influence foreign audiences, and affect adversary and enemy decision making (ADRP 3-0). An information theme is a unifying or dominant idea or image that expresses the purposes for an action. A message is a verbal, written, or electronic communication that supports an information theme focused on an audience. It supports a specific action or objective.

Actions, themes, and messages are inextricably linked. Commanders use inform and influence activities to ensure actions, themes, and messages compliment and reinforce each other and support operational objectives. They keep in mind that every action implies a message, and they avoid apparently contradictory actions, themes, or messages.

Throughout operations, commanders inform and influence audiences, both inside and outside of their organizations. Some commanders inform and influence through Soldier and leader engagements, conducting radio programs, command information programs, operations briefs, and unit Web site posts. Inform and influence activities assist commanders in creating shared understanding and purpose both inside and outside their organizations and among all affected audiences. This supports the commander's operational goals by synchronizing words and actions. (See Army doctrine on inform and influence activities for more information.)

B. Staff Tasks

Ref: ADRP 6-0, Mission Command (May '12), pp. 3-5 to 3-7.

The staff supports the commander and subordinate commanders in understanding situations, decision making, and implementing decisions throughout the conduct of operations. The staff does this through the four staff tasks:

- Conduct the operations process: plan, prepare, execute, and assess
- Conduct knowledge management and information management
- Conduct inform and influence activities
- Conduct cyber electromagnetic activities

1. Conduct the Operations Process: Plan, Prepare, Execute, and Assess

The operations process consists of the major activities of mission command conducted during operations: planning, preparing, executing and assessing operations. Commanders drive the operations process, while remaining focused on the major aspects of operations. Staffs conduct the operations process; they assist commanders in the details of planning, preparing, executing, and assessing.

Upon receipt of a mission, planning starts a cycle of the operations process that results in a plan or operation order to guide the force during execution. After the completion of the initial order, however, the commander and staff revise the plan based on changing circumstances. While units and Soldiers always prepare for potential operations, preparing for a specific operation begins during planning and continues through execution. Execution puts plans into action. During execution, staffs focus on concerted action to seize and retain operational initiative, build and maintain momentum, and exploit success. As the unit executes the current operation, the commander and staff are planning future operations based on assessments of progress. Assessment is continuous and affects the other three activities. Subordinate units of the same command may be conducting different operations process activities.

The continuous nature of the operations process allows commanders and staffs to make adjustments enabling agile and adaptive forces. Commanders, assisted by their staffs, integrate activities within the headquarters and across the force as they exercise mission command. Throughout the operations process, they develop an understanding and appreciation of their operational environment. They formulate a plan and provide purpose, direction, and guidance to the entire force. Commanders then adjust operations as changes to the operational environment occur. It is this cycle that enables commanders and forces to seize, retain, and exploit the initiative to gain a position of relative advantage over the enemy. (See ADRP 5-0 for a detailed explanation of the operations process.)

2. Conduct Knowledge Management and Information Management

Knowledge management is the process of enabling knowledge flow to enhance shared understanding, learning, and decision making. Knowledge management facilitates the transfer of knowledge between staffs, commanders, and forces. Knowledge management aligns people, processes, and tools within an organization to distribute knowledge and promote understanding. Commanders apply judgment to the information and knowledge provided to understand their operational environment and discern operational advantages. (See Army doctrine on knowledge management for more information.)

Commanders are constantly seeking to understand their operational environment in order to facilitate decision making. The staff uses information management to assist the commander in building and maintaining understanding. Information management is the science of using procedures and information systems to collect, process, store, display, disseminate, and protect data, information, and knowledge products. The staff studies the operational environment, identifies information gaps, and helps the commander develop and answer information requirements. Collected data are then organized and processed into information for development into and use as knowledge. Information becomes knowledge, and that knowledge also becomes a source of information. As this happens, new knowledge is created, shared, and acted upon. During the course of operations, knowledge constantly flows between individuals and organizations. Staffs help manage this constant cycle of exchange. (See Army doctrine on information management for more information.)

Staffs use information and knowledge management to provide commanders the information they need to create and maintain their understanding and make effective decisions. Information is disseminated, stored, and retrieved according to established information management practices. Information management practices allow all involved to build on each other's knowledge to further develop a shared understanding across the force. Knowledge management practices enable the transfer of knowledge between individuals and organizations. Knowledge transfer occurs both formally—through established processes and procedures—and informally—through collaboration and dialogue. Participants exchange perspectives along with information. They question each other's assumptions and exchange ideas. In this way, they create and maintain shared understanding and develop new approaches. Teams benefit, and forces enhance integration and synchronization.

3. Conduct Inform and Influence Activities

Throughout the operations process, staffs assist commanders in developing themes and messages to inform domestic audiences and influence foreign friendly, neutral, adversary, and enemy populations. They coordinate the activities and operations of information-related capabilities to integrate and synchronize all actions and messages into a cohesive effort. Staffs assist the commander in employing those capabilities to inform and influence foreign target audiences to shape the operational environment, exploit success, and protect friendly vulnerabilities. (See Army doctrine on inform and influence activities for more information.)

All assets and capabilities at a commander's disposal have the capacity to inform and influence to varying degrees. Some examples of resources commanders may use include combat camera, counter intelligence, maneuver and fires, and network operations. The primary information-related capabilities of inform and influence activities are:

- Public affairs
- Military information support operations
- Soldier and leader engagement

4. Conduct Cyber Electromagnetic Activities

Cyber electromagnetic activities are activities leveraged to seize, retain, and exploit an advantage over adversaries and enemies in both cyberspace and the electromagnetic spectrum, while simultaneously denying and degrading adversary and enemy use of the same and protecting the mission command system (ADRP 3-0).

To succeed in unified land operations, cyber electromagnetic activities must be integrated and synchronized across all command echelons and war fighting functions. Commanders, supported by their staff, integrate cyberspace operations, electromagnetic spectrum operations and electronic warfare. The electronic warfare working group or similar staff organization coordinates cyber electromagnetic activities. These activities may employ the same technologies, capabilities, and enablers to accomplish assigned tasks. Cyber electromagnetic activities also enable inform and influence activities, signals intelligence, and network operations. (See Army doctrine on cyber electromagnetic activities for more information.)

C. Additional Tasks

Ref: ADRP 6-0, Mission Command (May '12), pp. 3-5 to 3-7.

Commanders, assisted by their staffs, integrate five additional mission command warfighting function tasks. These are:

1. Conduct Military Deception

Commanders may use military deception to establish conditions favorable to success. Military deception is actions executed to deliberately mislead adversary military decision makers as to friendly military capabilities, intentions, and operations, thereby causing the adversary to take specific actions (or inactions) that will contribute to the accomplishment of the friendly mission (JP 3-13.4). Commanders use military deception to confuse an adversary, to deter hostile actions, and to increase the potential of successful friendly actions. It targets adversary decision makers and affects their decision making process. Military deception can enhance the likelihood of success by causing an adversary to take (or not to take) specific actions, not just to believe certain things.

2. Conduct Civil Affairs Operations

Commanders use civil affairs operations to engage the civil component of the operational environment. Military forces interact with the civilian populace during operations. A supportive civilian population can provide resources and information that facilitate friendly operations. A hostile civilian population can threaten the operations of deployed friendly forces. Commanders use civil affairs operations to enhance the relationship between military forces and civil authorities in areas where military forces are present. Civil affairs operations are usually conducted by civil affairs forces due to the complexities and demands for specialized capabilities. (See Army doctrine on civil affairs for more information.)

3. Install, Operate, and Maintain the Network

Commanders rely on technical networks to communicate information and control forces. Technical networks facilitate information flow by connecting information users and information producers and enable effective and efficient information flow. Technical networks help shape and influence operations by getting information to decision makers, with adequate context, enabling them to make better decisions. They also assist commanders in projecting their decisions across the force. (See Army doctrine on network operations for more information.)

4. Conduct Airspace Control

Commanders conduct airspace control to increase combat effectiveness. Airspace control promotes the safe, efficient, and flexible use of airspace with minimum restraint on airspace users, and includes the coordination, integration, and regulation of airspace to increase operational effectiveness. Effective airspace control reduces the risk of fratricide, enhances air defense operations, and permits greater flexibility of operations. (See Army doctrine on airspace control for more information.)

5. Conduct Information Protection

Information protection is active or passive measures used to safeguard and defend friendly information and information systems. It denies enemies, adversaries, and others the opportunity to exploit friendly information and information systems for their own purposes. It is accomplished through active and passive means designed to help protect the force and preserve combat power.

I. Mission Command Philosophy of Command

Ref: ADRP 6-0, Mission Command (May '12), chap. 2.

I. Principles of Mission Command

The mission command philosophy helps commanders counter the uncertainty of operations by reducing the amount of certainty needed to act. Commanders understand that some decisions must be made quickly and are better made at the point of action. Mission command is based on mutual trust and a shared understanding and purpose between commanders, subordinates, staffs, and unified action partners. It requires every Soldier to be prepared to assume responsibility, maintain unity of effort, take prudent action, and act resourcefully within the commander's intent.

Through leadership, commanders build teams. They develop and maintain mutual trust and a shared understanding throughout the force and with unified action partners. Commanders understand that subordinates and staffs require resources and a clear intent to guide their actions. They allow them the freedom of action to exercise disciplined initiative to adapt to changing situations. Because mission command decentralizes decision making authority and grants subordinates' significant freedom of action, it demands more of commanders at all levels and requires rigorous training and education. In exercising mission command, commanders are guided by six principles:

Principles of Mission Command

1. **Build Cohesive Teams Through Mutual Trust**
2. **Create Shared Understanding**
3. **Provide a Clear Commander's Intent**
4. **Exercise Disciplined Initiative**
5. **Use Mission Orders**
6. **Accept Prudent Risk**

Ref: ADP 6-0, Misson Command, pp. 2 to 5.

See following pages (pp. 2-14 to 2-15) for further discussion of the principles of mission command.

Principles of Mission Command

Ref: ADP 6-0, Mission Command (May '12), pp. 2 to 5.

1. Build Cohesive Teams Through Mutual Trust

Mutual trust is shared confidence among commanders, subordinates, and partners. Effective commanders build cohesive teams in an environment of mutual trust. There are few shortcuts to gaining the trust of others. Trust takes time and must be earned. Commanders earn trust by upholding the Army values and exercising leadership, consistent with the Army's leadership principles. (See the Army leadership publication for details on the leadership principles.)

Effective commanders build teams within their own organizations and with unified action partners through interpersonal relationships. Unified action partners are those military forces, governmental and nongovernmental organizations, and elements of the private sector with whom Army forces plan, coordinate, synchronize, and integrate during the conduct of operations (ADRP 3-0). Uniting all the diverse capabilities necessary to achieve success in operations requires collaborative and cooperative efforts that focus those capabilities toward a common goal. Where military forces typically demand unity of command, a challenge for building teams with unified action partners is to forge unity of effort. Unity of effort is coordination and cooperation toward common objectives, even if the participants are not necessarily part of the same command or organization—the product of successful unified action (JP 1).

2. Create Shared Understanding

A defining challenge for commanders and staffs is creating shared understanding of their operational environment, their operation's purpose, its problems, and approaches to solving them. Shared understanding and purpose form the basis for unity of effort and trust. Commanders and staffs actively build and maintain shared understanding within the force and with unified action partners by maintaining collaboration and dialogue throughout the operations process (planning, preparation, execution, and assessment). (See ADP 5-0 for a discussion of the operations process.)

Commanders use collaboration to establish human connections, build trust, and create and maintain shared understanding and purpose. Collaborative exchange helps commanders increase their situational understanding, resolve potential misunderstandings, and assess the progress of operations. Effective collaboration provides a forum. It allows dialogue in which participants exchange information, learn from one another, and create joint solutions. Establishing a culture of collaboration is difficult but necessary. Creating shared understanding of the issues, concerns, and abilities of commanders, subordinates, and unified action partners takes an investment of time and effort. Successful commanders talk with Soldiers, subordinate leaders, and unified action partners. Through collaboration and dialogue, participants share information and perspectives, question assumptions, and exchange ideas to help create and maintain a shared understanding and purpose.

3. Provide a Clear Commander's Intent

The commander's intent is a clear and concise expression of the purpose of the operation and the desired military end state that supports mission command, provides focus to the staff, and helps subordinate and supporting commanders act to achieve the commander's desired results without further orders, even when the operation does not unfold as planned (JP 3-0). Commanders establish their own commander's intent within the intent of their higher commander. The higher commander's intent provides the basis for unity of effort throughout the larger force.

Commanders articulate the overall reason for the operation so forces understand why it is being conducted. A well-crafted commander's intent conveys a clear image of the operation's purpose, key tasks, and the desired outcome. It expresses the broader purpose of the operation—beyond that of the mission statement. This helps subordinate commanders and Soldiers to gain insight into what is expected of them, what constraints apply, and, most important, why the mission is being undertaken. A clear commander's intent that lower-level leaders can understand is key to maintaining unity of effort. (See ADRP 5-0 for the format of the commander's intent.)

4. Exercise Disciplined Initiative

Disciplined initiative is action in the absence of orders, when existing orders no longer fit the situation, or when unforeseen opportunities or threats arise. Leaders and subordinates exercise disciplined initiative to create opportunities. Commanders rely on subordinates to act, and subordinates take action to develop the situation. This willingness to act helps develop and maintain operational initiative that sets or dictates the terms of action throughout an operation.

The commander's intent defines the limits within which subordinates may exercise initiative. It gives subordinates the confidence to apply their judgment in ambiguous and urgent situations because they know the mission's purpose, key task, and desired end state. They can take actions they think will best accomplish the mission. Using disciplined initiative, subordinates strive to solve many unanticipated problems. They perform the necessary coordination and take appropriate action when existing orders no longer fit the situation.

Commanders and subordinates are obligated to follow lawful orders. Commanders deviate from orders only when they are unlawful, needlessly risk the lives of Soldiers, or no longer fit the situation. Subordinates inform their superiors as soon as possible when they have deviated from orders. Adhering to applicable laws and regulations when exercising disciplined initiative builds credibility and legitimacy. Straying beyond legal boundaries undermines trust and jeopardizes tactical, operational, and strategic success.

5. Use Mission Orders

Mission orders are directives that emphasize to subordinates the results to be attained, not how they are to achieve them. Commanders use mission orders to provide direction and guidance that focus the forces' activities on the achievement of the main objective, set priorities, allocate resources, and influence the situation. They provide subordinates the maximum freedom of action in determining how best to accomplish missions. Mission orders seek to maximize individual initiative, while relying on lateral coordination between units and vertical coordination up and down the chain of command. The mission orders technique does not mean commanders do not supervise subordinates in execution. However, they do not micromanage. They intervene during execution only to direct changes, when necessary, to the concept of operations.

6. Accept Prudent Risk

Commanders accept prudent risk when making decisions because uncertainty exists in all military operations. Prudent risk is a deliberate exposure to potential injury or loss when the commander judges the outcome in terms of mission accomplishment as worth the cost. Opportunities come with risks. The willingness to accept prudent risk is often the key to exposing enemy weaknesses.

Making reasonable estimates and intentionally accepting prudent risk are fundamental to mission command. Commanders focus on creating opportunities rather than simply preventing defeat—even when preventing defeat appears safer. Reasonably estimating and intentionally accepting risk are not gambling. Gambling, in contrast to prudent risk taking, is staking success on a single event without considering the hazard to the force should the event not unfold as envisioned. Therefore, commanders avoid taking gambles. Commanders carefully determine risks, analyze and minimize as many hazards as possible, and then take prudent risks to exploit opportunities.

II. Art of Command

Ref: ADP 6-0, Mission Command, pp. 5 to 7.

Joint doctrine defines command as the authority that a commander in the armed forces lawfully exercises over subordinates by virtue of rank or assignment. Command includes the authority and responsibility for effectively using available resources and for planning the employment of, organizing, directing, coordinating, and controlling military forces for the accomplishment of assigned missions. It also includes responsibility for health, welfare, morale, and discipline of assigned personnel (JP 1). Army doctrine defines the art of command as the creative and skillful exercise of authority through timely decision making and leadership. As an art, command requires exercising judgment. Commanders constantly use their judgment for such things as delegating authority, making decisions, determining the appropriate degree of control, and allocating resources. Although certain facts such as troop-to-task ratios may influence a commander, they do not account for the human aspects of command. A commander's experience and training also influence decision making skills. Proficiency in the art of command stems from years of schooling, self-development, and operational and training experiences.

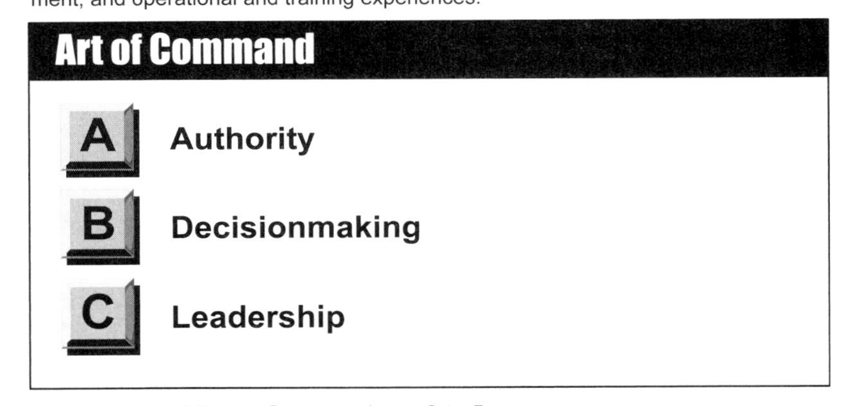

Ref: ADP 6-0, Misson Command, pp. 2 to 5.

As an art, command also requires providing leadership. Leadership is the process of influencing people by providing purpose, direction, and motivation to accomplish the mission and improve the organization. Humans communicate to convey information and thoughts. Although various formats exist to communicate information, successful commanders understand the immeasurable value of collaboration and dialogue. Collaboration and dialogue help commanders obtain human information not collected by their mission command system. Based on the situation and the audience (Soldiers, subordinate commanders, or unified action partners), commanders determine the appropriate communication and leadership style. (See the Army leadership publication for details on leadership style.) Commanders then organize their mission command system to support their decision making and facilitate communication.

A. Authority

Authority is the delegated power to judge, act, or command. Commanders have a legal authority to enforce orders under the Uniform Code of Military Justice. Commanders understand that operations affect and are affected by human interactions. As such, they seek to establish personal authority. Personal authority ultimately arises from the actions of the commander and the trust and confidence generated by those actions. Commanders earn respect and trust by upholding laws and Army values, applying Army leadership principles, and demonstrating tactical and technical expertise. In this way, commanders enhance their authority.

Commanders are legally responsible for their decisions and for the actions, accomplishments, and failures of their subordinates. All commanders have a responsibility to act within their higher commander's intent to achieve the desired end state. However, humans sometimes make mistakes. Commanders realize that subordinates may not accomplish all tasks initially and that errors may occur. Successful commanders allow subordinates to learn through their mistakes and develop experience. With such acceptance in the command climate, subordinates gain the experience required to operate on their own. However, commanders do not continually underwrite subordinates' mistakes resulting from a critical lack of judgment. Nor do they tolerate repeated errors of omission when subordinates fail to exercise initiative. The art of command lies in discriminating between mistakes to underwrite as teaching points from those that are unacceptable in a military leader.

B. Decisionmaking

Decisionmaking requires knowing if, when, and what to decide and understanding the consequences of any decision. Commanders first seek to understand the situation. As commanders and staffs receive information, they process it to develop meaning. Commanders and staffs then apply judgment to gain understanding. This understanding helps commanders and staffs develop effective plans, assess operations and make quality decisions. Commanders use experience, training, and study to inform their decisions. They consider the impact of leadership, operational complexity, and human factors when determining how to best use available resources to accomplish the mission. Success in operations demands timely and effective decisions based on applying judgment to available information and knowledge. They use their judgment to assess information, situations, or circumstances shrewdly and to draw feasible conclusions.

See p. 2-19 for further discussion of judgment.

C. Leadership

Through leadership, commanders influence their organizations to accomplish missions. They develop mutual trust, create shared understanding, and build cohesive teams. Successful commanders act decisively, within the higher commander's intent, and in the best interest of the organization.

Commanders use their presence to lead their forces effectively. They recognize that military operations take a toll on the moral, physical, and mental stamina of Soldiers. They seek to maintain a constant understanding of the status of their forces and adjust their leadership appropriately. They gather and communicate information and knowledge about the command's purpose, goals, and status. Establishing command presence makes the commander's knowledge and experience available to subordinates. Skilled commanders communicate tactical and technical knowledge that goes beyond plans and procedures. Command presence establishes a background for all plans and procedures so that subordinates can understand how and when to adapt them to achieve the commander's intent. In many instances, a leader's physical presence is necessary to lead effectively.

Commanders position themselves where they can command effectively without losing the ability to respond to changing situations. They seek to establish a positive command climate that facilitates team building, encourages initiative, and fosters collaboration, dialogue and mutual trust and understanding. Commanders understand the importance of human relationships in overcoming uncertainty and chaos and maintaining the focus of their forces. The art of command includes exploiting the dynamics of human relationships to the advantage of friendly forces and to the disadvantage of an enemy. Success depends at least as much on understanding the human aspects as it does on any numerical and technological superiority.

See pp. 2-23 to 2-26 for further discussion (application of the mission command philosophy).

Understanding

Ref: ADRP 6-0, Mission Command (May '12), pp. 2-7 to 2-9.

To achieve understanding, commanders and staffs process data to develop meaning. A cognitive hierarchy model depicts how data are transformed into understanding. At the lowest level, processing transforms data into information. Analysis then refines information into knowledge. Commanders and staffs then apply judgment to transform knowledge into situational understanding.

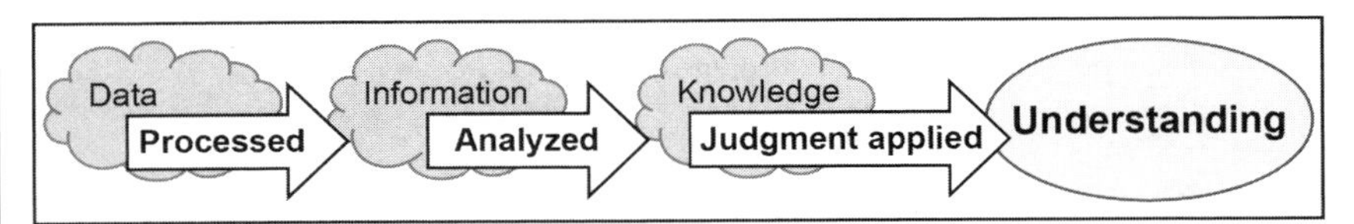

Data consist of unprocessed signals communicated between any nodes in an information system. It includes signals sensed from the environment, detected by a collector of any kind (human, mechanical, or electronic). Data can be quantified, stored, and organized in files and databases. However, to make data useful, people must process data into information.

Information is the meaning that a human assigns to data by means of the known conventions used in their representation (JP 3-13.1). Data become information, which also becomes a source of more data. Information alone rarely provides an adequate basis for deciding and acting. Effective mission command requires further developing information into knowledge so commanders can achieve understanding.

Knowledge is information analyzed to provide meaning and value or evaluated as to implications for an operation. Ultimately, knowledge is the result of individual cognition. Individuals learn through study, experience, practice, and human interaction, as they develop their expertise and skilled judgment. Individuals who develop knowledge determine how to preserve and share it for the benefit of others.

Understanding is knowledge that has been synthesized and had judgment applied to it to comprehend the situation's inner relationships. Judgment is based on experience, expertise, and intuition. Ideally, true understanding should be the basis for decisions. However, commanders and staffs realize that uncertainty and time preclude achieving perfect understanding before deciding and acting.

Analytic Decisionmaking

Analytic decisionmaking generates several alternative solutions, compares these solutions to a set of criteria, and selects the best course of action. It aims to produce the optimal solution by comparing options. It emphasizes analytic reasoning guided by experience, and commanders use it when time is available. This approach offers several advantages.

Intuitive Decisionmaking

Intuitive decisionmaking is reaching a conclusion through pattern recognition based on knowledge, judgment, experience, education, intelligence, boldness, perception, and character. Intuitive decision making is faster and more often done at the lowest levels of command. When using intuitive decision making, leaders should be aware of their own biases and how their current operational environment differs from past environments.

Commanders blend intuitive and analytic decision making to help them remain objective and make timely and effective decisions. Commanders avoid making decisions purely by intuition; they incorporate some analysis into their intuitive decisions. Combining both approaches provides a holistic perspective on the many factors that affect decisions. Commanders understand that decisions should be neither rushed nor over-thought.

Judgment

Ref: ADRP 6-0, Mission Command (May '12), pp. 2-9 to 2-10.

Commanders use judgment to assess information, situations, or circumstances shrewdly and to draw feasible conclusions. Through good judgment, commanders form sound opinions and make sensible decisions. They select the critical time and place to act, assign missions, manage risk, prioritize effort, allocate resources, and lead Soldiers. Commanders make decisions using judgment developed from experience, training, study, and creative and critical thinking. Experience contributes to judgment by providing a basis for rapidly identifying practical courses of actions and dismissing impractical ones. Commanders apply their judgment to:

A. Identify, Accept, and Mitigate Risk

Commanders use judgment when identifying risk, deciding what risk to accept, and mitigating accepted risk. They accept risk to create opportunities. They reduce risk by foresight and careful planning. Commanders use risk assessment and risk management to identify and mitigate risk. Risk management is a tool commanders can use to identify risk, assess risk, and develop mitigation and control measures to help manage risk. (See FM 5-19 for more information on risk management.)

Consideration of risk begins during planning, as commanders and staffs complete a risk assessment for each course of action and propose control measures. They use collaboration and dialogue. They integrate input from subordinates, staff, and appropriate organizations and partners. They determine how to manage identified risks. This includes delegating management of certain risks to subordinate commanders who will develop appropriate mitigation measures. Commanders then allocate the resources they deem appropriate to mitigate risks. Subordinates must also trust their leaders to underwrite their prudent risk taking and to reward their disciplined initiative. Successful commanders minimize risk and unify the effort by monitoring how well subordinates are using their authority and resources and exercising initiative.

B. Prioritize Resources

Commanders allocate resources to accomplish the mission. Allocating resources requires judgment because resources can be limited. Considerations for prioritizing resources include how to:

- Effectively accomplish the mission while conserving resources
- Protect the lives of Soldiers
- Apply the principles of mass and economy of force
- Posture the force for subsequent operations

C. Delegate Authority

Commanders use judgment when determining what authority to delegate. Commanders delegate authority verbally, in writing, or both. Examples of delegated authority are authority over an area of expertise or technical specialty, a geographic area, or specific kinds of actions. Commanders may limit delegated authority in time, or they may use a standing delegation.

Commanders delegate authority and set the level of their personal involvement in delegated tasks based on their assessment of the skill and experience of their subordinates. When delegating authority to subordinates, commanders do everything in their power to set the conditions for success. They allocate enough resources to subordinates so they can accomplish their missions. Resources include people, units, services, supplies, equipment, networks, information, and time. Delegation not only applies to subordinate commanders but also to members of the staff. Commanders rely on and expect initiative from staff officers as much as from subordinate commanders.

III. Science of Control

Control is the regulation of forces and war fighting functions to accomplish the mission in accordance with the commander's intent (ADP 6-0). Aided by staffs, commanders exercise control over assigned forces in their area of operations. Staffs coordinate, synchronize, and integrate actions, inform the commander, and exercise control for the commander. Control permits commanders to adjust operations to account for changing circumstances and direct the changes necessary to address the new situation. Commanders impose enough control to mass the effect of combat power at the decisive point in time while allowing subordinates the maximum freedom of action to accomplish assigned tasks.

The commander's mission command system, especially the staff, assists the commander with control. However, the commander remains the central figure. The science of control comprises:

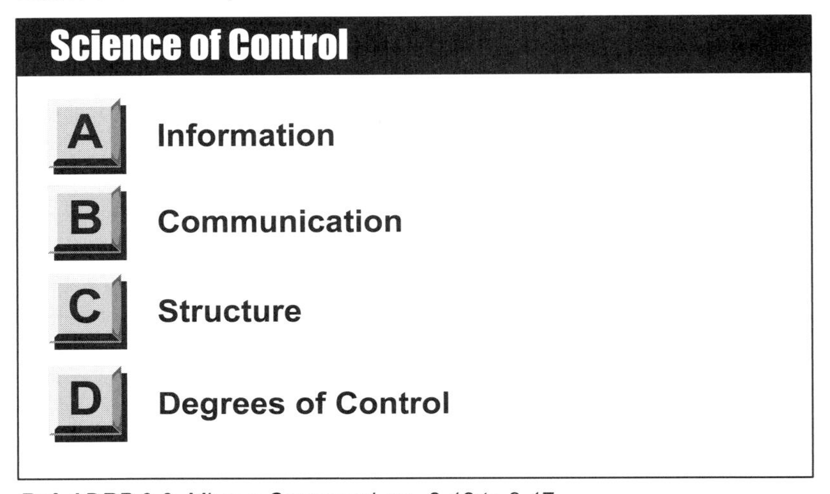

Ref: ADRP 6-0, Misson Command, pp. 2-12 to 2-17.

The science of control consists of systems and procedures used to improve the commander's understanding and support accomplishing missions (ADP 6-0). The science of control supports the art of command. In contrast to the art of command, the science of control is based on objectivity, facts, empirical methods, and analysis. Commanders and staffs use the science of control to overcome the physical and procedural constraints under which units operate. Units are bound by such factors as movement rates, fuel consumption, weapons effects, rules of engagement, and legal considerations. Commanders and staffs strive to understand aspects of operations they can analyze and measure, such as the physical capabilities and limitations of friendly and enemy organizations. Control requires a realistic appreciation for time and distance factors, including the time required to initiate certain actions.

A. Information

Ref: ADRP 6-0, Mission Command (May '12), pp. 2-13 to 2-14.

Commanders use the science of control to manage information. Information fuels understanding and decision making. Commanders establish information requirements to set priorities for collecting relevant information. An information requirement is any information element the commander and staff require to successfully conduct operations. Relevant information that answers information requirements is:

- **Accurate**: it conveys the true situation
- **Timely**: it is available in time to make decisions
- **Usable**: it is portrayed in common, easily understood formats and displays
- **Complete**: it provides all information necessary
- **Precise**: it contains sufficient detail
- **Reliable**: it is trustworthy and dependable

Commanders balance the art of command with the science of control as they assess information against these criteria. For example, in some situations, relevant information that is somewhat incomplete or imprecise may be better than no information at all, especially when time for execution is limited. However, effective commanders use the science of control to reduce the likelihood of receiving inaccurate, late, or unreliable information, which is of no value to the exercise of mission command.

Information can come in many forms, such as feedback and electronic means. Feedback comes from subordinates, higher headquarters, or adjacent, supporting, and supported forces and unified action partners. For feedback to be effective, a commander's mission command system must process it into knowledge, identifying any differences between the desired end state and the situation that exists. Commanders and staffs interpret information received to gain situational understanding and to exploit fleeting opportunities, respond to developing threats, modify plans, or reallocate resources.

Electronic means of communication have increased the access to and speed of finding information. However, they also have increased the volume of information and the potential for misinformation. Successful commanders are mindful of this when they configure their mission command system. Commanders determine information requirements and set information priorities. They avoid requesting too much information, which decreases the staff's chances of obtaining the right information. The quest for information is time-consuming; commanders who demand complete information place unreasonable burdens upon subordinates. Subordinates pressured to worry over every detail rarely have the desire to exercise initiative. At worst, excessive information demands corrupt the trust required for mission command.

Staffs provide commanders and subordinates information relevant to their operational environment and the progress of operations. They use the operational variables (political, military, economic, social, information, infrastructure, physical environment, and time—known as PMESII–PT) and the mission variables (mission, enemy, terrain and weather, troops and support available, time available and civil considerations—known as METT–TC) as major subject categories to group relevant information. Commanders and staffs develop a common operational picture (known as a COP), a single display of relevant information within a commander's area of interest tailored to the user's requirements and based on common data and information shared by more than one command. They choose any appropriate technique to develop and display the COP, such as graphical representations, verbal narratives, or written reports. Development of the COP is ongoing throughout operations. This tool supports developing knowledge and understanding.

B. Communication

Commanders and staffs disseminate and share information among people, elements, and places. Communication is more than the simple transmission of information. It is a means to exercise control over forces. Communication links information to decisions and decisions to action. No decision during operations can be executed without clear communication between commanders and subordinates. Communication among the parts of a command supports their coordinated action. Effective commanders do not take the importance of communication for granted. They use multidirectional communication and suitable communication media to achieve objectives. Commanders choose appropriate times, places, and means to communicate. They use face-to-face talks, written and verbal orders, estimates and plans, published memos, electronic mail, Web sites, social media, and newsletters.

The traditional view of communication within military organizations is that subordinates send commanders information, and commanders provide subordinates with decisions and instructions. This linear form of communication is inadequate for mission command. Communication has an importance far beyond exchanging information. Commanders and staffs communicate to learn, exchange ideas, and create and sustain shared understanding. Information needs to flow up and down the chain of command as well as laterally to adjacent units and organizations. Separate from the quality or meaning of information exchanged, communication strengthens bonds within a command. It is an important factor in building trust, cooperation, cohesion, and mutual understanding.

Effective commanders conduct face-to-face talks with their subordinates to ensure subordinates fully understand them. Humans communicate by what they say and do and by their manner of speaking and behaving. Nonverbal communication may include gestures, sighs, and body language. Commanders pay attention to verbal and nonverbal feedback to ascertain the effectiveness of their communication.

Commanders and staffs should communicate face-to-face whenever possible. This does not mean they do not keep records of information communicated or follow-up with written documentation. Records are important as a means of affirming understanding and for later study and critique. Records support understanding over time, whereas memory may distort or even omit elements of the information required or passed. (See DA Pam 25-403 for record keeping guidance.)

1. Channels

Information normally moves throughout a force along various transmission paths, or channels. Commanders and staffs transfer information horizontally as well as vertically. Establishing command and support relationships helps create channels that streamline information dissemination by ensuring the right information passes promptly to the right people. Three common channels are known as command, staff, and technical channels. Command channels are direct chain-of-command transmission paths. Commanders and authorized staff officers use command channels for command-related activities.

Staff channels are staff-to-staff transmission paths between headquarters and are used for control related activities. Staff channels transmit planning information, status reports, controlling instructions, and other information to support mission command. The intelligence and sustainment nets are examples of staff channels.

Technical channels are the transmission paths between two technically similar units or offices that perform a specialized technical function, requiring special expertise or control the performance of technical functions. Technical channels are typically used to control performance of technical functions. They are not used for conducting operations or supporting another unit's mission. An example is network control.

2. Feedback

Commanders use feedback to compare the actual situation to their visualization, and decide whether to adjust operations, and direct actions. Feedback takes many forms, including information, knowledge, experience, and wisdom. Feedback comes from many sources: subordinates, higher headquarters, or adjacent, supporting, and supported forces. It can arrive any time: before, during, or after operations. Feedback helps commanders and staffs gain understanding. For feedback to be effective, it must be processed it into knowledge, identifying any differences between the desired end state and the current situation. New information that conflicts with the expectations established during planning requires commanders and staffs to validate those expectations or revise them to reflect reality. This contributes to an accurate understanding that allows commanders to exploit fleeting opportunities, respond to developing situations, modify plans, or reallocate resources.

C. Structure

Organizational structure helps commanders exercise control. Structure refers to a defined organization that establishes relationships and guides interactions among elements. It also includes procedures for coordinating among an organization's groups and activities. The commander establishes control with a defined organization. Structure is both internal (such as a command post) and external (such as command and support relationships among subordinate forces). Commanders apply the doctrinal guidance provided in FM 6-0 for organizing Army command post operations and command and support relationships. The most basic organization in control is a hierarchy. In military terms, this relationship is between the commander and staff, and subordinate forces.

D. Degrees of Control

A key aspect of mission command is determining the appropriate degree of control to impose on subordinates. The appropriate degree of control varies with each situation and is not easy to determine. Different operations and phases of operations require tighter or more relaxed control over subordinate elements than other phases require.

See following pages (pp. 2-24 to 2-25) for further discussion and considerations for determining degree of control include:

- *Level of acceptable risk*
- *Delegation of authority and resources*
- *Ability to sustain the force*
- *Span of control*
- *Forms of control*

IV. Application of the Mission Command Philosophy

In mission command, the commander is the central figure. Guided by the principles of mission command, commanders skillfully balance of the art of command with the science of control. They exploit and enhance uniquely human skills while systematically regulating forces and war fighting functions. They understand and use human relationships to overcome uncertainty and chaos and maintain the focus of their forces. Collaboration and dialogue help commanders build mutual trust, create a shared understanding and purpose, and receive human information not collected by their mission command system. They consider the impact of leadership, operational complexity, and human factors when determining how to best use available resources to accomplish the mission. Applying the mission command philosophy helps commanders exercise authority skillfully and master the systems and procedures that help forces accomplish missions. They use the mission command warfighting function to help them integrate and synchronize operations.

D. Degrees of Control

Ref: ADRP 6-0, Mission Command (May '12), pp. 2-15 to 2-17.

A key aspect of mission command is determining the appropriate degree of control to impose on subordinates. The appropriate degree of control varies with each situation and is not easy to determine. Different operations and phases of operations require tighter or more relaxed control over subordinate elements than other phases require. An air assault's air movement and landing phases, for example, require precise control and synchronization. Its ground maneuver plan may require less detail. Successful commanders understand that swift action may be necessary to capitalize on fleeting opportunities. They centralize or decentralize control of operations as needed to ensure that units can adapt to changing situations.

As a rule, commanders use the mission orders technique for plans and orders. They limit information in a base plan or order to the minimum needed to synchronize combat power at the decisive time and place and allow subordinates as much freedom of action as possible. Commanders rely on subordinates to act within the commander's intent and concept of operations. The attachments to the plan or order contain details regarding the situation and instructions necessary for synchronization.

Commanders concentrate and synchronize many units to mass effects, and they centralize or decentralize control of operations as needed to ensure that units can adapt to changing situations. Commanders ensure they maintain enough control: the higher headquarters imposes enough control to maximize total combat power, while delegating appropriate authorities and resources to subordinates. Commanders and subordinates understand what risks the higher commander will accept and what risks will remain with the subordinate commander. Considerations include:

Degrees of Control Considerations

1. **Level of Acceptable Risk**
2. **Delegation of Authority and Resources**
3. **Ability to Sustain the Force**
4. **Span of Control**
5. **Forms of Control**

1. Level of Acceptable Risk

Although some risk is inherent in all operations, commanders consider the impact of centralizing or decentralizing control of authority and resources. Commanders must be aware of how much flexibility a higher commander will allow for subordinate initiative in decision making and execution. They weigh the benefit of decentralizing control against the risk that the higher headquarter may not be able to respond immediately to a subordinate unit's request for assistance. Commanders avoid over-control of authority and resources that may leave subordinate units lacking the ability to respond rapidly to emerging situations.

2. Delegation of Authority and Resources

The experience of subordinate commanders, their ability to make decisions, and the impact of subordinate decisions on the higher headquarters mission factor into any decision to delegate decision making authority. The degree of trust and confidence a commander has in subordinate commanders, as well as confidence in one's own abilities, weighs heavily in deciding delegation of authority. Additionally, the commander should be confident in subordinate commanders' training, education, experience, and staff expertise to be able to use effectively any resources allocated to them.

3. Ability to Sustain the Force

Sustaining the force involves equipping it with materiel (for individuals and units), maintaining Soldier readiness, and sustaining readiness for unified land operations. Commanders analyze their operational environment to understand what is needed and allocate resources. Commanders and staffs consider availability and adequacy of resources when deciding how and when to allocate assets to subordinate units. Effective commanders are prepared to request additional assets, as necessary, from higher headquarters and then task-organize to best utilize those assets.

4. Span of Control

Unit organization should ensure reasonable span of control—the number of subordinates or activities under a single commander. Generally, commanders can effectively command and effectively control two to five subordinate headquarters. A commander's span of control should not exceed that commander's capability to command effectively. The optimal number of subordinates depends on the situation.

Narrowing the span of control—that is, lessening the number of immediate subordinates—deepens the organization by adding layers of command. Conversely, eliminating echelons of command or "flattening" an organization widens the span of control. The aim is to flatten the organization to the extent compatible with reasonable spans of control. Commanders balance width and depth, so that the structure fits the situation. For example, higher-tempo operations, such as offensive operations, tend to favor wider spans of control.

An effective task organization enables the commander and subordinate commanders to command effectively. The commander establishes the span of control and an organizational structure that best fits the situation and supports mission command while maintaining operational adaptability.

5. Forms of Control

Two techniques for control are positive and procedural. Although commonly associated with airspace control, positive and procedural controls also apply to land operations. All military operations require both forms to offset the weaknesses of each. They complement each other and enhance operations.

Positive control is a technique for actively regulating forces that requires explicit coordination between commanders and subordinate leaders. Positive control requires active command participation.

Procedural control is a technique of regulating forces where actions are governed by written and oral instructions which do not require authorization to execute. Examples of procedural control include orders, regulations, policies, and doctrine. Once established, procedural control requires no intervention by the higher headquarters. Forces share a common understanding of the procedures and how to apply them in operations.

Control measures provide procedural control without requiring detailed explanations. A control measure is a means of regulating forces or war fighting functions. Control measures can be permissive (which allows something to happen) or restrictive (which limits how something is done). Some control measures are graphic. A graphic control measure is a symbol used on maps and displays to regulate forces and war fighting functions. (See FM 1-02 for illustrations of graphic control measures and rules for their use.)

Commanders use the minimum number of control measures necessary to control their forces. Commanders tailor their use of control measures to conform to the higher commander's intent. They also consider the mission, terrain, and amount of authority delegated to subordinates. Effectively employing control measures requires commanders and staffs to understand their purposes and ramifications, including the permissions or limitations imposed on subordinates' freedom of action and initiative. Each measure should have a specific purpose: mass the effects of combat power, synchronize subordinate forces' operations, or minimize the possibility of fratricide.

Command Presence

Ref: ADRP 6-0, Mission Command (May '12), pp. 2-11 to 2-12.

Command presence is creating a favorable impression in carriage, appearance, and professional and personal conduct. Commanders use their presence to gather and communicate information and knowledge and assess operations. Establishing command presence makes the commander's knowledge and experience available to subordinates. It does not require giving subordinates detailed instructions, nor does it include second-guessing subordinates' performance. Skilled commanders communicate tactical and technical knowledge that goes beyond plans and procedures. Command presence establishes a background for all plans and procedures so that subordinates can understand how and when to adapt them to achieve the commander's intent. Commanders can establish command presence in a variety of ways, including:

Command Presence

- **Briefings**
- **Back-briefs**
- **Rehearsals**
- **Leader's Reconnaissance**
- **Commander's Intent**
- **On-site Visits and Battlefield Circulation**
- **After Action Reviews**
- **Commander's Guidance**
- **Personal Example**

Directly engaging subordinates and staffs allows commanders to motivate Soldiers, build trust and confidence, exchange information, and assess operations. Commanders understand and use human relationships to overcome uncertainty and chaos and maintain the focus of their forces. They communicate in a variety of ways, adjusting their communication style to fit the situation and the audience. They communicate both formally and informally, through questions, discussions, conversations, and other direct or indirect communication.

In many instances, a leader's physical presence is necessary to lead effectively. Commanders position themselves where they can best command without losing the ability to respond to changing situations. Commanders carefully consider where they need to be, balancing the need to inspire Soldiers with maintaining an overall perspective of the entire operation. The commander's forward presence demonstrates a willingness to share danger and hardship. It also allows commanders to appraise for themselves a subordinate unit's condition, including its leaders' and Soldiers' morale. Forward presence allows commanders to sense the human dimension of conflict, particularly when fear and fatigue reduce effectiveness. Commanders cannot let the perceived advantages of improved information technology compromise their obligation to lead by example, face-to-face with Soldiers.

Command Climate

Commanders create their organization's tone—the characteristic atmosphere in which people work. This is known as the command climate. It is directly attributable to the leader's values, skills, and actions. A positive climate facilitates team building, encourages initiative, and fosters collaboration, dialogue, mutual trust, and shared understanding. Commanders shape the climate of the organization, no matter what the size.

Successful commanders recognize that all Soldiers can contribute to mission accomplishment. Commanders establish clear and realistic goals and communicate their goals openly. They establish and maintain communication between subordinates and leaders. They encourage subordinates to bring creative and innovative ideas to the forefront.

II. Command and Support Relationships

Ref: FM 6-0 (C2), Commander and Staff Organization and Operations (Apr '16), app. B.

Command and support relationships provide the basis for unity of command and unity of effort in operations. Command relationships affect Army force generation, force tailoring, and task organization. Commanders use Army support relationships when task-organizing Army forces. All command and support relationships fall within the framework of joint doctrine. *Note: JP 1 discusses joint command relationships and authorities.*

I. Chain of Command

The President and Secretary of Defense exercise authority and control of the armed forces through two distinct branches of the chain of command as described in JP-1. One branch runs from the President, through the Secretary of Defense, to the combatant commanders for missions and forces assigned to combatant commands. The other branch runs from the President through the Secretary of Defense to the secretaries of the military departments. This branch is used for purposes other than operational direction of forces assigned to the combatant commands. Each military department operates under the authority, direction, and control of the secretary of that military department. These secretaries exercise authority through their respective Service chiefs over Service forces not assigned to combatant commanders. The Service chiefs, except as otherwise prescribed by law, perform their duties under the authority, direction, and control of the secretaries to whom they are directly responsible.

The typical operational chain of command extends from the combatant commander to a joint task force commander, then to a functional component commander or a Service component commander. Joint task forces and functional component commands, such as a land component, comprise forces that are normally subordinate to a Service component command but have been placed under the operational control (OPCON) of the joint task force, and subsequently to a functional component commander. Conversely, the combatant commander may designate one of the Service component commanders as the joint task force commander or as a functional component commander. In some cases, the combatant commander may not establish a joint task force, retaining operational control over subordinate functional commands and Service components directly.

Army Service Component Command (ASCC)

Under joint doctrine, each joint force includes a Service component command that provides administrative and logistic support to Service forces under OPCON of that joint force. However, Army doctrine distinguishes between the Army component of a combatant command and Army components of subordinate joint forces. Under Army doctrine, Army Service component command (ASCC) refers to the Army component assigned to a combatant command. There is only one ASCC within a combatant command's area of responsibility.

ARFOR

The Army components of all other joint forces are called ARFORs. An ARFOR is the Army Service component headquarters for a joint task force or a joint and multinational force. It consists of the senior Army headquarters and its commander (when not designated as the joint force commander) and all Army forces that the combatant commander subordinates to the joint task force or places under the control of a multinational force commander. The ARFOR becomes the conduit for most Service-related issues and administrative support. The Army Service component command may function as an ARFOR headquarters when the combatant commander does not exercise command and control through subordinate joint force commanders.

The Secretary of the Army directs the flow of administrative control (ADCON). Administrative control for Army units within a combatant command normally extends from the Secretary of the Army through the ASCC, through an ARFOR, and then to Army units assigned or attached to an Army headquarters within that joint command. However, administrative control is not tied to the operational chain of command. The Secretary of the Army may redirect some or all Service responsibilities outside the normal ASCC channels. In similar fashion, the ASCC may distribute some administrative responsibilities outside the ARFOR. Their primary considerations are the effectiveness of Army forces and the care of Soldiers.

A. Combatant Commands

The Unified Command Plan establishes combatant commanders' missions and geographic responsibilities. Combatant commanders directly link operational military forces to the Secretary of Defense and the President. The Secretary of Defense deploys troops and exercises military power through the combatant commands.

Refer to JFODS4: The Joint Forces Operations & Doctrine SMARTbook for additional information on combatant commands, joint task forces and operational warfighting doctrine.

B. Joint Task Forces and Service Components

Joint task forces are the organizations most often used by a combatant commander for contingencies. Combatant commanders establish joint task forces and designate the joint force commanders for these commands. Those commanders exercise OPCON of all U.S. forces through functional component commands, Service components, subordinate joint task forces, or a combination of these. The senior Army officer assigned to a joint task force, other than the joint force commander and members of the joint task force staff, becomes the ARFOR commander. The ARFOR commander answers to the Secretary of the Army through the Army Service component command for most ADCON responsibilities.

Depending on the joint task force organization, the ARFOR commander may exercise OPCON of some or all Army forces assigned to the task force. For example, an Army corps headquarters may become a joint force land component within a large joint task force. The corps commander exercises OPCON of Army divisions and tactical control (TACON) of Marine Corps forces within the land component. As the senior Army headquarters, the corps becomes the ARFOR for not only the Army divisions but also all other Army units within the joint task force, including those not under OPCON of the corps. Unless modified by the Secretary of the Army or the ASCC, Service responsibilities continue through the ARFOR to the respective Army commanders. The corps has OPCON of the Army divisions and TACON of the Marine division. The corps does not have OPCON over the other Army units but does, as the ARFOR, exercise ADCON over them. The corps also assists the ASCC in controlling Army support to other Services and to any multinational forces as directed.

When an Army headquarters becomes the joint force land component as part of a joint task force, Army units subordinated to it are normally under OPCON. Marine Corps forces made available to a joint force land component command built around an Army headquarters are normally under TACON. The land component commander makes recommendations to the joint force commander on properly using attached, OPCON, or TACON assets; planning and coordinating land operations; and accomplishing such operational missions as assigned.

II. Task Organization

Ref: FM 6-0 (C2), Commander and Staff Organization and Operations (Apr '16), app. D.

Task-organizing is the act of designing an operating force, support staff, or sustainment package of specific size and composition to meet a unique task or mission (ADRP 3-0). Characteristics to examine when task-organizing the force include, but are not limited to, training, experience, equipment, sustainability, operational environment, (including enemy threat), and mobility. For Army forces, it includes allocating available assets to subordinate commanders and establishing their command and support relationships. Command and support relationships provide the basis for unity of command in operations. The assistant chief of staff, plans (G-5) or assistant chief of staff, operations (G-3 [S-3]) develops Annex A (Task Organization).

Fundamental Considerations

Military units consist of organic components. Organic parts of a unit are those forming an essential part of the unit and are listed in its table of organization and equipment (TOE). Commanders can alter organizations' organic unit relationships to better allocate assets to subordinate commanders. They also can establish temporary command and support relationships to facilitate exercising mission command.

Establishing clear command and support relationships is fundamental to organizing any operation. These relationships establish clear responsibilities and authorities between subordinate and supporting units. Some command and support relationships (for example, tactical control) limit the commander's authority to prescribe additional relationships. Knowing the inherent responsibilities of each command and support relationship allows commanders to effectively organize their forces and helps supporting commanders to understand their unit's role in the organizational structure.

Commanders designate command and support relationships to weight the decisive operation and support the concept of operations. Task organization also helps subordinate and supporting commanders support the commander's intent. These relationships carry with them varying responsibilities to the subordinate unit by the parent and gaining units.

Commanders consider two organizational principles when task-organizing forces:

- **When possible, commanders maintain cohesive mission teams.** They organize forces based on standing headquarters, their assigned forces, and habitual associations when possible. When not feasible and ad hoc organizations are created, commanders arrange time for training and establishing functional working relationships and procedures. Once commanders have organized and committed a force, they keep its task organization unless the benefits of a change clearly outweigh the disadvantages. Reorganizations may result in a loss of time, effort, and tempo. Sustainment considerations may also preclude quick reorganization.
- **Commanders carefully avoid exceeding the span of control capabilities of subordinates.** Span of control refers to the number of subordinate units under a single commander. This number is situation dependent and may vary. As a rule, commanders can effectively command two to six subordinate units. Allocating subordinate commanders more units gives them greater flexibility and increases options and combinations. However, increasing the number of subordinate units increases the number of decisions for commanders to make in a timely fashion. This slows down the reaction time among decisionmakers.

Refer to BSS5: The Battle Staff SMARTbook (Leading, Planning & Conducting Military Operations) for complete discussion of task organization to include fundamental considerations, Army command and support relationships, unit listing sequence, and outline format (sample).

III. Joint Command & Support Relationships

Ref: JP 1, Doctrine for the Armed Forces of the United States (Mar '13), chap. V.

Levels of Authority

The specific command relationship (COCOM, OPCON, TACON, and support) will define the authority a commander has over assigned or attached forces. An overview of command relationships is shown in Figure V-1, below.

Command Relationships Synopsis

Combatant Command (Command Authority)

(Unique to Combatant Commander)

- Planning, programming, budgeting, and execution process input
- Assignment of subordinate commanders
- Relationships with Department of Defense agencies
- Directive authority for logistics

Operational control when delegated

- Authoritative direction for all military operations and joint training
- Organize and employ commands and forces
- Assign command functions to subordinates
- Establish plans and requirements for intelligence, surveillance, and reconnaissance activities
- Suspend subordinate commanders from duty

Tactical control when delegated

Local direction and control of movements or maneuvers to accomplish mission

Support relationship when assigned

Aid, assist, protect, or sustain another organization

Ref: JP 1, Doctrine for the Armed Forces of the United States, fig. V-1, p. V-2.

Command Relationships Overview

- Forces, not command relationships, are transferred between commands. When forces are transferred, the command relationship the gaining commander will exercise (and the losing commander will relinquish) over those forces must be specified.
- When transfer of forces to a joint force will be permanent (or for an unknown but long period of time) the forces should be reassigned. Combatant commanders will exercise combatant command (command authority) and subordinate joint force commanders (JFCs), will exercise operational control (OPCON) over reassigned forces.
- When transfer of forces to a joint force will be temporary, the forces will be attached to the gaining command and JFCs, normally through the Service component commander, will exercise OPCON over the attached forces.
- Establishing authorities for subordinate unified commands and joint task forces direct the assignment or attachment of their forces to those subordinate commands as appropriate.

A. Combatant Command (COCOM) - Command Authority

COCOM provides full authority for a CCDR to perform those functions of command over assigned forces involving organizing and employing commands and forces, assigning tasks, designating objectives, and giving authoritative direction over all aspects of military operations, joint training (or in the case of USSOCOM, training of assigned forces), and logistics necessary to accomplish the missions assigned to the command. COCOM should be exercised through the commanders of subordinate organizations, normally JFCs, Service and/or functional component commanders.

B. Operational Control (OPCON)

OPCON is the command authority that may be exercised by commanders at any echelon at or below the level of CCMD and may be delegated within the command. It is the authority to perform those functions of command over subordinate forces involving organizing and employing commands and forces, assigning tasks, designating objectives, and giving authoritative direction over all aspects of military operations and joint training necessary to accomplish the mission.

C. Tactical Control (TACON)

TACON is an authority over assigned or attached forces or commands, or military capability or forces made available for tasking, that is limited to the detailed direction and control of movements and maneuvers within the operational area necessary to accomplish assigned missions or tasks assigned by the commander exercising OPCON or TACON of the attached force.

Support Relationships

Support is a command authority. A support relationship is established by a common superior commander between subordinate commanders when one organization should aid, protect, complement, or sustain another force. The support command relationship is used by SecDef to establish and prioritize support between and among CCDRs, and it is used by JFCs to establish support relationships between and among subordinate commanders.

A. General Support

That support which is given to the supported force as a whole rather than to a particular subdivision thereof.

B. Mutual Support

That support which units render each other against an enemy because of their assigned tasks, their position relative to each other and to the enemy, and their inherent capabilities.

C. Direct Support

A mission requiring a force to support another specific force and authorizing it to answer directly to the supported force's request for assistance.

D. Close Support

That action of the supporting force against targets or objectives that are sufficiently near the supported force as to require detailed integration or coordination of the supporting action with the fire, movement, or other actions of the supported force

Refer to JFODS4: The Joint Forces Operations & Doctrine SMARTbook (Guide to Joint, Multinational & Interorganizational Operations) for further discussion. Topics and chapters include joint doctrine fundamentals, joint operations, joint operation planning, joint logistics, joint task forces, information operations, multinational operations, and interorganizational coordination.

Mission Cmd (ADRP 6-0)

IV. Army Command & Support Relationships

Ref: FM 6-0 (C2), Commander and Staff Organization and Operations (Apr '16), app. B.

Army command and support relationships are similar but not identical to joint command authorities and relationships. Differences stem from the way Army forces task-organize internally and the need for a system of support relationships between Army forces. Another important difference is the requirement for Army commanders to handle the administrative support requirements that meet the needs of Soldiers

A. Command Relationships

Army command relationships define superior and subordinate relationships between unit commanders. By specifying a chain of command, command relationships unify effort and enable commanders to use subordinate forces with maximum flexibility. Army command relationships identify the degree of control of the gaining Army commander. The type of command relationship often relates to the expected longevity of the relationship between the headquarters involved and quickly identifies the degree of support that the gaining and losing Army commanders provide.

If relationship is:	Then inherent responsibilities:							
	Have command relationship with:	May be task-organized by: [1]	Unless modified, ADCON have responsibility through:	Are assigned position or AO by:	Provide liaison to:	Establish/ maintain communications with:	Have priorities established by:	Can impose on gaining unit further command or support relationship of:
Organic	All organic forces organized with the HQ	Organic HQ	Army HQ specified in organizing document	Organic HQ	N/A	N/A	Organic HQ	Attached; OPCON; TACON; GS; GSR; R; DS
Assigned	Combatant command	Gaining HQ	Gaining Army HQ	OPCON chain of command	As required by OPCON	As required by OPCON	ASCC or Service-assigned HQ	As required by OPCON HQ
Attached	Gaining unit	Gaining unit	Gaining Army HQ	Gaining unit	As required by gaining unit	Unit to which attached	Gaining unit	Attached; OPCON; TACON; GS; GSR; R; DS
OPCON	Gaining unit	Parent unit and gaining unit; gaining unit may pass OPCON to lower HQ[1]	Parent unit	Gaining unit	As required by gaining unit	As required by gaining unit and parent unit	Gaining unit	OPCON; TACON; GS; GSR; R; DS
TACON	Gaining unit	Parent unit	Parent unit	Gaining unit	As required by gaining unit	As required by gaining unit and parent unit	Gaining unit	TACON;GS GSR; R; DS

Note: [1] In NATO, the gaining unit may not task-organize a multinational force. (See TACON.)

ADCON administrative control
AO area of operations
ASCC Army Service component command
DS direct support
GS general support
GSR general support–reinforcing
HQ headquarters
N/A not applicable
NATO North Atlantic Treaty Organization
OPCON operational control
R reinforcing
TACON tactical control

Ref: FM 6-0 (C2), Commander and Staff Organization and Operations, table B-2, p. B-5.

B. Support Relationships

Table B-3 on the following page lists Army support relationships. Army support relationships are not a command authority and are more specific than the joint support relationships. Commanders establish support relationships when subordination of one unit to another is inappropriate. Commanders assign a support relationship when—

- The support is more effective if a commander with the requisite technical and tactical expertise controls the supporting unit, rather than the supported commander. The echelon of the supporting unit is the same as or higher than that of the supported unit. For example, the supporting unit may be a brigade, and the supported unit may be a battalion. It would be inappropriate for the brigade to be subordinated to the battalion, hence the use of an support relationship.
- The supporting unit supports several units simultaneously. The requirement to set support priorities to allocate resources to supported units exists. Assigning support relationships is one aspect of mission command.

If relationship is:	Then inherent responsibilities:							
	Have command relationship with:	May be task-organized by:	Receive sustainment from:	Are assigned position or an area of operations by:	Provide liaison to:	Establish/ maintain communications with:	Have priorities established by:	Can impose on gaining unit further command or support relationship by:
Direct support[1]	Parent unit	Parent unit	Parent unit	Supported unit	Supported unit	Parent unit; supported unit	Supported unit	See note[1]
Reinforcing	Parent unit	Parent unit	Parent unit	Reinforced unit	Reinforced unit	Parent unit; reinforced unit	Reinforced unit; then parent unit	Not applicable
General support–reinforcing	Parent unit	Parent unit	Parent unit	Parent unit	Reinforced unit and as required by parent unit	Reinforced unit and as required by parent unit	Parent unit; then reinforced unit	Not applicable
General support	Parent unit	Parent unit	Parent unit	Parent unit	As required by parent unit	As required by parent unit	Parent unit	Not applicable

Note: [1] Commanders of units in direct support may further assign support relationships between their subordinate units and elements of the supported unit after coordination with the supported commander.

Ref: FM 6-0 (C2), Commander and Staff Organization and Operations, table B-3, p. B-6.

Army support relationships allow supporting commanders to employ their units' capabilities to achieve results required by supported commanders. Support relationships are graduated from an exclusive supported and supporting relationship between two units—as in direct support—to a broad level of support extended to all units under the control of the higher headquarters—as in general support. Support relationships do not alter administrative control. Commanders specify and change support relationships through task organization.

Direct support is a support relationship requiring a force to support another specific force and authorizing it to answer directly to the supported force's request for assistance (ADRP 5-0). A unit assigned a direct support relationship retains its command relationship with its parent unit, but is positioned by and has priorities of support established by the supported unit.

General support is that support which is given to the supported force as a whole and not to any particular subdivision thereof (JP 3-09.3). Units assigned a GS relationship are positioned and have priorities established by their parent unit.

Reinforcing is a support relationship requiring a force to support another supporting unit (ADRP 5-0). Only like units (for example, artillery to artillery) can be given a reinforcing mission. A unit assigned a reinforcing support relationship retains its command relationship with its parent unit, but is positioned by the reinforced unit. A unit that is reinforcing has priorities of support established by the reinforced unit, then the parent unit.

General support-reinforcing is a support relationship assigned to a unit to support the force as a whole and to reinforce another similar-type unit (ADRP 5-0). A unit assigned a general support-reinforcing (GSR) support relationship is positioned and has priorities established by its parent unit and secondly by the reinforced unit.

V. Command Posts

Ref: FM 6-0 (C2), Commander and Staff Organization and Operations (Apr '16), chap. 1.

In operations, effective mission command requires continuous, close coordination, synchronization, and information sharing across staff sections. To promote this, commanders cross-functionally organize elements of staff sections in command posts (CPs) and CP cells. Additional staff integration occurs in meetings, including working groups and boards.

A. Command Posts

A command post is a unit headquarters where the commander and staff perform their activities. The headquarters' design of the modular force, combined with robust communications, gives commanders a flexible mission command structure consisting of a main CP, a tactical CP, and a command group for brigades, divisions, and corps. Combined arms battalions are also resourced with a combat trains CP and a field trains CP. Theater army headquarters are resourced with a main CP and a contingency CP.

- **Main Command Post (Main CP)**. The main command post is a facility containing the majority of the staff designed to control current operations, conduct detailed analysis, and plan future operations. The main CP is the unit's principal CP.
- **Tactical Command Post (Tactical CP).** The tactical command post is a facility containing a tailored portion of a unit headquarters designed to control portions of an operation for a limited time. Commanders employ the tactical CP as an extension of the main CP to help control the execution of an operation or a specific task, such as a gap crossing, a passage of lines, or an air assault operation.

B. Command Group

A command group consists of commander and selected staff members who assist the commander in controlling operations away from a command post. The command group is organized and equipped to suit the commander's decision making and leadership requirements. It does this while enabling the commander to accomplish critical mission command tasks anywhere in the area of operations.

C. Early Entry Command Post (EECP)

While not part of the unit's table of organization and equipment, commanders can establish an early-entry command post to assist them in controlling operations during the deployment phase of an operation. An early-entry command post is a lead element of a headquarters designed to control operations until the remaining portions of the headquarters are deployed and operational. The early-entry command post normally consists of personnel and equipment from the tactical CP with additional intelligence analysts, planners, and other staff officers from the main CP based on the situation.

CP Effectiveness and Survivability

Commanders consider the effectiveness and survivability of a CP when planning CP organization. In many cases these factors work against each other; therefore, neither can be optimized. Commanders make trade-offs to acceptably balance survivability and effectiveness.

Effectiveness factors include: CP design and fusion of command and staff efforts, standardization, continuity, deployability, and capacity and range. Survivability factors include: dispersion, size, redundancy, and mobility.

Refer to BSS5: The Battle Staff SMARTbook (Leading, Planning & Conducting Military Operations) for complete discussion of command post organization and operations to include command post organization considerations, command post operations (SOPs, CP battle drills, shift-change briefings, reports), effectiveness and survivablity factors, and command post cells (functional and integrating) and staff sections.

III. Army Airspace Command & Control

Ref: FM 3-52, Airspace Control (Oct '16), chap. 1.

Airspace is a component of an operational environment critical to successful Army or land operations. Airspace is not owned by individual subordinate organizations in the sense an assigned area of operations confers ownership of the ground. Airspace over an Army area of operations remains under the purview of the joint force commander (JFC). Other military and civilian organizations operating in the joint operations area have airspace requirements over an Army area of operations. These organizations may require airspace to—

- Conduct joint air operations
- Conduct area air defense
- Deliver joint fires
- Conduct civil air operations

Joint Air Operations

Normally, the JFC designates a joint force air component commander (JFACC) to synchronize the joint air effort. Components retain organic capabilities (sorties) to accomplish missions assigned by the JFC. Components also make capabilities (sorties), either JFC directed or excess, available to the JFC for tasking by the JFACC. Generally, Army capabilities (sorties) are never made available to nor directed by the JFC for tasking by the JFACC. The JFACC plans for and tasks only those joint capabilities (sorties) made available to the JFC for tasking by the JFACC. The JFACC has the authority to direct and employ these joint capabilities (sorties) for a common purpose based on the JFC's concept of operations and air apportionment decisions.

The responsibilities of the JFACC, the area air defense commander (AADC), and airspace control authority (ACA) are interrelated and the JFC normally assigns them to one individual for unity of effort. These responsibilities are normally assigned to the JFACC. Designating one Service component commander as the JFACC, AADC, and ACA often simplifies the coordination required to develop and execute fully integrated air operations.

Area Air Defense (AADC)

The AADC oversees defensive counterair (DCA) operations, which include both air and missile threats. The AADC identifies airspace coordinating measures (ACMs) that support and enhance DCA operations, identifies required airspace management systems, establishes procedures for systems to operate within the airspace, and incorporates them into the airspace control system.

Refer to JP 3-01 for more information on the AADC.

Joint Airspace Control

Competing airspace users balance the demands for and integrate their requirements for airspace. Airspace control is a process used to increase operational effectiveness by promoting the safe, efficient, and flexible use of airspace. To help balance the various airspace user demands, the JFC usually esignates an ACA responsible for establishing an airspace control system. An airspace control system is an arrangement of those organizations, personnel, policies, procedures, and facilities required to perform airspace control functions.

Refer to JP 3-52 for more information.

I. Airspace Coordinating Measures

Ref: FM 3-52, Airspace Control (Oct '16), app B (adapted from previous sources).

Methods of airspace control range from positive control of all air assets in an airspace control area to procedural control of air assets, or a combination of both.

Positive Control

Positive control relies on positive identification, tracking, and direction of aircraft within the airspace control area. It uses electronic means such as radar; sensors; identification, friend or foe (IFF) systems; selective identification feature (SIF) capabilities; digital data links; and other elements of the intelligence system and C2 network structures.

Procedural Control

Procedural control relies on a combination of mutually agreed and promulgated orders and procedures. These may include comprehensive AD identification procedures and ROE, aircraft identification maneuvers, fire support coordinating measures (FSCMs), and airspace control measures (ACMs). Service, joint, and multinational capabilities and requirements determine which method, or which elements of each method, that airspace control plans and systems use. Airspace control measures provide a variety of procedural measures of controlling airspace users and airspace.

Airspace Control Measures

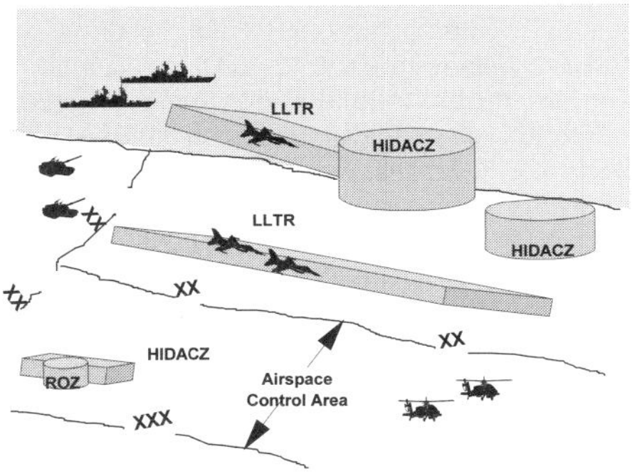

When established, airspace control measures accomplish one or more of the following:

- Reserve airspace for specific airspace users
- Restrict actions of airspace users
- Control actions of specific airspace users
- Require airspace users to accomplish specific actions

1. Air Control Point (ACP)

An easily identifiable point on the terrain or an electronic navigational aid used to provide necessary control during air movement. ACPs are generally designated at each point where the flight route or air corridor makes a definite change in direction and at any other point deemed necessary for timing or control of the operation.

2. Air Corridor

A restricted air route of travel specified for use by friendly aircraft and established to prevent friendly aircraft from being fired on by friendly forces. (Army) — Used to deconflict artillery firing positions with aviation traffic, including unmanned aerial vehicles.

3. Communications Checkpoint (CCP)

An air control point that requires serial leaders to report either to the aviation mission commander or the terminal control facility.

4. Downed Aircrew Pickup Point

A point to where aviators will attempt to evade and escape to be recovered by friendly forces.

5. High Density Airspace Control Zone (HIDACZ)

Airspace designated in an airspace control plan or airspace control order, in which there is a concentrated employment of numerous and varied weapons and airspace users. A high-density airspace control zone has defined dimensions, which usually coincide with geographical features or navigational aids. Access to a high-density airspace control zone is normally controlled by the maneuver commander. The maneuver commander can also direct a more restrictive weapons status within the high-density airspace control zone.

6. Low-Level Transit Route (LLTR)

A temporary corridor of defined dimensions established in the forward area to minimize the risk to friendly a/c from friendly ADA or ground forces.

7. Minimum-Risk Route (MRR)

A temporary corridor of defined dimensions recommended for use by high-speed, fixed-wing aircraft that presents the minimum known hazards to low-flying aircraft transiting the combat zone.

8. Restricted Operations Zone (ROZ)

A volume of airspace of defined dimensions designated for a specific mission. Entry into that zone is authorized only by the originating headquarters.

9. Standard Use Army Aircraft Flight Route (SAAFR)

Routes established below the coordinating altitude to facilitate the movement of Army aviation assets. Routes are normally located in the corps through brigade rear areas of operation and do not require approval by the airspace control authority.

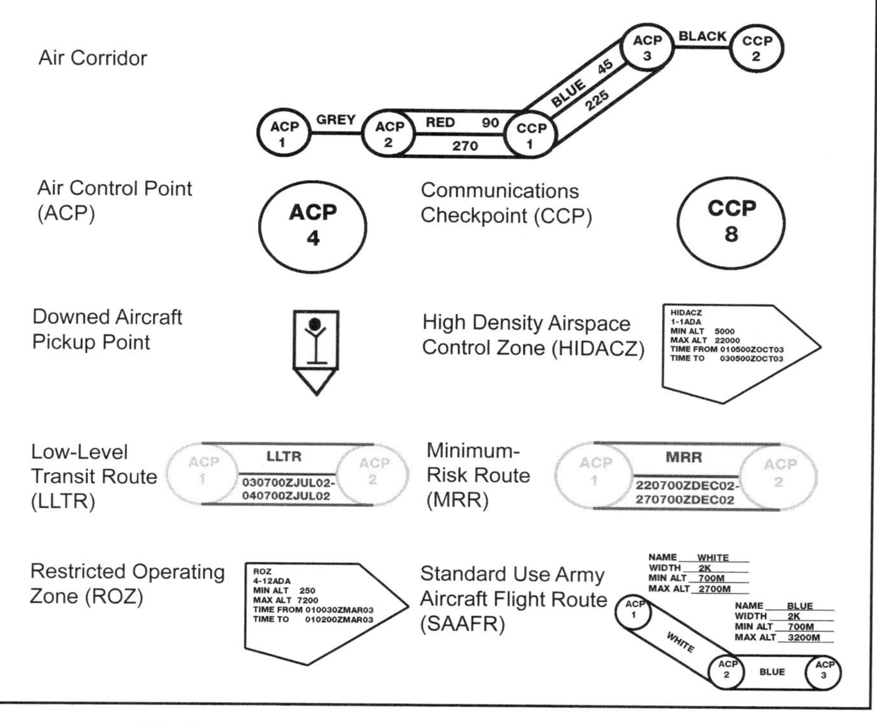

II. Key Positions and Responsibilities

There are four key positions critical to planning for and executing airspace control.

1. Joint Force Commander

The joint force commander has many responsibilities, to include the airspace control. For airspace control, the JFC specifically must—

- Include overall responsibility of airspace control and air defense in a joint theater
- Establish airspace control objectives and priorities for the joint force
- Oversee the planning and force integration activities that affect the TAGS
- Resolve matters on which the ACA is unable to obtain agreement
- Possibly retain airspace control responsibilities (or he may appoint an ACA)

The JFC may designate a JFACC as a single component commander for theater- or JOA-wide counterair operations. The JFACC will have the preponderance of air power. He also has the ability to provide C2 and produce and disseminate an ATO and ACO. He is normally appointed as the ACA and AADC. The JFC normally tasks the same person as the ACA, AADC, and JFACC to maintain the flexibility to effectively meet the enemy air threat and manage airspace control. Additional information on the selection and responsibilities of the JFACC can be found in JP 3-56.1.

2. Joint Force Air Component Commander (JFACC)

The JFACC may be sea or land based. JFACC responsibilities include—

- Developing a joint air operations plan to best support force objectives
- Recommending apportionment of the joint air effort to the JFC
- Providing centralized direction for the allocation and tasking capabilities and forces
- Controlling execution of joint operations as specified by the JFC
- Coordinating joint air operations with operations of other component commanders and forces assigned to or supporting the JFC
- Evaluating the results of joint air operations
- Functioning as the supported & supporting commander, as directed by the JFC

3. Airspace Control Authority (ACA)

The ACA is responsible for operating the airspace control system in the airspace control area. Centralized direction by the ACA does not imply assumption of operational or tactical control over any air assets. The ACA has broad responsibilities to include—

- Coordinate, integrate, and regulate the use of the airspace
- Establish broad policies and procedures for airspace control
- Establish airspace control system & integrating host-nation/multinational forces
- Develop airspace control plan (and implement through airspace control order)

4. Area Air Defense Commander (AADC)

The AADC is responsible for planning, coordinating, and integrating the joint area air defense plan. The AADC develops broad policies and procedures for air defense. The AADC has broad responsibilities to include—

- Developing and executing a plan to disseminate timely cueing of information and air and missile early warnings
- Planning, coordinating, and integrating joint air defense operations
- Developing and implementing identification and engagement procedures for air and missile threats
- Appointing a deputy AADC to assist the AADC in planning and coordinating air and missile defense operations

Chap 3

Movement & Maneuver Warfighting Function

Ref: ADRP 3-0, Operations (Nov '16), p. 5-4

The movement and maneuver warfighting function is the related tasks and systems that move and employ forces to achieve a position of relative advantage over the enemy and other threats. Direct fire and close combat are inherent in maneuver. The movement and maneuver warfighting function includes tasks associated with force projection related to gaining a position of relative advantage over the enemy. Movement is necessary to disperse and displace the force as a whole or in part when maneuvering. Maneuver is employment of forces in the operational area through movement in combination with fires to achieve a position of advantage in respect to the enemy (JP 3-0). It works through movement and with fires to achieve a position of relative advantage over the enemy to accomplish the mission and consolidate gains. Commanders use maneuver for massing the effects of combat power to achieve surprise, shock, and momentum. Effective maneuver requires close coordination with fires. Both tactical and operational maneuver require sustainment support. The movement and maneuver warfighting function includes the following tasks:

- Move.
- Maneuver.
- Employ direct fires.
- Occupy an area.
- Conduct mobility and countermobility.
- Conduct reconnaissance and surveillance.
- Employ battlefield obscuration.

The movement and maneuver warfighting function does not include administrative movements of personnel and materiel. These movements fall under the sustainment warfighting function.

In addition to the basic tactical concepts in ADRP 3-90, commanders consider the following when performing movement and maneuver warfighting function tasks:

- Various ways and means help maneuver forces attain positional advantage. For example, the planning of civil affairs operations may minimize civilian interference with operations and minimize the impact of military operations on the population.
- Successful movement and maneuver requires agility and versatility of thought, plans, operations, and organizations. It requires designating the main effort and then, if necessary, shifting the main effort and applying the principles of mass and economy of force.

Editor's note: For the purposes of The Army Operations & Doctrine SMARTbook, an overview of the following topics are represented as they relate to the movement and maneuver warfighting function:

I. Offense and Defense (Decisive Action)

Ref: ADP 3-90, Offense and Defense (Aug '12).

Tactics is the employment and ordered arrangement of forces in relation to each other (CJCSM 5120.01). Through tactics, commanders use combat power to accomplish missions. The tactical-level commander employs combat power in the conduct of engagements and battles. This section addresses the tactical level of war, the art and science of tactics, and hasty versus deliberate operations.

The Tactical Level of War

ADP 3-90 is the primary manual for offensive and defensive tasks at the tactical level. It does not provide doctrine for stability or defense support of civil authorities tasks. It is authoritative and provides guidance in the form of combat tested concepts and ideas for the employment of available means to win in combat. These tactics are not prescriptive in nature, and they require judgment in application.

The tactical level of war is the level of war at which battles and engagements are planned and executed to achieve military objectives assigned to tactical units or task forces (JP 3-0). Activities at this level focus on the ordered arrangement and maneuver of combat elements in relation to each other and to the enemy to achieve combat objectives. It is important to understand tactics within the context of the levels of war. The strategic and operational levels provide the context for tactical operations. Without this context, tactical operations are just a series of disconnected and unfocused actions. Strategic and operational success is a measure of how one or more battles link to winning a major operation or campaign. In turn, tactical success is a measure of how one or more engagements link to winning a battle.

See also pp. 3-17 to 3-22.

The Offense

The offense is the decisive form of war. While strategic, operational, or tactical considerations may require defending for a period of time, defeat of the enemy eventually requires shifting to the offense. Army forces strike the enemy using offensive action in times, places, or manners for which the enemy is not prepared to seize, retain, and exploit the operational initiative. Operational initiative is setting or dictating the terms of action throughout an operation (ADRP 3-0).

The main purpose of the offense is to defeat, destroy, or neutralize the enemy force. Additionally, commanders conduct offensive tasks to secure decisive terrain, to deprive the enemy of resources, to gain information, to deceive and divert the enemy, to hold the enemy in position, to disrupt his attack, and to set the conditions for future successful operations.

The Defense

While the offense is the most decisive type of combat operation, the defense is the stronger type. Army forces conduct defensive tasks as part of major operations and joint campaigns, while simultaneously conducting offensive and stability tasks as part of decisive action outside the United States.

Commanders choose to defend to create conditions for a counteroffensive that allows Army forces to regain the initiative. Other reasons for conducting a defense include to retain decisive terrain or deny a vital area to the enemy, to attrit or fix the enemy as a prelude to the offense, in response to surprise action by the enemy, or to increase the enemy's vulnerability by forcing the enemy to concentrate forces.

Tactical Enabling Tasks

Commanders direct tactical enabling tasks to support the conduct of decisive action. Tactical enabling tasks are usually shaping or sustaining. They may be decisive in the conduct of stability tasks. Tactical enabling tasks discussed in ADRP 3-90 include reconnaissance, security, troop movement, relief in place, passage of lines, encirclement operations, and urban operations. Stability ultimately aims to create a condition so the local populace regards the situation as legitimate, acceptable, and predictable.

Offense and Defense (Unifying Logic Chart)

Unified Land Operations

Seize, retain, and exploit the initiative to gain and maintain a position of relative advantage in sustained land operations in order to create the conditions for favorable conflict resolution.

Executed through...

Decisive Action

offensive | defensive | stability | DSCA

Offensive tasks	Defensive tasks
• Movement to contact - Search and attack - Cordon and search **• Attack** - Ambush - Counterattack - Demonstration - Spoiling attack - Feint - Raid **• Exploitation** **• Pursuit**	**• Area defense** **• Mobile defense** **• Retrograde operations** - Delay - Withdrawal - Retirement
Forms of maneuver • Envelopment • Flank attack • Frontal attack • Infiltration • Penetration • Turning movement	**Forms of the defense** • Defense of a linear obstacle • Perimeter defense • Reverse slope defense

Tactical enabling tasks

Tactical mission tasks

Ref: ADP 3-90, Offense and Defense, fig. 1, p. iv.

Refer to SUTS2: The Small Unit Tactics SMARTbook (Leading, Planning & Conducting Tactical Operations) for complete discussion of offensive and defensive operations. Related topics include tactical mission fundamentals, stability & counterinsurgency operations, tactical enabling operations, special purpose attacks, urban operations & fortifications, and patrols & patrolling.

II. Stability Operations (Decisive Action)

Ref: ADP 3-07, Stability (Aug '12).

Stability ultimately aims to create a condition so the local populace regards the situation as legitimate, acceptable, and predictable. These conditions consist of the level of violence; the functioning of governmental, economic, and societal institutions; and the general adherence to local laws, rules, and norms of behavior.

Sources of instability manifest themselves locally. First, instability stems from decreased support for the government based on what locals actually expect of their government. Second, instability grows from increased support for anti-government elements, which usually occurs when locals see spoilers as helping to solve the priority grievance. Lastly, instability stems from the undermining of the normal functioning of society where the emphasis must be on a return to the established norms.

Stabilization is a process in which personnel identify and mitigate underlying sources of instability to establish the conditions for long-term stability. While long-term development requires stability, stability does not require long-term development. Therefore, stability tasks focus on identifying and targeting the root causes of instability and by building the capacity of local institutions.

Primary Army Stability Tasks

Army units conduct five primary stability tasks. These tasks support efforts that encompass both military and nonmilitary efforts generally required to achieve stability. These tasks are similar to and nested with the joint functions and DOS stability sectors. Taken together, they provide a base for linking the execution of activities among the instruments of national and international power as part of unified action.

1. Establish Civil Security. Establishing civil security involves providing for the safety of the host nation and its population, including protection from internal and external threats. Establishing civil security provides needed space for host-nation and civil agencies and organizations to work toward sustained peace.

2. Establish Civil Control. Establishing civil control supports efforts to institute rule of law and stable, effective governance. Civil control relates to public order—the domain of the police and other law enforcement agencies, courts, prosecution services, and prisons (known as the Rule of Law sector).

3. Restore Essential Services. The restoration of essential services in a fragile environment is essential toward achieving stability. The basic functions of local governance stop during conflict and other disasters. Initially, military forces lead efforts to establish or restore the most basic civil services: the essential food, water, shelter, and medical support necessary to sustain the population until forces restore local civil services. Military forces follow the lead of other USG agencies, particularly United States Agency for International Development, in the long restoration of essential services.

4. Support to Governance. When a legitimate and functional host-nation government exists, military forces operating to support the state have a limited role. However, if the host-nation government cannot adequately perform its basic civil functions—whatever the reason—some degree of military support to governance may be necessary. Military efforts to support governance focus on restoring public administration and resuming public services.

5. Support to Economic and Infrastructure Development. Military efforts to support the economic sector are critical to sustainable economic development. The economic viability of a host nation often exhibits stress and ultimately fractures as conflict, disaster, and internal strife overwhelms the government. Signs of economic stress include rapid increases in inflation, uncontrolled escalation of public debt, and a general decline in the host nation's ability to provide for the well-being of its people. Economic problems inextricably connect to governance and security concerns. As one institution begins to fail, others likely follow.

Stability Underlying Logic

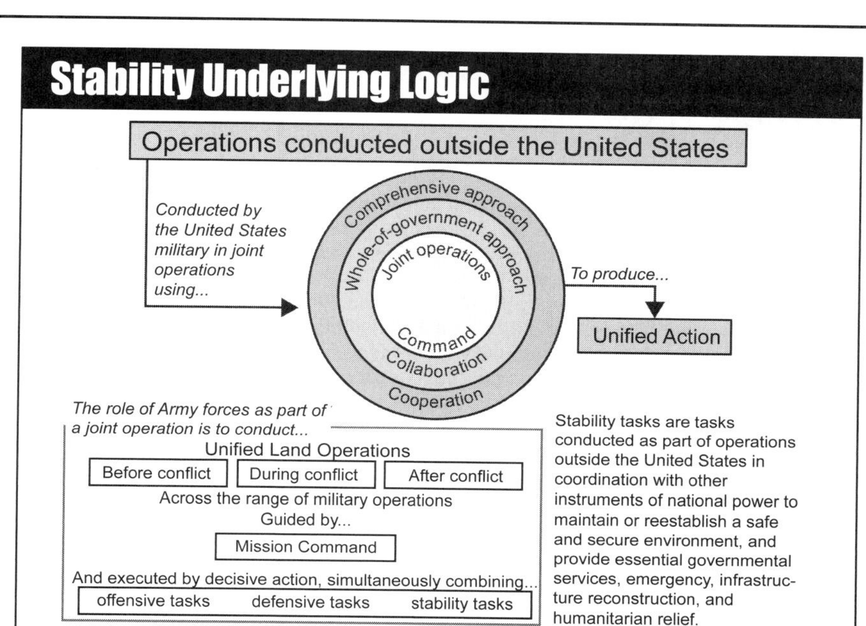

Stability tasks are tasks conducted as part of operations outside the United States in coordination with other instruments of national power to maintain or reestablish a safe and secure environment, and provide essential governmental services, emergency, infrastructure reconstruction, and humanitarian relief.

To do this the Army conducts the primary stability tasks integrated into the joint stability functions and the United States Government stability sectors to achieve the end state conditions...

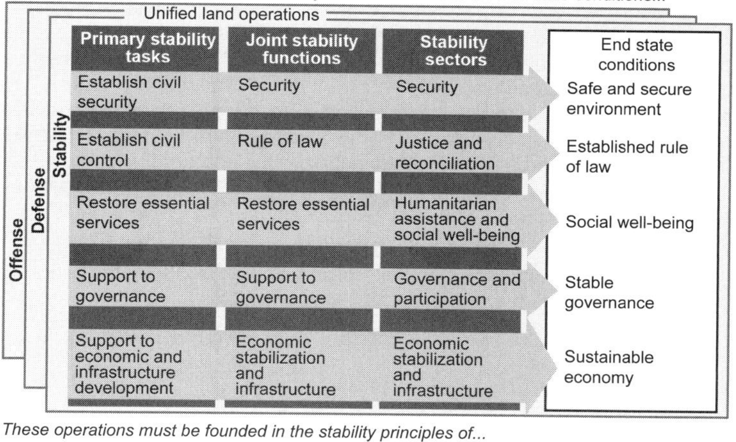

Primary stability tasks	Joint stability functions	Stability sectors	End state conditions
Establish civil security	Security	Security	Safe and secure environment
Establish civil control	Rule of law	Justice and reconciliation	Established rule of law
Restore essential services	Restore essential services	Humanitarian assistance and social well-being	Social well-being
Support to governance	Support to governance	Governance and participation	Stable governance
Support to economic and infrastructure development	Economic stabilization and infrastructure	Economic stabilization and infrastructure	Sustainable economy

These operations must be founded in the stability principles of...

Conflict termination

Unity of effort

Legitmacy and host-nation ownership

Building partner capacity

Ref: ADP 3-07, Stability, fig. 1, p. iii.

Refer to TAA2: Military Engagement, Security Cooperation & Stability SMARTbook (Foreign Train, Advise, & Assist) for further discussion. Topics include the Range of Military Operations (JP 3-0), Security Cooperation & Security Assistance (Train, Advise, & Assist), Stability Operations , Peace Operations, Counterinsurgency Operations, Civil-Military Operations, and more!

III. Defense Support of Civil Authorities (Decisive Action)

Ref: ADP 3-28, Defense Support of Civil Authorities (Jul '12).

In March 2011, The President of the United States signed Presidential Policy Directive 8, to strengthen "the security and resilience of the United States through systematic preparation for the threats that pose the greatest risk to the security of the Nation, including acts of terrorism, cyber attacks, pandemics, and catastrophic natural disasters." In support of this directive, the Department of Homeland Security, primarily through the Federal Emergency Management Agency (FEMA), maintains national doctrine for all aspects of incident management, defined as a national comprehensive approach to preventing, preparing for, responding to, and recovering from terrorist attacks, major disasters, and other emergencies. Incident management includes measures and activities performed at the local, state, and national levels and includes both crisis and consequence management activities (JP 3-28). ADP 3-0 states that Army forces operate as part of a larger national effort characterized as unified action—the synchronization, coordination, and/or integration of the activities of governmental and nongovernmental entities with military operations to achieve unity of effort (JP 1).

Army Support of Civil Authorities

Army forces support civil authorities by performing defense support of civil authorities tasks. Defense support of civil authorities is defined as support provided by United States Federal military forces, DoD [Department of Defense] civilians, DoD contract personnel, DoD component assets, and National Guard forces (when the Secretary of Defense, in coordination with the Governors of the States, elects and requests to use those forces in title 32, United States Code, status) in response to requests for assistance from civil authorities for domestic emergencies, law enforcement support, and other domestic activities, or from qualifying entities for special events. Also known as civil support (DODD 3025.18).

Military forces provide civil support at federal and state levels. Federal military forces are active Army, Marine Corps, Navy, and Air Force; mobilized Army, Marine Corps, Navy, and Air Force Reserve; and National Guard mobilized for federal service under title 10, United States Code (USC). State National Guard forces under state control perform DSCA tasks when serving under title 32, USC.

Primary Purposes of Army Support

While there are many potential missions for Soldiers as part of DSCA, the overarching purposes of all DSCA missions are, in the following order, to—

- Save lives
- Alleviate suffering
- Protect property

Primary Characteristics of Army Support

Army forces operating within the United States encounter very different operational environments than they face outside the Nation's boundaries. The support provided by Army forces depends on specific circumstances dictated by law. Soldiers and Army civilians need to understand domestic environments so they can employ the Army's capabilities efficiently, effectively, and legally.

While every domestic support mission is unique, four defining characteristics shape the actions of commanders and leaders in any mission. These characteristics are that—

- State and federal laws define how military forces support civil authorities
- Civil authorities are in charge, and military forces support them
- Military forces depart when civil authorities are able continue w/o military support
- Military forces must document costs of all direct and indirect support provided

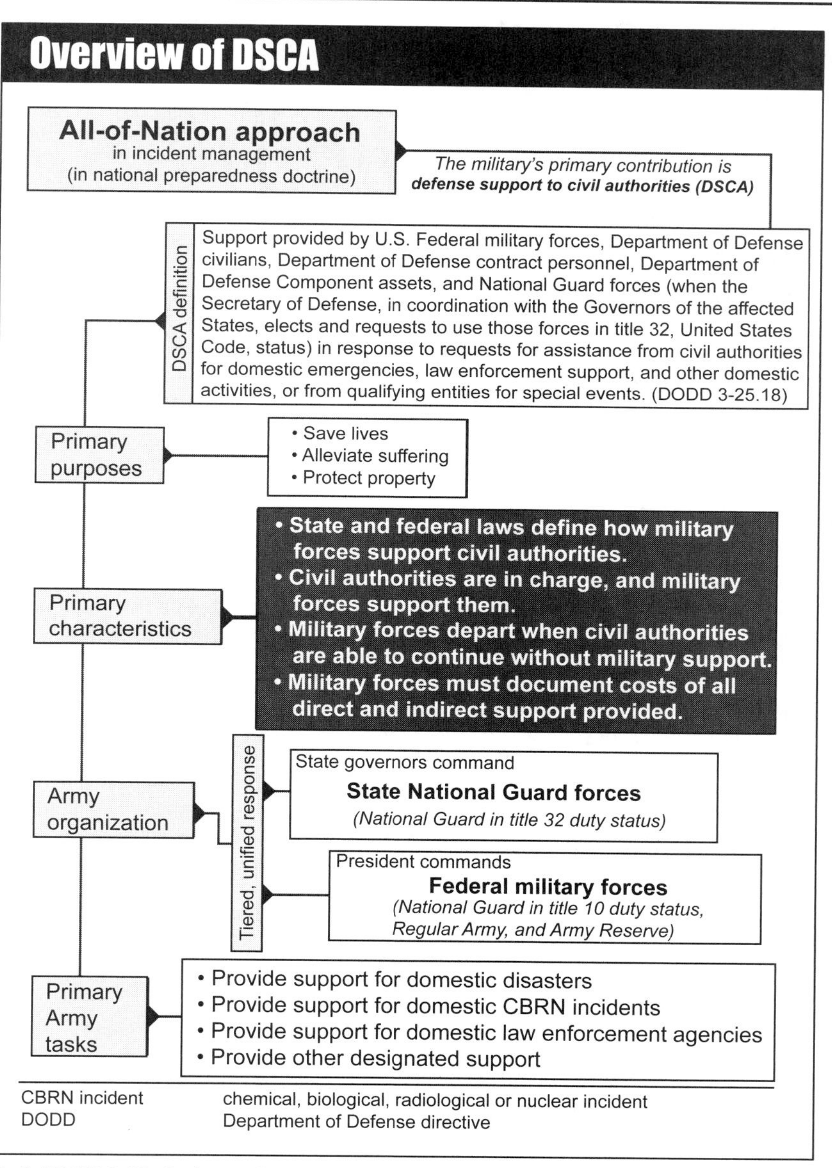

Ref: ADRP 3-28, Defense Support of Civil Authorities, fig. 1, p. iv.

Refer to HDS1: The Homeland Defense & DSCA SMARTbook. Topics include homeland defense, (JP 3-28), defense support of civil authorities (JP 3-28), Army support of civil authorities (ADRP 3-28), multi-service DSCA TTPs (ATP 3-28.1/MCWP 3-36.2), DSCA liaison officer toolkit (GTA 90-01-020), key legal and policy documents, and specific hazard and planning guidance.

IV. Special Operations

Ref: ADP 3-05, Special Operations (Aug '12).

Special operations are operations requiring unique modes of employment, tactical techniques, equipment, and training often conducted in hostile, denied, or politically sensitive environments and characterized by one or more of the following: time sensitive, clandestine, low visibility, conducted with and/or through indigenous forces, requiring regional expertise, and/or a high degree of risk (JP 3-05, Special Operations).

Army Special Operations Forces Critical Capabilities

Army special operations forces have a significant role in the successful outcome of unconventional warfare, counterterrorism, and counterinsurgency campaigns. Special operations forces provide a lethal, unilateral, or collaborative and indigenous counter-network capability against insurgent and terrorist groups, a means to assess and moderate population behavior by addressing local underlying causes, and a means to organize indigenous security and governmental structures. In both special warfare and surgical strike capabilities, Army special operations forces provide a population centric, intelligence-enabled capability that works with multinational partners and host nations to develop regional stability, enhance global security, and facilitate future operations.

Special Warfare

Special warfare is the execution of activities that involve a combination of lethal and nonlethal actions taken by a specially trained and educated force that has a deep understanding of cultures and foreign language, proficiency in small-unit tactics, and the ability to build and fight alongside indigenous combat formations in a permissive, uncertain, or hostile environment. Special warfare is an umbrella term that represents special operations forces conducting combinations of unconventional warfare, foreign internal defense, and/or **counterinsurgency** through and with indigenous forces or personnel in politically sensitive and/or hostile environments.

- **Unconventional warfare** is defined as activities conducted to enable a resistance movement or insurgency to coerce, disrupt, or overthrow a government or occupying power by operating through or with an underground, auxiliary, and guerrilla force in a denied area (JP 3-05).
- **Foreign internal defense** is participation by civilian and military agencies of a government in any of the action programs taken by another government or other designated organization to free and protect its society from subversion, lawlessness, insurgency, terrorism, and other threats to its security (JP 3-22).

Surgical Strike

Surgical strike is the execution of activities in a precise manner that employ special operations forces in hostile, denied, or politically sensitive environments to seize, destroy, capture, exploit, recover or damage designated targets, or influence threats.

Surgical strike activities include actions against critical operational or strategic targets, to include counterproliferation actions, counterterrorism actions, and hostage rescue and recovery operations.

- **Counterproliferation** actions prevent the threat and/or use of weapons of mass destruction against the United States, its forces, allies, and partners.
- **Counterterrorism** actions taken directly and indirectly against terrorist networks influence and render global and regional environments inhospitable to terrorist networks. Hostage rescue and recovery operations, which are sensitive crisis-response missions, include offensive measures taken to prevent, deter, preempt, and respond to terrorist threats and incidents, including recapture of U.S. facilities, installations, and sensitive material.

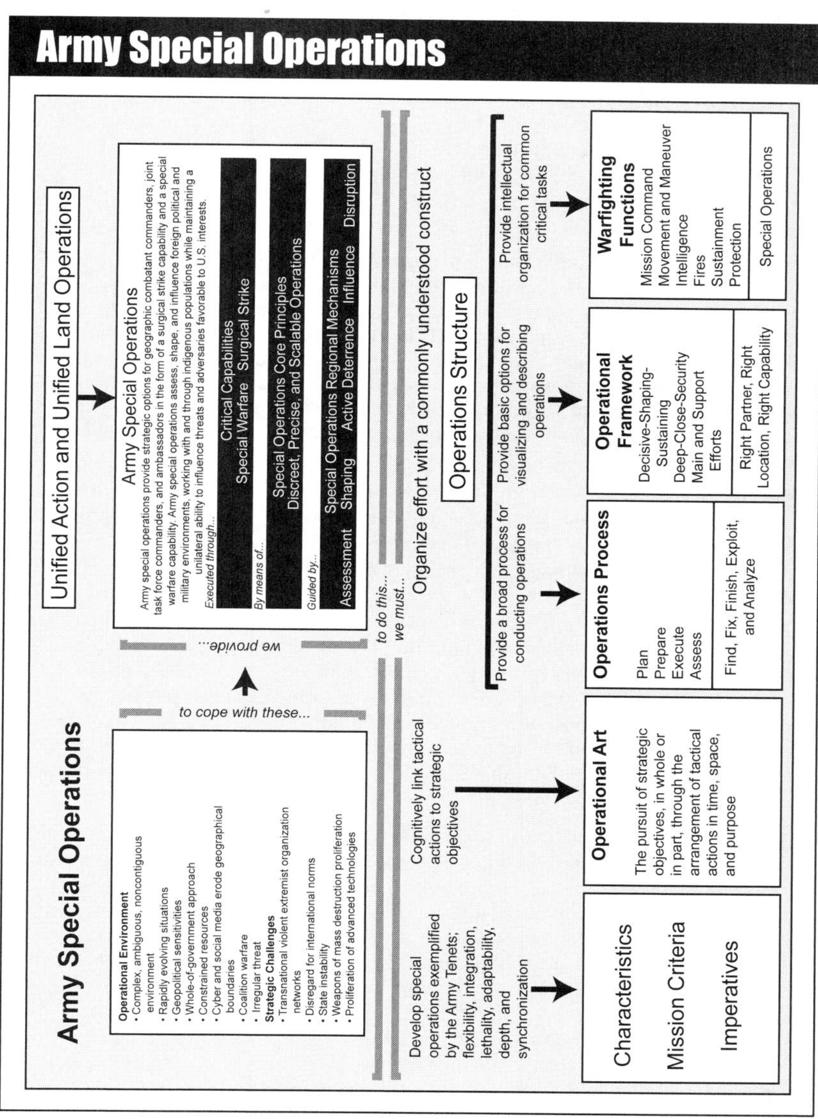

Ref: ADP 3-05, Special Operations, fig. 1, p. iii.

Refer to CTS1: The Counterterrorism & Hybrid Threat SMARTbook for related discussion. Counterterrorism is one of the core tasks of the US special operations forces (SOF), and their role and additive capability is to conduct offensive measures within Department of Defense's (DOD's) overall combating terrorism (CbT) efforts.

V. Deployment & Redeployment

Ref: FM 3-35, Army Deployment and Redeployment (Apr '10).

FM 3-35 is the Army's authoritative doctrine for planning, organizing, executing, and supporting deployment and redeployment. It represents the culmination of our efforts to consolidate all deployment doctrine (FM 100-17, FM 100-17-3, FM 3-35.4, FM 100-17-5, and FM 4-01.011) into a single manual to align Army deployment doctrine with joint deployment doctrine.

Force Projection

Force projection is the military element of national power that systemically and rapidly moves military forces in response to requirements of full spectrum operations. It is a demonstrated ability to alert, mobilize, rapidly deploy, and operate effectively anywhere in the world. The Army, as a key member of the joint team, must be ready for global force projection with an appropriate mix of combat forces together with support and sustainment units. Moreover, the world situation demands that the Army project its power at an unprecedented pace. The flexible and rapidly deployable Army forces with sufficient depth and strength to sustain multiple, simultaneous operations contributes to defusing the crisis.

Deployment

Deployment is composed of activities required to prepare and move forces, supplies, and equipment to a theater. This involves the force as it task organizes, tailors itself for movement based on the mission, concept of operations, available lift, and other resources.

The Joint deployment process is divided into four phases -- deployment planning; predeployment activities; movement; and JRSOI. The terminology used to describe the Army deployment phases is in synch with the Joint process. The Joint process includes a planning phase at the outset whereas the Army considers planning to be woven through all the phases. Moreover, the movement phase in the Army process is discussed in two segments – fort to port and port to port. The Army relies on U.S. Transportation Command (USTRANSCOM) to provide the strategic lift to and from the port of embarkation (POE).

Army Force Generation (ARFORGEN)

ARFORGEN is the structured progression of increased unit readiness over time resulting in recurring periods of availability of trained, ready, and cohesive units prepared for operational deployment in support of civil authorities and combatant commander requirements. ARFORGEN uses personnel, equipment, and training to generate forces to meet current and future requirements of combatant commanders. These cyclical readiness process forces commanders to recognize that all units are not ready all the time.

- **Reset**. Units returning from operations or have experienced significant organizational changes are placed in the reset phase
- **Train/Ready.** Units determined to be at a ready level are capable of beginning their mission preparation and collective training with other operational headquarters
- **Available.** Units are capable of conducting a mission under any combatant commander

Refer to SMFLS4: The Sustainment & Multifunctional Logistics SMARTbook (Warfighter's Guide to Logistics, Personnel Services, & Health Services Support) for complete discussion of deployment; reception, staging, onward movement and integration (RSOI); and redeployment operations.

Chap 3

I. Tactical Echelons

Ref: ADRP 3-90, Offense and Defense (Aug '12), pp. 2-13 to 2-17.

The Army echelons its broad array of capabilities to perform diverse functions. These functions vary with the type of unit and, particularly at operational echelons, with the organization of the theater, the nature of the conflict, and the number of friendly forces committed to the effort.

Tactical Echelons

- **A** Fire Team
- **B** Crew
- **C** Squad
- **D** Section
- **E** Platoon
- **F** Companies, Batteries, Troops and Detachments
- **G** Battalions and Squadrons
- **H** Brigades, Regiments, and Groups
- **I** Division

At each echelon, the commander task organizes available capabilities to accomplish the mission. The commander's purpose in task organization is to maximize subordinate commanders' abilities to generate a combined arms effect consistent with the concept of operations. Commanders and staffs work to ensure the distribution of capabilities to the appropriate components of the force to weight the decisive operation. The relationships between units within and supporting an echelon are described in terms of command and support relationships.

See pp. 2-27 to 2-34 for discussion of command and support relationships.

A. Fire Team

A fire team is a small military unit. A fire team generally consists of four or fewer soldiers and is usually grouped by two or three teams into a squad or section. The concept of the fire team is based on the need for tactical flexibility. A fire team is capable of autonomous operations as part of its next larger unit, such as a squad or section. It is usually led by a sergeant.

B. Crew

A crew consists of all personnel operating a particular system. This system might be a weapons system, such as a tank or machinegun. The system might also be a vehicle, such as a helicopter, or a sensor system, such as a target acquisition radar. The rank of the senior crew member can vary widely from a junior noncommissioned officer to a commissioned or warrant officer based on the system.

C. Squad

A squad is a small military unit typically containing two or more fire teams. It typically contains a dozen Soldiers or less. In some cases the crew of a system may also be designated as a squad. Squads are usually led by a staff sergeant.

D. Section

A section is an Army unit smaller than a platoon and larger than a squad. A section may consist of the crews of two or more Army systems, such as a tank section, or several fire teams.

E. Platoon

A platoon is a subdivision of a company or troop consisting of two or more squads or sections. A platoon is normally led by a lieutenant. Platoons tend to contain roughly 30 Soldiers, but in some cases they contain significantly more or less than that number.

F. Companies, Batteries, Troops, and Detachments

A company is a unit consisting of two or more platoons, usually of the same type, with a headquarters and a limited capacity for self-support. A troop is a company-size unit in a reconnaissance organization. A battery is a company-size unit in a fires or air defense artillery battalion. A company normally consists of more than 75 soldiers. Some aviation and armor companies are exceptions to this rule. Companies and air defense and artillery batteries are the basic elements of battalions. Companies, batteries, and troops may also be assigned as separate units of brigades and larger organizations. Some companies, such as special forces companies, have subordinate detachments, instead of platoons, which are organized and trained to operate independently for extended periods. A detachment is a tactical element organized on either a temporary or permanent basis for special duties.

Company-size combat units can fight in mass or by subordinate platoons. Reconnaissance troops frequently operate with their platoons in separate areas. In combined arms battalions, companies fight either as integral units or as task-organized teams reinforced with close-combat platoons of the same or different types. A company team is a combined arms organization formed by attaching one or more nonorganic armor, mechanized infantry, Stryker, or infantry platoons to an armor, mechanized infantry, Stryker, or infantry company, either in exchange for, or in addition to, its organic platoons. These company teams can include other supporting squads or platoons, such as engineers. Company teams are task-organized for specific missions. Such teams can match capabilities to missions with greater precision than units using only organic platoons. However, the attachment of different units at the company level demands thorough training to achieve the maximum complementary effects. Whenever possible, platoons and detachments should train together before they are committed.

Military Units & Echelons

Ref: ADRP 1-02, Operational Terms and Military Symbols (Feb '15), chap. 3.

ADRP 1-02 provides a single standard for developing and depicting hand-drawn and computer-generated military symbols for situation maps, overlays, and annotated aerial photographs for all types of military operations. It is the Army proponent publication for all military symbols and complies with Department of Defense (DOD) Military-Standard (MIL-STD) 2525C, Common Warfighting Symbology.

A military symbol is a graphic representation of a unit, equipment, installation, activity, control measure, or tactical task relevant to military operations that is used for planning or to represent the common operational picture on a map, display or overlay. Military symbols are governed by the rules in Military Standard (MIL-STD) 2525C.

Unit

A unit is a military element whose structure is prescribed by a competent authority, such as a table of organization and equipment; specifically, part of an organization (JP 1-02).

Echelons

An echelon is a separate level of command (JP 1-02). In addition, there is also a separate echelon known as a command. A command is a unit or units, an organization, or an area under the command of one individual. It does not correspond to any of the other echelons. The table below shows the field 6 amplifiers for Army echelons and commands.

Full-Frame Icons for Units

Full-frame icons may reflect the main function of the symbol or may reflect modifying information.

Function (Hstorical derivation in italics)	Icon	Example
Air defense (radar dome)		
Antitank or antiarmor (upside down V)		
Armored cavalry		
Armored infantry (mechanized infantry)		
Corps support		
Headquarters or headquarters element		
Infantry (crossed straps)		
Medical (Geneva cross)		
Reconnaissance (cavalry) (cavalry bandoleer)		
Signal (lightning flash)		
Supply		
Theater or echelons above corps support		

Echelon Amplifiers (Field 6)

An echelon is a separate level of command. In addition, there is also a separate echelon known as a command.

Echelon	Amplifier
Team/crew	Ø
Squad	●
Section	●●
Platoon/detachment	●●●
Company/battery/troop	I
Battalion/squadron	II
Regiment/group	III
Brigade	X
Division	XX
Corps	XXX
Army	XXXX
Army group	XXXXX
Theater	XXXXXX
Nonechelon	**Amplifier**
Command	++

Refer to BSS5: The Battle Staff SMARTbook (Leading, Planning & Conducting Military Operations) for a full chapter on operational terms and graphics from ADRP 1-02. Sections include operational terms and acronyms; military symbology basics; units, individuals and organizations; equipment, installations and activities; control measure symbols; and tactical mission tasks.

Fires batteries are the basic firing units of fires battalions. They are organized with firing platoons, a headquarters, and limited support sections. They may fire and displace together or by platoons. Normally, batteries fight as part of their parent battalions, but the commander can establish a command or support relationship to other batteries or other units. In rare cases batteries or even platoons respond directly to a maneuver battalion or company. If required, multiple launch rocket system (MLRS) batteries are equipped to operate independently from battalion control.

Air defense artillery batteries operate as the fighting elements of air defense artillery battalions. Air defense batteries provide fires for protection against lower and upper tier aerial threats. Various air defense systems provide short to long range capabilities. They are employed using the principles of mass, mix, mobility, and integration.

Combat engineer companies control a mixture of different types of engineer platoons. Each type of BCT has an organic engineer company. In armored BCTs and infantry BCTs they are part of the brigade special troops battalion. The engineer company in a Stryker BCT is a separate company. Combat engineer companies attached to engineer battalions may be employed in a variety of tasks, or they may be placed in some type of support relationship to BCTs or functional or multifunctional support brigades.

Most functional and multifunctional support brigades contain attached separate companies with greater self-sustainment capabilities than normally found in comparable size maneuver organizations. However, they may receive unit-level sustainment support on an area basis. Such functional and multifunctional support companies vary widely in size, employment, and assignment.

G. Battalions and Squadrons

A battalion (or a reconnaissance squadron) is a unit consisting of two or more company-, battery-, or troop-size units and a headquarters. Most battalions range in size between 500 and 800 Soldiers, although some sustainment battalions are larger. Most maneuver battalions are organized by branch, arm, or service and, in addition to their line companies, contain a headquarters company. Combined arms battalions are exceptions to this rule, in that they contain two mechanized infantry companies and two armor companies. Typically, battalions have three to five companies in addition to their headquarters.

A BCT commander can task organize subordinate maneuver battalions with other maneuver and functional and multifunctional support companies to form task forces for special missions. A battalion task force is a maneuver battalion-size unit consisting of a battalion headquarters, at least one assigned company-size element, and at least one attached company-size element from another maneuver or support unit (functional or multifunctional). Combined arms battalions are permanently organized battalion task forces and have their own graphic symbol. Task organization increases the capability of maneuver battalions. Fires battalions may control batteries of any kind from other fires battalions through an established support relationship. The commander can reinforce engineer battalions with the same or different types of engineer companies and platoons to form engineer task forces.

Functional and multifunctional support and sustainment battalions vary widely in type and organization. They may perform functional services for a larger supported unit within that unit's AO. All battalions are capable of short-term, limited self-defense.

H. Brigades, Regiments, and Groups

A brigade is a unit consisting of two or more battalions and a headquarters company or detachment. A brigade normally contains between 2,500 to 5,000 Soldiers. Its capacity for independent action varies by its type. Division commanders use armored, infantry, or Stryker brigade combat teams, supported by multifunctional support brigades—fires brigades, combat aviation brigades, maneuver enhancement brigades (MEBs), battlefield surveillance brigades, and sustainment brigades—and functional brigades, such as air and missile defense brigades, engineer brigades, civil affairs brigades, and military police brigades, to accomplish their assigned missions. Sustainment brigades are normally assigned to the theater sustainment command and provide support to other Army units, usually on an area basis.

A brigade combat team is a combined arms organization consisting of a brigade headquarters, at least two maneuver battalions, and necessary supporting functional capabilities. BCTs are the largest fixed tactical units in the Army. However, additional battalions and companies may be attached to them or their organic battalions, and companies can be detached from them as part of force tailoring at the strategic and operational levels and task organization at the tactical level as required by the mission variables of METT-TC. Infantry and armored BCTs normally include a fires battalion, a brigade support battalion, and a brigade special troops battalion in addition to their two maneuver battalions and a reconnaissance squadron. Stryker BCTs do not have a brigade special troops battalion. They have separate functional companies. They also have a third maneuver battalion and a fires battalion. BCTs combine the efforts of their battalions and companies to fight engagements and perform tactical tasks within division-level battles and major operations. Their chief tactical responsibility is synchronizing the plans and actions of their subordinate units to accomplish assigned tasks for a division headquarters.

Army Regulation (AR) 600-82 defines a regiment as a single or a group of like-type combat arms or training units authorized a regimental color. (The term "combat arms" is a rescinded doctrinal term, but it still exists in Army regulations.) Traditionally, a regiment was a fixed tactical unit with organic combat battalions and supply and support organizations. The Army currently retains only one traditional tactical regiment, the 75th Rangers. All of the Army's other regiments have no tactical function. Instead, they are intended to perpetuate regimental history, esprit de corps, and traditions for Soldiers affiliated with a regiment. Many of the Army's branches contain only a single regiment, such as the Corps of Engineers and the Military Police Corps. Each maneuver battalion or squadron carries an association with a parent regiment. In some BCTs and brigades several numbered battalions carrying the same regimental association serve together, and tend to consider themselves part of the traditional regiment, when in fact they are independent battalions serving a brigade, rather than a regimental headquarters. For example, some BCTs, such as the 3rd Cavalry Regiment, bear the title of cavalry regiments for historical purposes.

Groups are brigade-size organizations that, as a result of Army modularity, will continue to be rarely used outside the Army's special operations forces. The Army's modular design deactivates group headquarters in favor of activating additional brigade headquarters. Exceptions to this are explosive ordnance disposal (EOD), criminal investigation division (CID), regional support groups, military information support, and special forces. Traditionally, a group headquarters could be established under a brigade as an intermediate headquarters for two or more functional and multifunctional support and sustainment battalions when the span of control of the brigade exceeded seven battalion-size subordinate units. Group headquarters were extensively organized as building blocks in large sustainment organizations and functional commands, such as theater army sustainment commands and engineer commands.

I. Division

A division is an Army echelon of command above brigade and below corps. It is a tactical headquarters which employs a combination of brigade combat teams, multifunctional brigades, and functional brigades in land operations. The division headquarters is a self-contained organization with a command group and a fully functional staff that requires no staff support from subordinate units to provide functional staff capabilities for its primary role. The organization consists of a headquarters and headquarters battalion, which provides administrative and sustainment support to the division headquarters in garrison and when deployed for operations.

Movement & Maneuver

The division headquarters may command or be supported by a variable number of subordinate BCTs, multifunctional brigades, and functional brigades, depending upon the role of the headquarters and the mission variables of METT-TC. Typically, a division headquarters would have between two and five subordinate BCTs, plus a tailored set of subordinate multifunctional brigades. These multifunctional brigades may include a fires brigade and one or more combat aviation brigades, a battlefield surveillance brigade, and a maneuver enhancement brigade. The functional brigades consist of military police, engineer, air and missile defense, and military intelligence brigades. BCTs and multifunctional brigades normally have command relationships with the division headquarters (assigned, attached, operational control [OPCON] or tactical control [TACON]). Functional brigades may have a command relationship with the division headquarters, or may have a support relationship (direct support or general support), or may support friendly forces on an area basis. Sustainment and medical units supporting the division headquarters and the forces task organized under it are assigned to the theater sustainment command and will normally provide support on an area basis, although under unusual circumstances they may be placed OPCON or TACON to the division headquarters, when the mission variables of METT-TC make a command relationship more practical and effective.

The division headquarters provides a flexible mission command capability and utility in all operational environments. The division headquarters may be used in other roles, including as the senior Army headquarters, joint force land component, or joint task force headquarters in a joint operations area for smaller scale operations. However, when performing these roles, the division requires significant Army and joint augmentation.

The mission variables of METT-TC determine the optimal size and mix of capabilities of the forces task organized under each division headquarters. The size, composition and capabilities of the forces task organized under the division headquarters may vary between divisions involved in the same joint campaign, and may change from one phase of that campaign to another. Operations focused on destruction of a conventional enemy military force (offense and defense tasks) may require a mix of forces and capabilities that is quite different from those required for an operation focused on protection of civil populations (stability tasks). For conventional military operations, the division should have one of each type of multifunctional and functional brigade in order to have available all of the capabilities required to conduct combined arms operations.

The division normally operates as a tactical headquarters under the operational control of an Army corps, ARFOR, or joint force land component commander. As a tactical echelon of mission command, the division headquarters arranges the multiple tactical actions of its subordinates in time, space and purpose to achieve significant military objectives. The division headquarters coordinates and synchronizes the tactical actions of subordinate brigades and directs them toward a common purpose or higher order objective. The division headquarters leverages joint force capabilities and conducts shaping operations within its area of operations in order to establish favorable conditions for the success of its main effort or decisive operation. The division allocates resources, designates the main effort, forecasts operational requirements, and establishes the priorities of support within its task organized forces. Sustainment and medical forces and functional units (military police, engineer, air and missile defense, and military intelligence) provide support in accordance with the priorities established by the supported division commander. The historical designations of the division headquarters, such as the 1st Cavalry Division and 101st Airborne Division, do not necessarily reflect the capabilities of the subordinate forces task organized under them.

Chap 3

II. Tactical Mission Tasks (ADRP 3-90)

Ref: ADRP 3-90, Offense & Defense (Aug '12), chap. 1 and 9.

I. The Tactical Level of War

Through tactics, commanders use combat power to accomplish missions. The tactical-level commander employs combat power to accomplish assigned missions.

The tactical level of war is the level of war at which battles and engagements are planned and executed to achieve military objectives assigned to tactical units or task forces (JP 3-0). Activities at this level focus on the ordered arrangement and maneuver of combat elements in relation to each other and to the enemy to achieve combat objectives. It is important to understand tactics within the context of the levels of war. The strategic and operational levels provide the context for tactical operations. Without this context, tactical operations are reduced to a series of disconnected and unfocused actions.

Movement & Maneuver

Tactical operations always require judgment and adaptation to the unique circumstances of a specific situation. Techniques and procedures are established patterns that can be applied repeatedly with little or no judgment in a variety of circumstances. Tactics, techniques, and procedures (TTP) provide commanders and staffs with a set of tools to use in developing the solution to a tactical problem.

See pp. 3-2 to 3-3 for related discussion and an overview of the offense and defense.

Individuals, Crews, and Small Units

Individuals, crews, and small units act at the tactical level. At times, their actions may produce strategic or operational effects. However, this does not mean these elements are acting at the strategic or operational level. Actions are not strategic unless they contribute directly to achieving the strategic end state. Similarly, actions are considered operational only if they are directly related to operational movement or the sequencing of battles and engagements. The level at which an action occurs is determined by the perspective of the echelon in terms of planning, preparation, and execution.

Battles, Engagements and Small-Unit Actions

Tactics is the employment and ordered arrangement of forces in relation to each other. Through tactics, commanders use combat power to accomplish missions. The tactical-level commander uses combat power in battles, engagements, and small-unit actions. A battle consists of a set of related engagements that lasts longer and involves larger forces than an engagement. Battles can affect the course of a campaign or major operation. An engagement is a tactical conflict, usually between opposing, lower echelons maneuver forces (JP 1-02). Engagements are typically conducted at brigade level and below. They are usually short, executed in terms of minutes, hours, or days.

II. The Science and Art of Tactics

The tactician must understand and master the science and the art of tactics, two distinctly different yet inseparable concepts. Commanders and leaders at all echelons and supporting commissioned, warrant, and noncommissioned staff officers must be tacticians to lead their soldiers in the conduct of full spectrum operations.

A. The Science

The science of tactics encompasses the understanding of those military aspects of tactics—capabilities, techniques, and procedures—that can be measured and codified. The science of tactics includes the physical capabilities of friendly and enemy organizations and systems, such as determining how long it takes a division to move a certain distance. It also includes techniques and procedures used to accomplish specific tasks, such as the tactical terms and control graphics that comprise the language of tactics. While not easy, the science of tactics is fairly straightforward. Much of what is contained in this manual is the science of tactics—techniques and procedures for employing the various elements of the combined arms team to achieve greater effects.

Mastery of the science of tactics is necessary for the tactician to understand the physical and procedural constraints under which he must work. These constraints include the effects of terrain, time, space, and weather on friendly and enemy forces. However—because combat is an intensely human activity—the solution to tactical problems cannot be reduced to a formula. This realization necessitates the study of the art of tactics.

B. The Art

The art of tactics consists of three interrelated aspects: the creative and flexible array of means to accomplish assigned missions, decision making under conditions of uncertainty when faced with an intelligent enemy, and understanding the human dimension—the effects of combat on soldiers. An art, as opposed to a science, requires exercising intuitive faculties that cannot be learned solely by study. The tactician must temper his study and evolve his skill through a variety of relevant, practical experiences. The more experience the tactician gains from practice under a variety of circumstances, the greater his mastery of the art of tactics.

Military professionals invoke the art of tactics to solve tactical problems within his commander's intent by choosing from interrelated options, including—

- Types and forms of operations, forms of maneuver, and tactical mission tasks
- Task organization of available forces, to include allocating scarce resources
- Arrangement and choice of control measures
- Tempo of the operation
- Risks the commander is willing to take

III. Tactical Mission Tasks

A tactical mission task is a specific activity performed by a unit while executing a form of tactical operation or form of maneuver. A tactical mission task may be expressed as either an action by a **friendly force** or **effects on an enemy force** (FM 7-15). The tactical mission tasks describe the results or effects the commander wants to achieve.

Not all tactical mission tasks have symbols. Some tactical mission task symbols will include unit symbols, and the tactical mission task "delay until a specified time" will use an amplifier. However, no modifiers are used with tactical mission task symbols. Tactical mission task symbols are used in course of action sketches, synchronization matrixes, and maneuver sketches. They do not replace any part of the operation order.

See following pages (pp. 3-20 to 3-21) for tactical mission tasks.

Tactical Doctrinal Taxonomy

Ref: ADRP 3-90, Offense and Defense (Aug '12), fig. 2-1, p. 2-3.

The following shows the Army's tactical doctrinal taxonomy for the four elements of decisive action (in accordance with ADRP 3-0) and their subordinate tasks. The commander conducts tactical enabling tasks to assist the planning, preparation, and execution of any of the four elements of decisive action. Tactical enabling tasks are never decisive operations in the context of the conduct of offensive and defensive tasks. (They are also never decisive during the conduct of stability tasks.) The commander uses tactical shaping tasks to assist in conducting combat operations with reduced risk.

Elements of Decisive Action (and subordinate tasks)

Offensive Tasks

Movement to Contact
- Search and attack
- Cordon and search

Attack
- Ambush*
- Counterattack*
- Demonstration*
- Spoiling attack*
- Feint*
- Raid*
- **Also known as special purpose attacks*

Exploitation

Pursuit

Forms of Maneuver
- Envelopment
- Frontal attack
- Infiltration
- Penetration
- Turning Movement

Defensive Tasks

Area defense

Mobile defense

Retrograde operations
- Delay
- Withdrawal
- Retirement

Forms of the Defense
- Defense of linear obstacle
- Perimeter defense
- Reverse slope defense

Stability Tasks
- Civil security
- Civil control
- Restore essential services
- Support to governance
- Support to economic and infrastructure development

Defense Support to Civil Authorities
- Provide support for domestic disasters
- Provide support for domestic CBRN incidents
- Provide support for domestic law enforcement agencies
- Provide other designated support

Tactical Enabling Tasks

Reconnaissance Operations
- Zone
- Area
- Route
- Recon in force

Security Operations
- Screen
- Guard
- Cover
- Area (also route & convoy)
- Local

Troop Movement
- Administrative movement
- Approach march
- Road march

Encirclement Operations

Passage of Lines

Relief in Place

Mobility Operations
- Breaching operations
- Clearing operations (area and route)
- Gap-crossing operations
- Combat roads and trails
- Forward airfields and landing zones
- Traffic operations

Tactical Mission Tasks

Actions by Friendly Forces
- Attack-by-Fire
- Breach
- Bypass
- Clear
- Control
- Counterreconnaissance
- Disengage
- Exfiltrate
- Follow and Assume
- Follow and Support
- Occupy
- Reduce
- Retain
- Secure
- Seize
- Support-by-Fire

Effects on Enemy Force
- Block
- Canalize
- Contain
- Defeat
- Destroy
- Disrupt
- Fix
- Interdict
- Isolate
- Neutralize
- Suppress
- Turn

A. Effects on Enemy Forces

Term	Graphic	Description
Block		*Block* is a tactical mission task that denies the enemy access to an area or prevents his advance in a direction or along an avenue of approach.
		Block is also an engineer obstacle effect that integrates fire planning and obstacle effort to stop an attacker along a specific avenue of approach or prevent him from passing through an engagement area.
Canalize		*Canalize* is a tactical mission task in which the commander restricts enemy movement to a narrow zone by exploiting terrain coupled with the use of obstacles, fires, or friendly maneuver.
Contain		*Contain* is a tactical mission task that requires the commander to stop, hold, or surround enemy forces or to cause them to center their activity on a given front and prevent them from withdrawing any part of their forces for use elsewhere.
Defeat	*No graphic*	*Defeat* occurs when an enemy has temporarily or permanently lost the physical means or the will to fight. The defeated force is unwilling or unable to pursue his COA, and can no longer interfere to a significant degree. Results from the use of force or the threat of its use.
Destroy		*Destroy* is a tactical mission task that physically renders an enemy force combat-ineffective until it is reconstituted. Alternatively, to destroy a combat system is to damage it so badly that it cannot perform any function or be restored to a usable condition without being entirely rebuilt.
Disrupt		*Disrupt* is a tactical mission task in which a commander integrates direct and indirect fires, terrain, and obstacles to upset an enemy's formation or tempo, interrupt his timetable, or cause his forces to commit prematurely or attack in a piecemeal fashion.
		Disrupt is also an engineer obstacle effect that focuses fire planning and obstacle effort to cause the enemy to break up his formation and tempo, interrupt his timetable, commit breaching assets prematurely, and attack in a piecemeal effort.
Fix		*Fix* is a tactical mission task where a commander prevents the enemy from moving any part of his force from a specific location for a specific period. Fixing an enemy force does not mean destroying it. The friendly force has to prevent the enemy from moving in any direction.
		Fix is also an engineer obstacle effect that focuses fire planning and obstacle effort to slow an attacker's movement within a specified area, normally an engagement area.
Interdict		*Interdict* is a tactical mission task where the commander prevents, disrupts, or delays the enemy's use of an area or route. Interdiction is a shaping operation conducted to complement and reinforce other ongoing offensive or defensive
Isolate		*Isolate* is a tactical mission task that requires a unit to seal off-both physically and psychologically-an enemy from his sources of support, deny him freedom of movement, and prevent him from having contact with other enemy forces.
Neutralize		*Neutralize* is a tactical mission task that results in rendering enemy personnel or materiel incapable of interfering with a particular operation.
Turn		*Turn* is a tactical mission task that involves forcing an enemy element from one avenue of approach or movement corridor to another.
		Turn is also a tactical obstacle effect that integrates fire planning and obstacle effort to divert an enemy formation from one avenue of approach to an adjacent avenue of approach or into an engagement area.

B. Actions by Friendly Forces

Task	Graphic	Description
Attack by Fire		*Attack-by-fire* is a tactical mission task in which a commander uses direct fires, supported by indirect fires, to engage an enemy without closing with him to destroy, suppress, fix, or deceive him.
Breach	B	*Breach* is a tactical mission task in which the unit employs all available means to break through or secure a passage through an enemy defense, obstacle, minefield, or fortification.
Bypass	B	Bypass is a tactical mission task in which the commander directs his unit to maneuver around an obstacle, position, or enemy force to maintain the momentum of the operation while deliberately avoiding combat with an enemy force.
Clear	C	*Clear* is a tactical mission task that requires the commander to remove all enemy forces and eliminate organized resistance within an assigned area.
Control	*No graphic*	*Control* is a tactical mission task that requires the commander to maintain physical influence over a specified area to prevent its use by an enemy or to create conditions for successful friendly operations.
Counterrecon	*No graphic*	*Counterreconnaissance* is a tactical mission task that encompasses all measures taken by a commander to counter enemy reconnaissance and surveillance efforts.
Disengage	*No graphic*	*Disengage* is a tactical mission task where a commander has his unit break contact with the enemy to allow the conduct of another mission or to avoid decisive engagement.
Exfiltrate	*No graphic*	*Exfiltrate* is a tactical mission task where a commander removes soldiers or units from areas under enemy control by stealth, deception, surprise, or clandestine means.
Follow and Assume		*Follow and assume* is a tactical mission task in which a second committed force follows a force conducting an offensive operation and is prepared to continue the mission if the lead force is fixed, attritted, or unable to continue. The follow-and-assume force is not a reserve but is committed to accomplish specific tasks.
Follow and Support		*Follow and support* is a tactical mission task in which a committed force follows and supports a lead force conducting an offensive operation. The follow-and-support force is not a reserve but is a force committed to specific tasks.
Occupy	O	*Occupy* is a tactical mission task that involves moving a friendly force into an area so that it can control that area. Both the force's movement to and occupation of the area occur without enemy opposition.
Reduce	*No graphic*	*Reduce* is a tactical mission task that involves the destruction of an encircled or bypassed enemy force.
Retain	R	*Retain* is a tactical mission task in which the cdr ensures that a terrain feature controlled by a friendly force remains free of enemy occupation or use. The commander assigning this task must specify the area to retain and the duration of the retention, which is time- or event-driven.
Secure	S	*Secure* is a tactical mission task that involves preventing a unit, facility, or geographical location from being damaged or destroyed as a result of enemy action. This task normally involves conducting area security operations.
Seize	S	*Seize* is a tactical mission task that involves taking possession of a designated area by using overwhelming force. An enemy force can no longer place direct fire on an objective that has been seized.
Support by Fire		*Support-by-fire* is a tactical mission task in which a maneuver force moves to a position where it can engage the enemy by direct fire in support of another maneuvering force. The primary objective of the support force is normally to fix and suppress the enemy so he cannot effectively fire on the maneuvering force.

C. Mission Symbols

Mission	Symbol	Description
Counterattack (dashed axis)	CATK	A form of attack by part or all of a defending force against an enemy attacking force, with the general objective of denying the enemy his goal in attacking (FM 3-0).
Cover	C C	A form of security operation whose primary task is to protect the main body by fighting to gain time while also observing and reporting information and preventing enemy ground observation of and direct fire against the main body.
Delay	D	A form of retrograde in which a force under pressure trades space for time by slowing down the enemy's momentum and inflicting maximum damage on the enemy without, in principle, becoming decisively engaged (JP 1-02, see delaying operation).
Guard	G G	A form of security operations whose primary task is to protect the main body by fighting to gain time while also observing and reporting information and preventing enemy ground observation of and direct fire against the main body. Units conducting a guard mission cannot operate independently because they rely upon fires and combat support assets of the main body.
Penetrate		A form of maneuver in which an attacking force seeks to rupture enemy defenses on a narrow front to disrupt the defensive system (FM 3-0).
Relief in Place	RIP	A tactical enabling operation in which, by the direction of higher authority, all or part of a unit is replaced in an area by the incoming unit.
Retirement	R	A form of retrograde [JP 1-02 uses *operation*] in which a force out of contact with the enemy moves away from the enemy (JP 1-02).
Screen	S S	A form of security operations that primarily provides early warning to the protected force.
Withdraw	W	A planned operation in which a force in contact disengages from an enemy force (JP 1-02) [The Army considers it a form of retrograde.]

Chap 3

III. Mobility and Countermobility

Ref: ADRP 3-90, Offense and Defense (Aug '12), pp. 3-10 to 3-14 and FM 3-34, Engineer Operations (Apr '16).

I. Mobility

Mobility is a quality or capability of military forces which permits them to move from place to place while retaining the ability to fulfill their primary mission (JP 3-17). Mobility operations are those combined arms activities that mitigate the effects of natural and man-made obstacles to enable freedom of movement and maneuver (ATTP 3-90.4).

Movement & Maneuver

Mobility Operations Primary Tasks

- **Breaching Operations**
- **Clearing Operations (Areas and Routes)**
- **Gap-Crossing Operations**
- **Combat Roads and Trails**
- **Forward Airfields and Landing Zones**
- **Traffic Operations**

Offensive Considerations

Mobility is necessary for successful offensive actions. Its major focus is to enable friendly forces to move and maneuver freely on the battlefield. The commander seeks the capability to move, exploit, and pursue the enemy across a wide front. When attacking, the commander concentrates the effects of combat power at selected locations. This may require the unit to improve or construct combat trails through areas where routes do not exist. The surprise achieved by attacking through an area believed to be impassable may justify the effort expended in constructing these trails. The force bypasses existing obstacles and minefields identified before starting the offensive operation instead of breaching them whenever possible. Units mark bypassed minefields whenever the mission variables of METT-TC allow.

Maintaining the momentum of the offense requires the attacking force to quickly pass through obstacles as it encounters them. There is a deliberate effort to capture bridges, beach and port exits, and other enemy reserved obstacles intact. The preferred method of fighting through a defended obstacle is employing a hasty (in-stride) breach, because it avoids the loss of time and momentum associated with conducting a deliberate breach.

Rivers and other gaps remain major obstacles despite advances in high-mobility weapon systems and extensive aviation support. Wet gap crossings are among the most critical, complex, and vulnerable combined arms operations. A crossing is conducted as a hasty crossing and as a continuation of the attack whenever possible because the time needed to prepare for a gap crossing allows the enemy more time to strengthen the defense. The size of the gap, as well as the enemy and friendly situations, will dictate the specific tactics, techniques, and procedures used in conducting the crossing. Functional engineer brigades contain the majority of tactical bridging assets. Military police and CBRN assets may also be required.

Assured Mobility

Ref: ADRP 3-90, Offense and Defense (Aug '12), pp. 3-12 to 3-13.

Assured mobility is a framework of processes, actions, and capabilities that assure the ability of a force to deploy, move, and maneuver where and when desired, without interruption or delay, to achieve the mission. The assured mobility fundamentals predict, detect, prevent, avoid, neutralize, and protect support the assured mobility framework. This framework is one means of enabling a force to achieve the commander's intent. Assured mobility emphasizes the conduct of proactive mobility, countermobility, and protection tasks in an integrated manner so as to increase the probability of mission accomplishment. While focused primarily on movement and maneuver, the assured mobility concept links to each warfighting function and both enables and is enabled by those functions.

Refer to ATTP 3-90.4, Combined Arms Mobility Operations for further discussion.

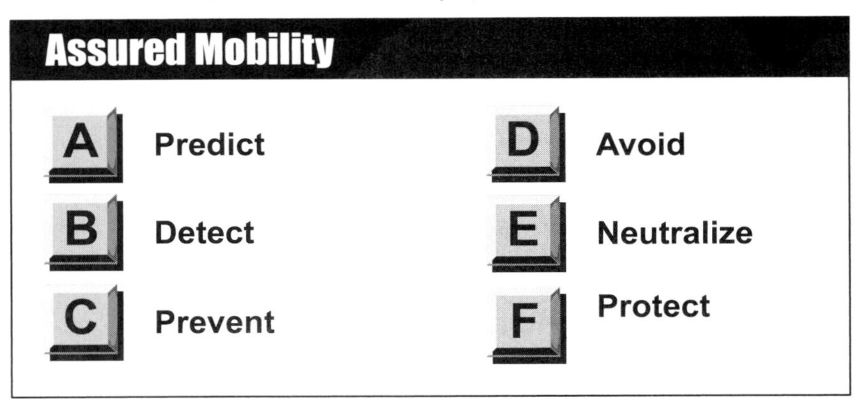

A. Predict

Commanders and staffs must accurately predict potential obstacles to force mobility by analyzing the enemy's capabilities and tactics, techniques, and procedures. This involves understanding how the enemy will evolve in reaction to friendly force countermeasures. It also involves understanding how the effects of terrain and the effects of the population, such as vehicular traffic and dislocated civilians, will impact force mobility. This helps build the mobility portion of the common operational picture and facilitates decisionmaking.

B. Detect

Commanders and staffs use intelligence products and information collection assets to identify the location of natural and manmade obstacles and potential means the enemy can use to create obstacles. Commanders employ available information collection assets to detect enemy obstacle preparations and also identify areas where there are no or only limited obstacles to ground movement and maneuver. This knowledge can be obtained through sustained surveillance of an area. Commanders identify both actual and potential obstacles and propose solutions and alternate COAs to minimize or eliminate their potential impact.

C. Prevent

Commanders and staffs apply this fundamental by preventing civilian interference with operations and denying the enemy's ability to influence friendly mobility. This is accomplished by forces acting proactively to elicit local populace support, or at least non-interference, and to eliminate enemy countermobility capabilities before those capabilities can emplace or activate obstacles, and by mitigating the factors that result in natural obstacles to friendly force movement and maneuver. This may include the employment of information-related capabilities to decrease uncertainty among the population to build support for or acceptance of operations.

Prevention may also consist of aggressive action to destroy enemy assets and capabilities before they can be used to create obstacles. In recent operations this included disrupting terrorist bomb-making cells by all available means, such as cutting off their funding, eliminating safe houses where bombs can be constructed, jamming frequencies to prevent remote detonators from being triggered, and either capturing or killing members of these cells. Forces also apply this fundamental by conducting countermobility operations to shape enemy movement and maneuver that may affect friendly movement and maneuver. This includes denying the enemy the ability and opportunity to attack critical infrastructure that supports mobility, such as airfields, roads, and bridges; or that could result in an obstacle; or have an obstacle effect if destroyed, such as dams and industrial chemical production and storage facilities.

D. Avoid

If prevention fails, the commander will move or maneuver forces to avoid impediments to mobility, if this is viable within the scheme of maneuver. If detection efforts can tell the commander where the enemy has not been, this frees up the unit to maneuver rapidly through those areas, even if they are not the most favorable movement routes.

E. Neutralize

Commanders and staffs plan to neutralize, reduce, or overcome obstacles and impediments as soon as possible to allow unrestricted movement of forces. The specific tactics, techniques, and procedures employed will depend on the mission variables of METT-TC, the rules of engagement, and where along the range of military operations the unit finds itself. For example, a small unit involved in major operations encountering surface-laid mines on a road in an urban area might attempt to destroy the mines in place using organic methods, such as aimed rifle or machinegun fire, after only minimal checks to reduce the danger to local civilians and accepting collateral damage to civilian buildings before proceeding on with its mission. That same unit encountering the same situation during the conduct of a peace-keeping operation would more likely secure the site, evacuate civilians from the area, and call for an explosive ordnance disposal team to disarm the mines in place to preclude any collateral damage.

F. Protect

Commanders and staffs plan and implement survivability and other protection measures that will prevent observation of the maneuvering force and thereby reduce the enemy's ability to engage or otherwise interfere with that force. This includes the use of combat formations and movement techniques. It may involve the use of electronic warfare systems—such as counter-radio controlled improvised explosive device electronic warfare systems, mine plows and rollers, and modifications to the rules of engagement. This may also include the conduct of countermobility missions to deny the enemy the capability to maneuver in certain directions and thereby provide additional protection to friendly maneuvering forces. It can also be as simple as altering patrol routes.

While engineers are the principal staff integrators for assured mobility, other staff sections play critical roles in ensuring the effective application and integration of mobility, countermobility, and protection tasks. In the case of amphibious operations, this would include naval forces that are responsible for assured mobility from amphibious shipping to beach and landing zone exits. These critical roles include providing information on threats to the routes. The senior engineer staff officer's role within assured mobility is similar to the role of the assistant chief of staff, intelligence (G-2) or the intelligence staff officer's (S-2s) integrating role with intelligence preparation of the battlefield. Ultimately, assured mobility is the commander's responsibility.

Refer to engineer doctrine on assured mobility for more information.

Clearing operations are conducted to eliminate the enemy's obstacle effort or residual obstacles within an assigned area or along a specified route. A clearing operation is a mobility operation, and, as with most mobility operations, it is typically performed by a combined arms force built around an engineer-based clearing force. A clearing operation could be conducted as a single mission to open or reopen a route or area, or it may be conducted on a recurring basis in support of efforts to defeat a sustained threat to a critical route.

Defensive Considerations

During the defense, mobility tasks include maintaining routes, coordinating gaps in existing obstacles, and supporting counterattacks. Engineers also open helicopter landing zones and tactical landing strips for fixed-wing aircraft. Maintaining and improving routes and creating bypass or alternate routes at critical points are major engineering tasks because movement routes are subjected to fires from enemy artillery and aircraft systems. These enemy fires may necessitate deploying engineer equipment, such as assault bridging and bulldozers, forward. The commander can also evacuate dislocated civilians or restrict their movements to routes not required by friendly forces to avoid detracting from the mobility of the defending force. The commander can do this provided the action is coordinated with the host nation or the appropriate civil-military operations units and fulfills the commander's responsibilities to dislocated civilians under the law of armed conflict.

The commander's priority of mobility support is first to routes used by counterattacking forces, then to routes used by main body forces displacing to subsequent positions. This mainly involves reducing obstacles and improving or constructing combat roads and trails to allow tactical support vehicles to accompany moving combat vehicles. The commander coordinates carefully to ensure that units leave lanes or gaps in their obstacles that allow for the repositioning of main body units and the commitment of the counterattack force. CBRN reconnaissance systems also contribute to the force's mobility in a contaminated environment.

II. Countermobility

Countermobility operations are those combined arms activities that use or enhance the effects of natural and manmade obstacles to deny an adversary freedom of movement and maneuver (FM 3-34). Countermobility operations help isolate the battlefield and protect attacking forces from enemy counterattack, even though force mobility in offensive actions normally has first priority. Obstacles provide security for friendly forces as the fight progresses into the depth of the enemy's defenses. They provide flank protection and deny the enemy counterattack routes. They assist friendly forces in defeating the enemy in detail and can be vital in reducing the amount of forces required to secure a given area. Further, they can permit the concentration of forces by allowing a relatively small force to defend a large AO. The commander ensures the use of obstacles is integrated with fires and fully synchronized with the concept of operations to avoid hindering the attacking force's mobility.

During visualization, the commander identifies avenues of approach that offer natural flank protection to an attacking force, such as rivers or ridgelines. Staff running estimates support this process. Flanks are protected by destroying bridges, emplacing minefields, and by using scatterable munitions to interdict roads and trails. Swamps, canals, lakes, forests, and escarpments are natural terrain features that can be quickly reinforced for flank security.

Offensive Considerations

Countermobility operations during the offense must stress rapid emplacement and flexibility. Engineer support must keep pace with advancing maneuver forces and be prepared to emplace obstacles alongside them. Obstacles are employed to maximize the effects of restrictive terrain, such as choke points, or deny the usefulness of key terrain, since time and resources will not permit developing the terrain's full defensive potential. The commander first considers likely enemy reactions and then plans how to block enemy avenues of approach or withdrawal with obstacles. The commander also plans the use of obstacles to contain bypassed enemy elements and prevent the enemy from withdrawing. The plan includes obstacles to use upon identification of the enemy's counterattack. Speed and interdiction capabilities are vital characteristics of the obstacles employed. The commander directs the planning for air- and artillery-delivered munitions on enemy counterattack routes. The fire support system delivers these munitions in front of or on top of enemy lead elements once they commit to one of the routes. Rapid cratering devices and surface minefields provide other excellent capabilities.

Control of minefields and obstacles and accurate reporting to all units are vital. Obstacles will hinder both friendly and enemy maneuver. Control of obstacle initiation is necessary to prevent the premature activation of minefields and obstacles.

Refer to FM 90-7 and FM 5-102 for information on obstacle integration and FM 3-34.210 for information on mine warfare.

Defensive Considerations

Countermobility operations help isolate the battlefield and protect friendly forces from enemy attacks. The commander normally concentrates engineer efforts on countering the enemy's mobility. A defending force typically requires large quantities of Class IV and V materiel and specialized equipment to construct fighting and survivability positions and obstacles. With limited assets, the commander must establish priorities among countermobility, mobility, and survivability efforts. The commander ensures that the unit staff synchronizes these efforts with the unit's sustainment plans.

The commander may plan to canalize the enemy force into a salient. In this case, the commander takes advantage of the enemy force's forward orientation by fixing the enemy and then delivering a blow to the enemy's flank or rear. As the enemy's attacking force assumes a defensive posture, the defending commander rapidly coordinates and concentrates all defending fires against unprepared and unsupported segments of the attacking enemy force. The unit may deliver these fires simultaneously or sequentially.

When planning obstacles, commanders and staffs consider not only current operations but also future operations. The commander should design obstacles for current operations so they do not hinder planned future operations. Any commander authorized to employ obstacles can designate certain obstacles to shape the battlefield as high-priority reserve obstacles. The commander assigns responsibility for preparation to a subordinate unit but retains authority for ordering their completion. One example of a reserve obstacle is a highway bridge over a major river. Such obstacles receive the highest priority in preparation and, if ordered, execution by the designated subordinate unit.

A commander integrates reinforcing obstacles with existing obstacles to improve the natural restrictive nature of the terrain to halt or slow enemy movement, canalize enemy movement into engagement areas, and protect friendly positions and maneuver. The commander may choose to employ scatterable mines, if allowed by the rules of engagement. Obstacles must be integrated with fires to be effective. This requires the ability to deliver effective fires well beyond the obstacle's location. When possible, units conceal obstacles from hostile observation. They coordinate obstacle plans with adjacent units and conform to the obstacle zone or belts of superior echelons.

Effective obstacles block, turn, fix, disrupt, or force the enemy to attempt to breach them. The defender tries to predict enemy points of breach based on terrain and probable enemy objectives. The defending force develops means to counter enemy breach attempts, such as pre-coordinated fires. The attacker will try to conceal the time and location of the breach. The defending commander's plan addresses how to counter such a breach attempt, to include reestablishing the obstacle by using scatterable mines and other techniques.

Improvement to defensive positions is continuous. Given time and resources, the defending force constructs additional obstacle systems in-depth, paying special attention to its assailable flanks and rear. The rear is especially vulnerable if there are noncontiguous areas of operations or nontraditional threats. Obstacle systems can provide additional protection from enemy attacks by forcing the enemy to spend time and resources to breach or bypass them. This gives the defending force more time to engage enemy forces attempting to execute a breach or bypass.

The commander designates the unit responsible for establishing and securing each obstacle. The commander may retain execution authority for some obstacles or restrict the use of some types of obstacles to allow other battlefield activities to occur. The commander allows subordinate commanders some flexibility in selecting the exact positioning of obstacles. However, all units must know which gaps or lanes—through obstacles and crossing sites—to keep open for movements, as well as the firing and self-destruct times of scatterable mines to prevent delays in movement. Commanders must be specific and clear in their orders for executing reserve obstacles and closing lanes. As each lane closes, the closing unit reports the lane's closure to the higher, subordinate, and adjacent headquarters to preclude displacing units from moving into areas with unmarked or abandoned obstacles.

Tactical and protective obstacles are constructed primarily at company level and below. Small-unit commanders ensure that observation and fires cover all obstacles to hinder breaching. Deliberate protective obstacles are common around fixed sites. Protective obstacles are a key enabler of survivability operations. They are tied in with FPFs and provide the friendly force with close-in protection. Commanders at all echelons track defensive preparations, such as establishing Class IV and V supply points and start or completion times of obstacle belts and groups. The commander plans how the unit will restore obstacles that the enemy has breached. The commander uses artillery, air, or ground systems to reseed minefields.

III. Engineer Support to Unified Land Operations

Army engineer support to operations encompasses a wide range of tasks that require many capabilities. Commanders use engineers throughout unified land operations across the range of military operations. They use them primarily to assure mobility, enhance protection, enable force projection and logistics, and build partner capacity and develop infrastructure. This chapter describes engineer tasks, the lines of engineer support, and engineer support to the warfighting functions.

A. Engineer Tasks

Engineer tasks provide the freedom of action as the objective. Engineer tasks that affect terrain deal with obstacles (including gaps), bridges, roads, trails, airfields, fighting positions, protective positions, deception, and a wide variety of other structures and facilities (base camps, aerial ports, seaports, utilities, buildings). Engineers affect these by clearing, reducing, emplacing, building, repairing, maintaining, camouflaging, protecting, conserving, or modifying them in some way through tasks (obstacle reduction, route clearance, technical rescue, infrastructure and environmental assessments, geospatial engineering).

Regardless of category, engineer tasks have different purposes in different situations. For example, a task to clear explosive hazards from a road that is designated as a direction of attack may have the purpose of assured mobility. Two days later, that same road may be designated as a main supply route, and a task to clear explosive hazards from the road may have the purpose of protecting critical assets or enabling logistics. The same task is involved, but with different purposes. In addition to the different purposes that an engineer task can have at different times, engineer support often involves simultaneous tasks with different purposes that support different warfighting functions.

Engineer Framework

Ref: FM 3-34, Engineer Operations (Apr '16).

The Engineer Regiment exists to provide the freedom of action for land power by mitigating the effects of terrain. FM 3-34 (2016) explains how (not what) to think about exploiting the capabilities of the Engineer Regiment in support of military operations. This version updates the engineer doctrinal framework (see introductory figure-1) that provides the intellectual underpinnings for the Engineer Regiment and better articulates its purpose and activities. It describes how engineers combine the skills and organizations of the three interrelated engineer disciplines (combat, general, and geospatial engineering) to provide support that helps ground force commanders—

- Assure mobility.
- Enhance protection.
- Enable force projection and logistics.
- Build partner capacity and develop infrastructure among populations and nations.

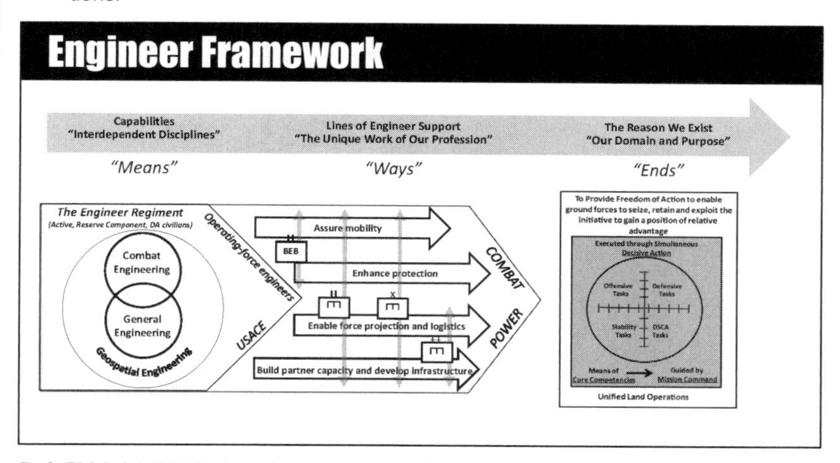

Ref: FM 3-34 (2016), Introductory figure-1. Engineer framework.

Engineer Organization

The Army organizes Soldiers and equipment into a variety of organizations, each with particular capabilities. Engineer units are organized based on the engineer disciplines. The Engineer Regiment is composed of organizations that are represented in the operating and generating forces. These organizations operate concurrently with one another to support combatant commanders (CCDRs) and unified action partners. The Engineer Regiment Active Army organizations include USACE and Army military engineer units within the combatant commands and Army commands. Approximately three-fourths of Army military engineer units are in the Reserve Component. The Reserve Component provides two theater engineer command (TEC) headquarters, including a wide range of specialized capabilities in its Army National Guard and U.S. Army Reserve components. At the center of this is the Office of the Chief of Engineers. The Chief of Engineers integrates capabilities and supports the planning, preparing, executing, and assessing of joint operations. The Regiment is experienced at providing interagency support and in leveraging nonmilitary and nongovernmental engineer assets to support mission accomplishment.

Engineer Planning (Integration with MDMP)

Ref: FM 3-34, Engineer Operations (Apr '16), pp. 3-7 to 3-16.

The engineer staff officer uses the running estimate as a logical thought process and as an extension of the military decisionmaking process.

Military Decisionmaking Process	Engineer Staff Running Estimate
Mission analysis: • Analyze the higher headquarters plan or order. • Perform the initial IPB. • Determine the specified, implied, and essential tasks. • Review the available assets, and identify resource shortfalls. • Determine the constraints. • Identify the critical facts, and develop assumptions. • Begin the composite risk assessment. • Determine the CCIR and EEFI. • Develop the information collection plan. • Update the plan for the use of available time. • Develop the initial information themes and messages. • Develop the proposed mission statement. • Present the mission analysis briefing. • Develop and issue the initial commander's intent. • Develop and issue the initial planning guidance. • Develop the COA evaluation criteria. • Issue the warning order.	Analyze the mission: • Analyze the higher headquarters orders. - Commander's intent. - Mission. - Concept of operation. - Timeline. - Area of operations. • Conduct the IPB, and develop the engineer staff running estimate. - Terrain and weather analysis. - Enemy mission and M/CM/S capabilities. - Friendly mission and M/CM/S capabilities. • Analyze the engineer mission. - Specified M/CM/S tasks. - Implied M/CM/S tasks. - Available assets. - Limitations. - Risk as applied to engineering capabilities. - Time analysis. - Essential tasks for M/CM/S. - Restated mission. • Conduct the risk assessment. - Safety. - Environment (EBS/OEHSA). • Determine the terrain and mobility restraints, obstacle intelligence, threat engineering capabilities, and critical infrastructure. • Recommend the CCIR. • Integrate the engineer reconnaissance effort.
COA development	Develop the scheme of engineer operations. • Analyze the relative combat power. • Refine the essential tasks for M/CM/S. • Identify the engineer missions and the allocation of forces and assets. • Determine the engineer priority of effort and support. • Refine the commander's guidance for M/CM/S operations. • Apply the engineer employment considerations. • Integrate engineer support into the maneuver COA. (See ATTP 5-0.1 for additional information on the scheme of engineer operations.)
COA analysis	War-game and refine the engineer plan.
COA comparison	Recommend a COA.
COA approval	Finalize the engineer plan.
Orders production, dissemination, and transition	Create the input to the basic operation order. • Scheme of engineer operations. • Essential tasks for M/CM/S. • Subunit instructions. • Coordinating instructions. Create the engineer annex and appendixes.

Ref: FM 3-34 (2016), table 3-1. MDMP and the engineer staff running estimate.

Engineer planners evaluate the sustainment significance of each phase of the operation during the entire planning process. They create a clear and concise concept of support that integrates the commander's intent and concept of operation. This includes analyzing the mission; developing, analyzing, war-gaming, and recommending a COA; and executing the plan. The table below lists some engineer planning considerations.

MDMP Steps	Engineer Considerations
Receipt of the mission	• Receive higher headquarters plans, orders, and construction directives. • Understand the commander's intent and time constraints. • Request geospatial information about the area of operations. • Establish engineer-related boards as appropriate.
Mission analysis	• Analyze the available information on existing obstacles or limitations. Evaluate terrain, climate, and threat capabilities to determine the potential impact on M/CM/S. • Develop the essential tasks for M/CM/S. • Identify the available information on routes and key facilities. Evaluate LOC, aerial port of debarkation, and seaport of debarkation requirements. • Determine the availability of construction and other engineering materials. • Review the availability of engineering capabilities, to include Army, joint, multinational, HN, and contracted support. • Determine the bed-down requirements for the supported force. Review theater construction standards and base camp master planning documentation. Review unified facilities criteria, as required. • Review the existing geospatial data on potential sites, conduct site reconnaissance (if possible) and environmental baseline surveys (if appropriated), and determine the threat (to include environmental considerations and explosive hazards). • Obtain the necessary geologic, hydrologic, and climatic data. • Determine the level of interagency cooperation required. • Determine the funding sources, as required. • Determine the terrain and mobility restraints, obstacle intelligence, threat engineering capabilities, and critical infrastructure. Recommend the commander's critical information requirements. • Integrate the reconnaissance effort.
COA development	• Identify the priority engineer requirements, including essential tasks for M/CM/S developed during mission analysis. • Integrate engineer support into COA development. • Recommend an appropriate level of protection effort for each COA based on the expected threat. • Produce construction designs that meet the commander's intent. (Use the Theater Construction Management System when the project is of sufficient size and scope.) • Determine alternate construction locations, methods, means, materials, and timelines to give the commander options. • Determine real-property and real estate requirements.
COA analysis	• War-game and refine the engineer plan. • Use the critical path method to determine the length of different COAs and the ability to crash the project.
COA comparison	• Determine the most feasible, acceptable, and suitable methods of completing the engineering effort.
COA approval	• Determine and compare the risks of each engineering COA. • Gain approval of the essential tasks for M/CM/S and construction management, safety, security, logistics, and environmental plans, as required.
Orders production, dissemination, and transition	• Produce construction directives, as required. • Provide input to the appropriate plans and orders. • Ensure that resources are properly allocated. • Coordinate combined arms rehearsals, as appropriate. • Conduct construction prebriefings. • Conduct preinspections and construction meetings. • Synchronize the construction plan with local and adjacent units. • Implement protection construction standards, including requirements for security fencing, lighting, barriers, and guard posts. • Conduct quality assurance and midproject inspections. • Participate in engineer-related boards. • Maintain as built and red line drawings. • Project turnover activities.

Ref: FM 3-34 (2016), table 3-2. Engineer considerations in the MDMP.

B. Lines of Engineer Support

Fundamental to engineer support to operations is the ability to anticipate and analyze the problem and understand the operational environment. Based on this understanding and the analysis of the problem, engineer planners select and apply the right engineer discipline and unit type to perform required individual and collective tasks. They must think in combinations of disciplines, which integrate and synchronize tasks in concert with the warfighting functions to generate combat power. Finally, they establish the necessary command and support relationship for these combinations. The lines of engineer support are the framework that underpins how engineers think in combinations, and these lines provide the connection between capabilities and tasks.

Commanders use lines of engineer support to synchronize engineer tasks with the rest of the combined arms force and to integrate them into the overall operation throughout the operations process. Lines of engineer support are categories of engineer tasks and capabilities that are grouped by purpose for specific operations. Lines of engineer support assist commanders and staffs with the capabilities of the three engineer disciplines throughout the Engineer Regiment and align activities according to purpose. The engineer disciplines are capabilities, based on knowledge and skills, that are organized in units. These units are organized based on the disciplines that are executed through individual and collective tasks. The combination of these tasks for a specific purpose, in the context of decisive action, achieves the lines of engineer support.

The three engineer disciplines encompass tasks along the lines of engineer support. The combat engineering discipline, due to its support to maneuver forces in close combat, is primarily focused on tasks that assure mobility and enhance protection. The general engineering and geospatial engineering disciplines performs tasks along all four lines of engineer support.

See facing page for a discussion of the three engineer disciplines.

1. Assure Mobility

The assure mobility line of engineer support orchestrates the combat, general, and geospatial engineering capabilities in combination to allow a commander to gain and maintain a position of advantage against an enemy (mobility operations) and deny the enemy the freedom of action to attain a position of advantage (countermobility operations). These tasks primarily focus on support to the movement and maneuver warfighting function, to include support to special operations forces. Although normally associated with organic combat engineers, general engineers may be task-organized to support this line of engineer support. This line of engineer support does not include engineer tasks intended to support the administrative movements of personnel and materiel. Such tasks are normally intended to enable logistics.

The assure mobility line of engineer support is achieved through the assured mobility framework described in ATTP 3-90.4.

Countermobility operations are those combined arms activities that use or enhance the effects of natural and man-made obstacles to deny an adversary freedom of movement and maneuver. The primary purpose of countermobility is to slow or divert the enemy, to increase time for target acquisition, and to increase weapon effectiveness. The advent of rapidly emplaced, remotely controlled, networked munitions enables engineers to conduct effective countermobility operations as part

2. Enhance Protection

This line of engineer support is the combination of the three engineer disciplines to support the preservation of the force so that the commander can apply maximum combat power. This line of engineer support consists largely of survivability operations, but can also include selected mobility tasks (construction of perimeter roads),

C. Engineer Disciplines

Ref: FM 3-34, Engineer Operations (Apr '16), pp. 1-1 to 1-2.

The engineer disciplines are areas of expertise within the Engineer Regiment. Each discipline is focused on capabilities that support, or are supported by, the other disciplines. Within these disciplines are personnel and equipment that provide unique technical knowledge, services, and capabilities that make engineers a member of the Army profession.

Ground forces conduct operations on, in, or above the terrain. The ground forces are affected by the terrain, and they often affect it. Engineer operations are unique because, whatever the intended purpose, engineer operations are directly aimed at affecting the terrain or at improving the understanding of the terrain. In this context, terrain includes natural and man-made terrain features (obstacles, roads, trails, bridges, airfields, ports, base camps). As a result, terrain is central to the three engineer disciplines. Combat and general engineering are focused on affecting the terrain, while geospatial engineering is focused on improving the understanding of the terrain.

Regardless of the disciplines, engineers must be prepared to conduct missions in close combat. Combat engineering is the only discipline that is trained and equipped to support movement and maneuver while in close combat. The general and geospatial engineering disciplines are not organized to move within combined arms formations or to apply fire and maneuver. The general and geospatial engineering disciplines have small arms and a limited number of crew-served weapons that are capable of engaging in close combat with fire and movement, primarily in a defensive role.

Combat Engineering

Combat engineering is the engineering capabilities and activities that closely support the maneuver of land combat forces consisting of three types: mobility, countermobility, and survivability (JP 3-34). This engineer discipline focuses on affecting terrain while in close support to maneuver. Combat engineering is integral to the ability of combined arms units to maneuver. Combat engineers enhance force momentum by shaping the physical environment to make the most efficient use of the space and time necessary to generate mass and speed while denying the enemy maneuver. By enhancing the unit ability to maneuver, combat engineers accelerate the concentration of combat power, increasing the velocity and tempo of the force to exploit critical enemy vulnerabilities. By reinforcing the natural restrictions of the physical environment, combat engineers limit the enemy ability to generate tempo and velocity. These limitations increase enemy reaction time and degrade its will to fight.

General Engineering

General engineering is the engineering capabilities and activities, other than combat engineering, that modify, maintain, or protect the physical environment (JP 3-34). This engineer discipline is primarily focused on providing construction support. It is the most diverse of the three engineer disciplines and is typically the largest percentage of engineer support that is provided to an operation, except in offensive and defensive operations at the tactical level when combat engineering will typically be predominant. It occurs throughout the area of operations, at all levels of war, and during every type of military operation. It may include the employment of all military occupational specialties within the Engineer Regiment.

Geospatial Engineering

Geospatial engineering is the engineering capabilities and activities that contribute to a clear understanding of the physical environment by providing geospatial information and services to commanders and staffs (JP 3-34). Geospatial engineers generate geospatial products and provide services to enable informed running estimates and decisionmaking. It is the art and science of applying geospatial information to enable an understanding of the physical environment as it affects terrain for military operations.

D. Tasks Supporting Decisive Action

Ref: FM 3-34, Engineer Operations (Apr '16), pp. 2-15 to 2-22.

Decisive action requires simultaneous combinations of offense, defense, and stability or DSCA tasks. ADRP 3-0 lists the tasks associated with each element and the purposes of each task. Each task has numerous associated subordinate tasks. Engineering capabilities are organized by the engineer disciplines and are synchronized in the application through the warfighting functions.

Offensive Tasks

Engineer support to the offense includes the simultaneous application of combat, general, and geospatial engineering disciplines through synchronizing warfighting functions and throughout the depth of the area of operations. Combat engineering in support of maneuver forces is the primary focus of engineers involved in the conduct of offensive tasks; however, the three disciplines are applied simultaneously to some degree. The primary focus will be support that enables movement and maneuver.

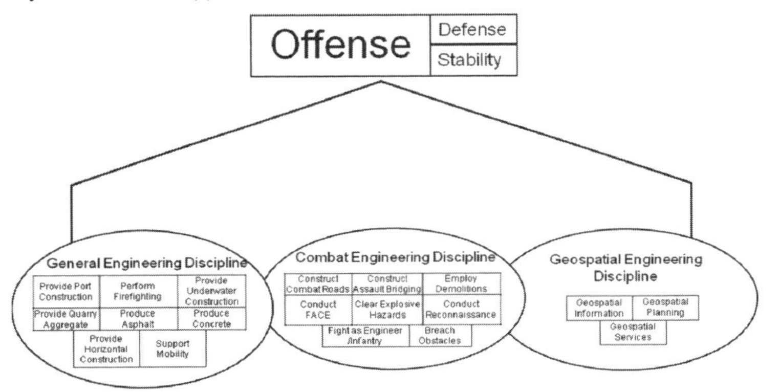

Ref: FM 3-34 (2016), fig. 2-2. Notional engineer support to offensive tasks.

Defensive Tasks

Engineer support to the defense includes the simultaneous application of combat, general, and geospatial engineering capabilities through synchronizing warfighting functions throughout the depth of the area of operations. Combat engineering in close support of maneuver forces is the primary focus in defensive operations; however, the three disciplines are applied simultaneously to some degree.

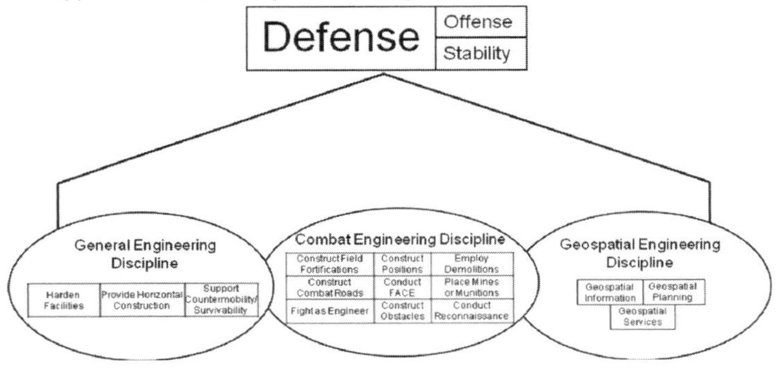

Ref: FM 3-34 (2016), fig. 2-3. Notional engineer support to defensive tasks.

Stability Tasks

Stability operations consist of five primary tasks—civil security, civil control, essential services restoration, support to governance, and support to economic and infrastructure development. The primary tasks are discussed in detail in ADRP 3-07. Engineer support to stability operations includes the simultaneous application of combat, general, and geospatial engineering capabilities through synchronizing warfighting functions and throughout the depth of the area of operations.

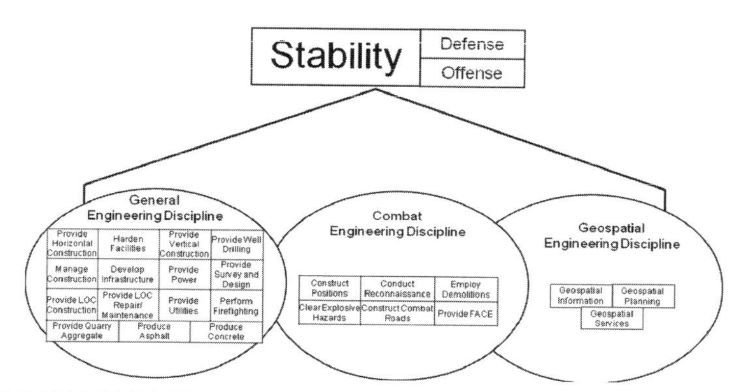

Ref: FM 3-34 (2016), fig. 2-4. Notional engineer support to stability tasks.

Defense Support of Civil Authorities (DSCA)

DSCA includes operations that address the consequences of natural or man-made disasters, accidents, and incidents within the United States and its territories. Army forces conduct DSCA when the size and scope of events exceed the capabilities or capacities of domestic civilian agencies. The Army National Guard is often the first military force to respond on behalf of state authorities.

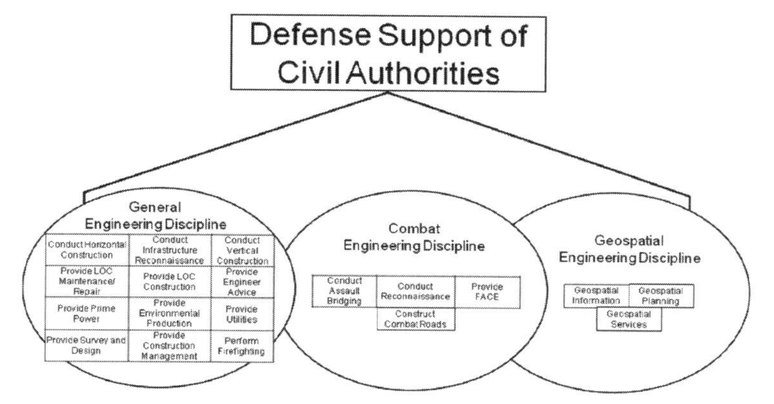

Ref: FM 3-34 (2016), fig. 2-5. Notional engineer support to DSCA tasks.

Engineering in DSCA may include the simultaneous application of combat, general, and geospatial engineering capabilities through synchronizing the warfighting functions throughout the area of operations. General engineering support for the restoration of essential services is the primary engineer focus in DSCA. Engineer support may also be required for Army forces providing mission command, protection, and sustainment to government agencies until they can function normally.

countermobility tasks (emplacement of protective obstacles), and explosive-hazards operations tasks. It also includes survivability and other protection tasks performed or supported by engineers.

Refer to ATP 3-37.34 and ADRP 3-37 for additional information.

3. Enable Force Projection and Logistics

Tasks in the enable force projection and logistics line of engineer support free combat engineers to support maneuver forces, establish and maintain the infrastructure necessary to support follow-on forces, and sustain military operations to enable force projection and logistics to continue after hostile action and to provide recommendations for the site selection of facilities, joint fires, and protection. Engineers combine capabilities from the three engineer disciplines to enable force projection and logistics. Primarily through the general engineering discipline, these capabilities are applied to enhance strategic through tactical movements. Tasks in this line of engineer support sustain military operations in-theater.

4. Build Partner Capacity and Develop Infrastructure

Engineers combine capabilities from across the three disciplines to support the build partner capacity and develop infrastructure line of engineer support, which are vital to stability and counterinsurgency tasks that do not align with a specific phase of operations. This line consists primarily of building, repairing, and maintaining various infrastructure facilities; providing essential services and, ultimately, building partner capacity to codevelop HN capabilities to perform such tasks. Linkages to stability are predominant in this line. Most infrastructure development takes place during Phase 0 (shape), Phase I (deter), Phase IV (stabilize), and Phase V (enable civil authority).

E. Engineer Support to Warfighting Functions

Unified land operations require the continuous generation and application of combat power, often for protracted periods. Combat power is the total means of destructive, constructive, and information capabilities that a military unit and formation can apply at a given time (ADRP 3-0). Army forces generate combat power by converting potential into effective action.

Engineer Support to Warfighting Functions

Leadership
Movement and Maneuver
Protection
Mission Command
Sustainment
Intelligence
Fires
Information

Engineer Functional Areas as Related to Combat Power

Combat power enables decisive actions

Enabling Capabilities	Movement and Maneuver	Intelligence	Fires	Sustainment	Mission Command	Protection
Mobility	P	S		S		S
Countermobility	P		S	S	S	S
Survivability	S		S	S		P
General Engineering	S			P	S	S
Geospatial	S	S	S	S	P	S

Enabling capabilities synchronized through their relationship to the warfighting functions.

P — Primary functional relationship
S — Secondary functional relationship

Ref: FM 3-34 (2016), fig. 2-1. Engineer application of combat power.

Chap 4

Intelligence Warfighting Function

Ref: ADP 2-0, Intelligence (Aug '12), ADRP 2-0, Intelligence (Aug '12), chap. 2 and ADRP 3-0, Operations (Nov '16), pp. 5-4 to 5-5.

The intelligence warfighting function is the related tasks and systems that facilitate understanding the enemy, terrain, weather, civil considerations, and other significant aspects of the operations environment.Specifically, other significant aspects of an operational environment include threats, adversaries, and operational variables, depending on the nature of operations. The intelligence warfighting function synchronizes information collection with the primary tactical tasks of reconnaissance, surveillance, security, and intelligence operations. Intelligence is driven by commanders and is more than just collection. Developing intelligence is a continuous process that involves analyzing information from all sources and conducting operations to develop the situation. The Army executes intelligence, surveillance, and reconnaissance through the operations and intelligence processes, with an emphasis on intelligence analysis and leveraging the larger intelligence enterprise, and information collection. The intelligence warfighting function includes the following tasks:

- Provide support to force generation.
- Provide support to situational understanding.
- Conduct information collection.
- Provide intelligence support to targeting and information capabilities.

For the intelligence warfighting function, setting the theater refers to executing the tasks needed to prepare for intelligence support to all echelons of a deployed force within a theater of operations. There are three core tasks. First, the G-2 or S-2 staff establishes and builds an intelligence architecture. Second, the G-2 or S-2 staff builds the knowledge needed to understand an operational environment through coordination and collaboration with regionally aligned forces, using the military intelligence brigade or theater as the anchor point. Building the knowledge to understand an operational environment includes connecting the intelligence architecture to and feeding the information systems. Last, the G-2 or S-2 staff supports the engagement that develops context and builds relationships through the successful conduct of intelligence operations; intelligence analysis; and intelligence processing, exploitation, and dissemination.

The commander drives intelligence, intelligence facilitates operations, and operations are supportive of intelligence; this relationship is continuous. Commanders' considerations for the intelligence warfighting function include—

- **Reducing operational uncertainty.** Intelligence does not eliminate uncertainty entirely. Commanders determine prudent risks inherent in any operation.
- **Determining the appropriate balance between the time allotted for collection and operational necessity.** It takes time to collect information and then develop that information into detailed and precise intelligence products.
- **Prioritizing finite resources and capabilities.**
- **Resourcing and prioritizing the intelligence warfighting function** appropriately to have enough network capability and access to meet the commander's needs.
- **Employing organic and supporting collection assets** as well as planning, coordinating, and articulating requirements to leverage the entire intelligence enterprise.

I. Intelligence Overview

Ref: ADP 2-0, Intelligence (Aug '12), preface and pp. 1 to 2.

The fundamentals of intelligence have migrated to two new publications—ADP 2-0 and Army Doctrine Reference Publication (ADRP) 2-0, Intelligence. The doctrinal constructs introduced in this publication are further explained in ADRP 2-0.

The Purpose of Intelligence *(see p. 4-6)*

Intelligence is the product resulting from the collection, processing, integration, evaluation, analysis, and interpretation of available information concerning foreign nations, hostile or potentially hostile forces or elements, or areas of actual or potential operations. The term is also applied to the activity that results in the product and to the organizations engaged in such activity (JP 2-0).

Intelligence is a continuous process that directly supports the operations process through understanding the commander's information requirements, analyzing information from all sources, and conducting operations to develop the situation. Intelligence is also a function that facilitates situational understanding and supports decisionmaking. This publication discusses intelligence as a function rather than intelligence as a product.

As a function, intelligence is inherently joint, interagency, intergovernmental, and multinational and leverages the intelligence enterprise. The Army focuses its intelligence effort through the intelligence warfighting function. The intelligence warfighting function systematically answers requirements to support unified land operations. This effort provides information and intelligence to all of the warfighting functions and directly supports the exercise of mission command throughout the conduct of operations.

Intelligence in Unified Land Operations

The Army synchronizes its intelligence efforts with unified action partners to achieve unity of effort and to meet the commander's intent. Intelligence unity of effort is critical to accomplish the mission. Unified action partners are important to intelligence in all operations. Multinational and interagency partners provide cultural awareness, as well as unique perspectives and capabilities that reinforce and complement Army intelligence capabilities. Using appropriate procedures and established policy, Army intelligence leaders provide information and intelligence support to multinational forces. The G-2/S-2 staff leverages the intelligence enterprise to answer the commander's requirements.

The Army executes intelligence, surveillance, and reconnaissance (ISR) through the operations and intelligence processes (with an emphasis on intelligence analysis and leveraging the larger intelligence enterprise) and information collection. Consistent with joint doctrine, intelligence, surveillance, and reconnaissance is an activity that synchronizes and integrates the planning and operation of sensors, assets, and processing, exploitation, and dissemination systems in direct support of current and future operations. This is an integrated intelligence and operations function (JP 2-01).

Army forces often bring unique intelligence capabilities to unified action. The intelligence warfighting function provides the commander with intelligence to plan, prepare, execute, and assess operations. The two most important aspects of intelligence are enabling mission command and providing support to commanders and decisionmakers. Mission command includes both the philosophy and the warfighting function. The mission command philosophy guides the intelligence warfighting function by emphasizing broad mission-type orders, individual initiative within the commander's intent, and leaders who can anticipate and adapt quickly to changing conditions. The mission command warfighting function integrates the elements of combat power across all of the warfighting functions. In order to ensure effective intelligence support, commanders and staffs must understand the interrelationship of mission command, the intelligence warfighting function, and fundamental intelligence doctrine. Timely, relevant, and accurate intelligence and predictive assessments help the commander maintain operational flexibility, exercise mission command, and mitigate risk.

Intelligence Logic Map

Joint Intelligence

The product resulting from the collection, processing, integration, evaluation, analysis, and interpretation of available information concerning foreign nations, hostile or potentially hostile forces or elements, or areas of actual or potential operations. The term is also applied to the activity which results in the product and to the organizations engaged in such activity.

- As a function, intelligence is inherently joint, interagency, intergovernmental, and multinational.
- Unified action partners provide cultural awareness as well as unique perspectives and capabilities that reinforce and complement Army intelligence capabilities.

Unified Land Operations

How the Army seizes, retains and exploits the intiative to gain and maintain a position of relative advantage in sustained land operations in order to create the conditions for favorable conflict resolution.

Intelligence in Unified Land Operations

The Army synchronizes its intelligence efforts with unified action partners to achieve unity of effort and to meet the commander's intent. Intelligence unity of effort is critical to accomplish the mission. Intelligence reduces operational uncertainty --

By facilitating...

Commanders' and Decisionmakers' Situational Understanding

Executed through the...

Intelligence Warfighting Function

The related tasks and systems that facilitate understanding of the enemy, terrain, and civil considerations.

Tasks:

- Support to force generation.
- Support to situational understanding.
- Conduct information collection.
- Support to targeting and information capabilities.

Which leverages the...

Intelligence Enterprise

- Intelligence community.
- Intelligence architecture.
- Intelligence professionals.

Guided by...

Mission Command

Intelligence Enterprise

To do this

The Army conducts the intelligence warfighting function through these fundamental doctrinal constructs

Basic activities and tasks used to describe the intelligence warfighting function and leverage the intelligence enterprise

Core Competencies

- Intelligence synchronization
- Intelligence operations
- Intelligence analysis

A broad process for supporting operations

Intelligence Process

- Plan and direct
- Collect
- Produce
- Disseminate

- Analyze
- Assess

The basic "building blocks" that together constitute the intelligence effort

Intelligence Capabilities

- All-source intelligence
- Single-source intelligence
  - Intelligence disciplines
    - Counterintelligence
    - Geospatial intelligence
    - Human intelligence
    - Measurement and signature intelligence
    - Open-source intelligence
    - Signals intelligence
    - Technical intelligence
  - Complementary intelligence capabilities
    - Biometrics-enabled intelligence
    - Cyber-enabled intelligence
    - Document and media exploitation
    - Forensic-enabled intelligence
  - Processing, exploitation, and dissemination (PED)

Ref: ADP 2-0, Intelligence, fig. 1, p. iii to iv.

II. ADRP 2-0, Intelligence (Aug '12)

Ref: ADRP 2-0, Intelligence (Aug '12), introduction.

The fundamentals of intelligence have migrated to two new publications—ADP 2-0 and Army Doctrine Reference Publication (ADRP) 2-0, Intelligence. The doctrinal constructs introduced in this publication are further explained in ADRP 2-0.

ADRP 2-0 includes the following changes:

Chapter 1

- Introduces the following concepts into intelligence doctrine:
  - Unified action
  - Unified land operations
  - Unified action partners
  - Intelligence support to decisive action tasks
  - Defense support of civil authorities
- Provides an overview of U.S. Army Intelligence and Security Command capabilities
- Incorporates doctrine on information collection established in FM 3-55
- Adopts the joint definition of intelligence, surveillance, and reconnaissance (ISR)

Chapter 2

- Introduces the following intelligence core competencies as the basic activities and tasks the Army uses to describe and drive the intelligence warfighting function and leverage the intelligence enterprise:
  - Intelligence synchronization
  - Intelligence operations
  - Intelligence analysis
- Introduces fusion centers into intelligence doctrine and describes the importance of fusion centers in achieving greater efficiency between the intelligence enterprise and mission command
- Updates the discussion on the purpose of intelligence. This information was addressed in FM 2-0, chapter 1
- Revises the discussion of the intelligence warfighting function found in FM 2-0, chapter 1, to more clearly account for threats, terrain and weather, and civil considerations
- Revises the discussion of the intelligence enterprise found in FM 2-0, chapter 1. The discussion describes a communications-enabled intelligence network and an Army intelligence interaction with the intelligence community and enterprise in more detail
- Revises the discussion of the intelligence community found in FM 2-0, chapter 1, to accurately reflect changes within joint doctrine

Chapter 3

- Contains a short discussion of the joint intelligence process and how it differs from the Army intelligence process
- Modifies the intelligence process steps and continuing activities, replaces the plan step and prepare step with the plan and direct step, makes disseminate a step instead of a continuing activity, and deletes generate intelligence knowledge as a discrete continuing activity
- Incorporates an overview of planning considerations into the plan and direct step of the intelligence process

Chapter 4

- Introduces the concept of Army intelligence capabilities
- Introduces a discussion of single-source intelligence:
  - Describes single-source intelligence as consisting of the seven intelligence disciplines (4-2) plus the complementary intelligence capabilities
  - Introduces cyber-enabled intelligence as a complementary intelligence capability that provides the ability to collect information and produce unique intelligence
  - Introduces forensic-enabled intelligence as a complementary intelligence capability that can support intelligence collection and analysis
  - Introduces the concept of processing, exploitation, and dissemination (PED) as a single-source intelligence activity
  - Categorizes biometrics-enabled intelligence (4-9) and document and media exploitation as complementary intelligence capabilities
- Modifies the description of all-source intelligence. It is no longer considered an Army intelligence discipline, but an approach for developing intelligence. Army doctrine now recognizes only the seven joint intelligence disciplines
- Removes imagery intelligence as a separate intelligence discipline. Imagery intelligence now falls under geospatial intelligence

Chapter 5

- Introduces and describes intelligence support to the Army design methodology
- Moves the discussion of intelligence support to the military decisionmaking process section to chapter 5, formerly addressed under all-source intelligence in FM 2-0
- Adds an overview of intelligence support to targeting
- Revises the discussion of the types of intelligence products
  - Significantly expands the discussion of the intelligence estimate
  - Incorporates material on the intelligence summary from FM 2-0, appendix A

Chapter 6

Updates the discussion of force projection from chapter 2 of FM 2-0.

New, Rescinded, and Modified Terms

ADRP 2-0 introduces four new Army terms, and rescinds and modifies several other terms.

Introductory table-1. New Army terms

Term	Remarks
fusion	Adds Army definition to existing joint term (see paragraph 4-4).
intelligence analysis	As a core competency (see paragraphs 2-61 through 2-64).
intelligence operations	As a core competency and as part of information collection (see paragraphs 2-55 through 2-60).
intelligence synchronization	As a core competency (see paragraphs 2-53 and 2-54).

Introductory table-2. Rescinded Army terms

Term	Remarks
broadcast dissemination	Rescinded.

Introductory table-3. Modified Army terms

Term	Remarks
all-source intelligence	Modified the definition—supersedes the definition in FM 2-0.
analysis	Retained based on common English usage. No longer formally defined.
biometrics-enabled intelligence	Modified the definition; new proponent publication—supersedes the definition in TC 2-22.82.
intelligence reach	Modified the definition—supersedes the definition in FM 2-0.
intelligence requirement	Adopts the joint definition.
open-source intelligence	Adopts the joint definition.

III. The Purpose of Intelligence

The purpose of intelligence is to support commanders and staffs in gaining situational understanding of threats, terrain and weather, and civil considerations. Intelligence supports the planning, preparing, execution, and assessment of operations. The most important role of intelligence is to support commanders and decisionmakers. The Army generates intelligence through the intelligence warfighting function.

Intelligence leaders ensure that the intelligence warfighting function operates effectively and efficiently. They are the commander's primary advisors on employing information collection assets and driving information collection. Additionally, intelligence analysts support their commanders with analysis and production of timely, relevant, accurate, and predictive assessments and products tailored to the commander's specific needs.

Facilitating Understanding

Conducting (planning, preparing, executing, and assessing) military operations requires intelligence products regarding threats and relevant aspects of the operational environment. These intelligence products enable commanders to identify potential COAs, plan operations, employ forces effectively, employ effective tactics and techniques, and implement protection.

A. Threats and Hazards

Although threats are a fundamental part of an operational environment for any operation, they are discussed separately here simply for emphasis. A threat is any combination of actors, entities, or forces that have the capability and intent to harm United States forces, United States national interests, or the homeland (ADRP 3-0). Threats may include individuals, groups of individuals (organized or not organized), paramilitary or military forces, nation-states, or national alliances. The intelligence warfighting function analyzes nation-states, organizations, people, or groups to determine their ability to damage or destroy life, vital resources, and institutions, or prevent mission accomplishment. Threats are sometimes categorized as traditional, irregular, disruptive, and catastrophic. While helpful in generally describing the nature of the threat, these categories do not precisely describe the threat's goals, organizations, and methods of operating.

See p. 1-9 for further discussion of threat categories.

Intelligence provides a deep understanding of the threat and how the threat can affect mission accomplishment, which is essential to conducting operations. Commanders and staffs must understand how current and potential threats organize, equip, train, employ, and control their forces. Therefore, the intelligence warfighting function must continually identify, monitor, and assess threats as they adapt and change over time.

Hazards are conditions or natural phenomena able to damage or destroy life, vital resources, and institutions, or prevent mission accomplishment. Understanding hazards and their effects on operations allows the commander to better understand the terrain, weather, and various other factors that best support the mission. It also helps the commander visualize potential impacts on operations. Successful interpretation of the environment aids in correctly applying threat COAs within a given geographical region. Hazards include disease, extreme weather phenomena, solar flares, and areas contaminated by toxic materials.

B. Terrain and Weather

Terrain aspects and weather conditions are inseparable, directly influence each other, and impact military operations based on the mission variables (METT-TC). Terrain analysis involves the study and interpretation of natural and manmade features of an area, their effects on military operations, and the effects of weather

IV. Intelligence Support to Commanders and Decisionmakers

Ref: ADP 2-0, Intelligence (Aug '12), pp. 2 to 3.

Commanders provide guidance and resources to support unique requirements of the staffs and subordinate commanders. Although commanders drive operations, as the principal decisionmakers, their relationship with their staffs must be one of close interaction and trust. This relationship must encourage initiative within the scope of the commander's intent. Independent thought and timely actions by staffs are vital to mission command.

Commanders provide guidance and continuous feedback throughout operations by—

- Providing direction
- Stating clear, concise commander's critical information requirements (CCIRs)
- Synchronizing the intelligence warfighting function
- Participating in planning
- Collaborating with the G-2/S-2 during the execution of operations

Teamwork within and between staffs produces integration essential to effective mission command and synchronized operations. While all staff sections have clearly defined functional responsibilities, they cannot work efficiently without complete cooperation and coordination among all sections and cells. Key staff synchronization and integration occur during—

- **Intelligence preparation of the battlefield (IPB).** The G-2/S-2 leads the IPB effort with the entire staff's participation during planning.
- **Army design methodology, the military decisionmaking process, and the rapid decisionmaking and synchronization process.** Intelligence provides important input that helps frame operational problems and drives decisionmaking processes.
- **Information collection.** The G-2/S-2 staff provides the analysis, supporting products, and draft plan necessary for the G-3/S-3 to task the information collection plan.
- **Targeting.** Intelligence is an inherent part of the targeting process and facilitates the execution of the decide, detect, deliver, and assess functions.
- **Assessments.** The G-2/S-2 staff collaborates closely with the rest of the staff to ensure timely and accurate assessments occur throughout operations.

The staff performs many different activities as a part of the intelligence warfighting function. This effort is extremely intensive during planning and execution. After the commander establishes CCIRs, the staff focuses the intelligence warfighting function on priority intelligence requirements and other requirements. The staff assesses the situation and refines or adds new requirements, as needed, and quickly retasks units and assets. It is critical for the staff to plan for and use well-developed procedures and flexible planning to track emerging targets, adapt to changing operational requirements, and meet the requirement for combat assessment.

Refer to ADRP 2-0, chap. 5, for further discussion.

Refer to BSS5:The Battle Staff SMARTbook (Leading, Planning & Conducting Military Operations) for further discussion of the intelligence warfighting function as it relates to the operations process -- to include warfighting function tasks, intelligence core competencies, the intelligence process, and types of intelligence products.

V. Reconnaissance & Surveillance

Ref: ADRP 3-90, Offense and Defense (Aug '12), pp. 5-1 to 5-3.

Reconnaissance operations are those operations undertaken to obtain, by visual observation or other detection methods, information about the activities and resources of an enemy or potential enemy, or to secure data concerning the meteorological, hydrographical or geographical characteristics and the indigenous population of a particular area. Reconnaissance primarily relies on the human dynamic rather than technical means. Reconnaissance is performed before, during, and after other operations to provide information used in the intelligence preparation of the battlefield (IPB) process, as well as by the commander in order to formulate, confirm, or modify his course of action (COA).

Forms of Reconnaissance

The four forms of reconnaissance are route, zone, area, and reconnaissance in force.

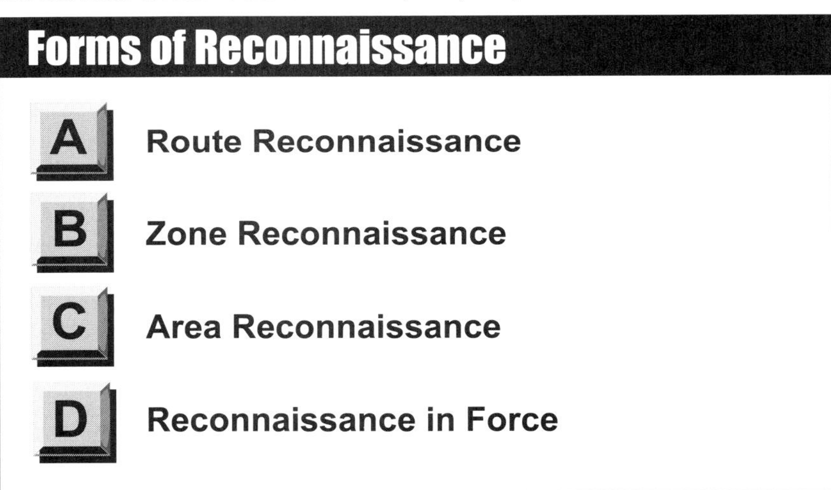

Ref: FM 3-90, Tactics, chap. 13.

Surveillance and reconnaissance missions are a principal means of information collection. A key difference between surveillance missions and reconnaissance is that surveillance is systematic, usually passive in collection of information, and may be continuous; while reconnaissance may be limited in duration of the assigned mission, is active in collection of information, and usually includes human participation. Reconnaissance employs many tactics, techniques, and procedures (TTP) throughout the course of the mission, one of which may include an extended period of surveillance.

A. Route Reconnaissance

Route reconnaissance is a form of reconnaissance that focuses along a specific line of communication, such as a road, railway, or cross-country mobility corridor. It provides new or updated information on route conditions, such as obstacles and bridge classifications, and enemy and civilian activity along the route. A route reconnaissance includes not only the route itself, but also all terrain along the route from which the enemy could influence the friendly force's movement.

B. Zone Reconnaissance

Zone reconnaissance is a form of reconnaissance that involves a directed effort to obtain detailed information on all routes, obstacles, terrain, and enemy forces within a zone defined by boundaries. It is appropriate when the enemy situation is vague, existing knowledge of the terrain is limited, or combat operations have altered the terrain. A zone reconnaissance may include several route or area reconnaissance missions assigned to subordinate units.

C. Area Reconnaissance

Area reconnaissance is a form of reconnaissance that focuses on obtaining detailed information about the terrain or enemy activity within a prescribed area. This area may include a town, a ridgeline, woods, an airhead, or any other feature critical to operations. The area may consist of a single point, such as a bridge or an installation. Areas are normally smaller than zones and are not usually contiguous to other friendly areas targeted for reconnaissance.

D. Reconnaissance in Force

A reconnaissance in force is a deliberate combat operation designed to discover or test the enemy's strength, dispositions, and reactions or to obtain other information. Battalion-size task forces or larger organizations usually conduct a reconnaissance in force (RIF) mission. A commander assigns a RIF mission when the enemy is known to be operating within an area and the commander cannot obtain adequate intelligence by any other means. A unit may also conduct a RIF in restrictive-type terrain where the enemy is likely to ambush smaller reconnaissance forces. A RIF is an aggressive reconnaissance, conducted as an offensive operation with clearly stated reconnaissance objectives. The overall goal of a RIF is to determine enemy weaknesses that can be exploited.

Every Soldier is a Sensor (ES2) Program

The Army established the every Soldier is a sensor (ES2) program, which is accomplished through Soldier surveillance and reconnaissance. The Soldier surveillance and reconnaissance AUTL task is designed to help units more effectively collect useful information in their AO. This task is critical because units often operate in an AO characterized by violence, uncertainty, and complex threats.

Refer to FM 2-91.6 for a detailed discussion about Soldier surveillance and reconnaissance.

Reconnaissance Objective

The commander orients his reconnaissance assets by identifying a reconnaissance objective within the area of operation (AO). The reconnaissance objective is a terrain feature, geographic area, or an enemy force about which the commander wants to obtain additional information. The reconnaissance objective clarifies the intent of the reconnaissance effort by specifying the most important result to obtain from the reconnaissance effort. The commander assigns a reconnaissance objective based on his priority information requirements (PIR) resulting from the IPB process and the reconnaissance asset's capabilities and limitations. The reconnaissance objective can be information about a specific geographical location, such as the cross-country trafficability, a specific enemy activity to be confirmed or denied, or a specific enemy unit to be located and tracked.

Refer to SUTS2: The Small Unit Tactics SMARTbook (Leading, Planning & Conducting Tactical Operations) for discussion of reconnaissance and security operations. Related topics include tactical mission fundamentals, offensive and defensive operations, stability & counterinsurgency operations, tactical enabling operations, special purpose attacks, tactical environments, and patrols & patrolling.

and climate on these features. Terrain analysis is a continuous process. Analyzing military aspects of terrain includes collection, analysis, evaluation, and interpretation of geographical information on natural and manmade features of the terrain. Then analysts combine other relevant factors with the terrain and weather to predict their effects on military operations.

Weather analysis evaluates forecasted weather effects on operations and various systems. Analysts should evaluate the effects of each military aspect of weather. However, just as in terrain analysis, they should focus on the aspects that have the most bearing on operations and decisionmaking. The evaluation of each aspect should begin with operational climatology and current weather forecasts. Analysts fine-tune the evaluation to determine the effects based on specific weather sensitivity thresholds for friendly and threat forces and systems.

C. Civil Considerations & Sociocultural Understanding

Civil considerations may be expressed using the joint systems perspective, the operational variables, or the mission variables. Intelligence analysts leverage information from many different sources, including open sources, to provide predictive intelligence and facilitate a broad understanding of the operational environment.

Analysts can draw relevant information from analysis of the operational environment using the operational variables (PMESII-PT). However, upon receipt of the mission, Army forces use ASCOPE (areas, structures, capabilities, organizations, people, and events) characteristics to describe civil considerations as part of the mission variables (METT-TC) during IPB. Additionally, a human terrain system team can provide detailed information and analysis pertaining to the sociocultural factors involved in the operation.

For additional information on ASCOPE and IPB, refer to FM 2-01.3.

Culture is a key factor in understanding all of the ASCOPE characteristics. Understanding a culture has become an increasingly important competency for Soldiers. Culture is the shared beliefs, values, customs, behaviors, and artifacts members of a society use to cope with the world and each other. Individuals belong to multiple groups through birth, assimilation, or achievement. Individuals' groups influence their beliefs, values, attitudes, and perceptions. As such, culture is internalized—it is habitual, taken for granted, and perceived as natural by people in the society.

Cultures—

- Influence people's range of action and ideas including what to do and not do, how to do or not do it, and with whom to do it or not do it.
- Include the circumstances for shifting and changing rules.
- Influence how people make judgments about what is right and wrong and how to assess what is important and unimportant.
- Affect how people categorize and deal with issues that do not fit into existing categories
- Provide the framework for rational thoughts and decisions. However, what one culture considers rational may not be rational to another culture.

Refer to FM 3-24 for a discussion of sociocultural analysis.

Army leaders seek to understand the situation in terms of the local cultures while avoiding their own cultural biases. Understanding other cultures applies to all operations, not only those dominated by stability.

VI. Intelligence Warfighting Function Tasks

The intelligence warfighting function facilitates support to the commander and staff through a broad range of supporting Army Universal Tasks List (AUTL) tasks. These tasks are interrelated, require the participation of the commander and staff, and are often conducted simultaneously. The intelligence warfighting function tasks facilitate the commander's visualization and understanding of the threat and other relevant aspects of the operational environment.

Intelligence Warfighting Function Tasks

Intelligence tasks ►	Commander's focus ►	Commander's decisions
Support to force generation • Provide intelligence readiness. • Establish an intelligence architecture. • Provide intelligence overwatch. • Generate intelligence knowledge. • Tailor the intelligence force.	Orient on contingencies.	Should the unit's level of readiness be increased? Should the operation plan be implemented?
Support to situational understanding • Perform intelligence preparation of the battlefield. • Perform situation development. • Provide intelligence support to protection. • Provide tactical intelligence overwatch. • Conduct police intelligence operations. • Provide intelligence support to civil affairs activities.	Plan an operation. Prepare. Execute. Assess. Secure the force. Determine 2d and 3d effects on operations and the populace.	Which course of action will be implemented? Which enemy actions are expected? What mitigation strategies should be developed and implemented to reduce the potential impact of operations on the population?
Conduct information collection • Plan requirements and assess collection. • Task and direct collection. • Execute collection.	Plan an operation. Prepare. Execute. Assess.	Which decision points, high-payoff targets (HPTs), and high-value targets (HVTs) are linked to the threat's actions? Are the assets available and in position to collect on the decision points, HPTs, and HVTs? Have the assets been repositioned for branches or sequels?
Support to targeting and information capabilities • Provide intelligence support to targeting. • Provide intelligence support to inform and influence activities. • Provide intelligence support to cyber electromagnetic activities. • Provide intelligence support to combat assessment.	Use lethal or nonlethal effects against targets. Destroy, suppress, disrupt, or neutralize targets. Reposition intelligence or attack assets.	Are the unit's lethal and nonlethal effects and maneuver effective? Which targets should be re-engaged? Are the unit's inform and influence activities effective?

Ref: ADRP 2-0, Intelligence, table 2-1, p. 2-3.

There are intelligence-related AUTL tasks beyond the four most significant intelligence warfighting tasks shown above. Soldiers, systems, and units from all branches conduct intelligence-related AUTL tasks. Every Soldier, as a part of a small unit, is a potential information collector. Soldiers develop a special awareness simply due to exposure to events occurring in the AO, and have the opportunity to collect and report information based on their observations and interactions with the local population. The increased awareness that Soldiers develop through personal contact and observation is a critical element of the unit's ability to understand the operational environment more fully.

VII. Characteristics of Effective Intelligence

Ref: ADRP 2-0, Intelligence (Aug '12), pp. 2-1 to 2-2.

The effectiveness of intelligence is measured against the relevant information quality criteria:

Accuracy

Intelligence gives commanders an accurate, balanced, complete, and objective picture of the threat and other aspects of the operational environment. To the extent possible, intelligence should accurately identify threat intentions, capabilities, limitations, and dispositions. It should be derived from multiple sources and disciplines to minimize the possibility of deception or misinterpretation. Alternative or contradictory assessments should be presented, when necessary, to ensure balance and unbiased intelligence.

Timeliness

Intelligence provided early supports operations and prevents surprise from threat actions. It must flow continuously to the commander before, during, and after an operation. Intelligence organizations, databases, and products must be available to develop estimates, make decisions, and plan operations.

Usability

Intelligence must be in the correct data file specifications for databasing and display. Usability facilitates further analysis, production of intelligence, integration of the product across the staff, and use within operations.

Completeness

Intelligence briefings and products convey all of the necessary components to be as complete as possible.

Precision

Intelligence briefings and products provide the required level of detail and complexity to answer the requirements.

Reliability

Intelligence evaluates and determines the extent to which the collected information and the information being used in intelligence briefings and products are trustworthy, uncorrupted, and undistorted. Any concerns with the reliability of intelligence must be stated up front.

Besides the relevant information quality criteria, intelligence must meet three additional criteria:

Relevant

Intelligence supports the commander's requirements.

Predictive

Intelligence informs the commander about what the threat can do (threat capabilities, emphasizing the most dangerous threat COA) and is most likely to do (the most likely threat COA). The G-2/S-2 staff should anticipate the commander's intelligence needs.

Tailored

Intelligence is shared and disseminated in the format requested by the commander, subordinate commanders, and staffs. It should support and satisfy the commander's priorities. The G-2/S-2 staff presents clear, concise intelligence that meets the commander's preferences, facilitates situational understanding, and is usable for decision-making or other action.

I. The Intelligence Process

Ref: ADRP 2-0, Intelligence (Aug '12), chap. 3.

Commanders use the operations process to drive the planning necessary to understand, visualize, and describe their operational environment; make and articulate decisions; and direct, lead, and assess military operations. Commanders successfully accomplish the operations process by using information and intelligence. The design and structure of the intelligence process support commanders by providing intelligence needed to support mission command and the commander's situational understanding. The commander provides guidance and focus by defining operational priorities and establishing decision points and CCIRs.

The Joint Intelligence Process

The joint intelligence process provides the basis for common intelligence terminology and procedures. (JP 2-0.) It consists of six interrelated categories of intelligence operations:

- Planning and direction
- Collection
- Processing and exploitation
- Analysis and production
- Dissemination and integration
- Evaluation and feedback

The Army Intelligence Process

Due to the unique characteristics of Army operations, the Army intelligence process differs from the joint process in a few subtle ways while accounting for each category of the joint intelligence process. The Army intelligence process consists of four steps (plan and direct, collect, produce, and disseminate) and two continuing activities (analyze and assess).

The Army views the intelligence process as a model that describes how the intelligence warfighting function facilitates situational understanding and supports decisionmaking. This process provides a common framework for Army professionals to guide their thoughts, discussions, plans, and assessments.

Commander's guidance drives the intelligence process. The intelligence process generates information, products, and knowledge about threats, terrain and weather, and civil considerations for the commander and staff. The intelligence process supports all of the activities of the operations process (plan, prepare, execute, and assess). The intelligence process can be conducted multiple times to support each activity of the operations process. The intelligence process, although designed similarly to the operations process, includes unique aspects and activities:

- The plan and direct step of the intelligence process closely corresponds with the plan activity of the operations process
- The collect, produce, and disseminate steps of the intelligence process together correspond to the execute activity of the operations process
- Assess, which is continuous, is part of the overall assessment activity of the operations process

Intelligence support to operations requires input from the entire intelligence enterprise. This support is coordinated through the G-2/S-2 staff at each echelon by using the intelligence process.

The G-2/S-2 produces intelligence for the commander as part of a collaborative process. The commander drives the G-2's/S-2's intelligence production effort by establishing intelligence and information requirements with clearly defined goals and criteria. Differing unit missions and operational environments dictate numerous and varied production requirements to the G-2/S-2 and staff.

The G-2/S-2 and staff provide intelligence products that enable the commander to—

- Plan operations and employ maneuver forces effectively
- Recognize potential COAs
- Conduct mission preparation
- Employ effective tactics, techniques, and procedures
- Take appropriate security measures
- Focus information collection
- Conduct effective targeting

Commander's Guidance

Commanders drive the intelligence process by both providing commander's guidance and approving CCIRs. While it is not part of the intelligence process, commander's guidance is one of the primary mechanisms used to focus the intelligence process. While issuing their guidance, commanders should limit the number of CCIRs so the staff can focus its efforts and allocate sufficient resources. Each commander dictates which intelligence products are required, when they are required, and in what format.

I. Intelligence Process Steps

Just as the activities of the operations process overlap and recur as the mission demands, so do the steps of the intelligence process.

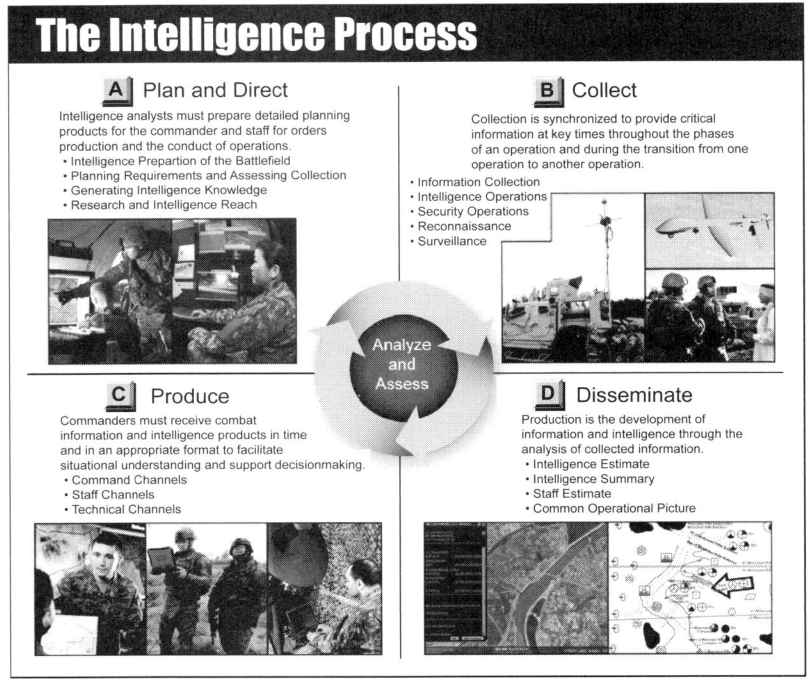

Ref: ADRP 2-0, Intelligence, fig. 3-1, p. 3-2.

Intelligence Planning Considerations

Ref: ADRP 2-0, Intelligence (Aug '12), pp. 3-4 to 3-5.

The intelligence warfighting function is designed to systematically answer intelligence requirements. Commanders focus the effort by clearly articulating intent, stating requirements, prioritizing targets, and assessing effectiveness as operations progress. However, commanders and staffs must have realistic expectations of intelligence capabilities to drive intelligence. The following are intelligence warfighting function planning considerations:

- Intelligence reduces uncertainty; it does not eliminate it. The commander has to determine the presence and degree of risk involved in conducting a particular operation. The time available to plan and prepare is directly related to the risk. Usually, the more time allotted to planning and preparation, the lower the risk. One of the commander's considerations is determining the appropriate balance between the time allotted for collection and operational necessity for executing an operation. It takes time to collect information and develop it into detailed and precise intelligence products.
- The intelligence warfighting function comprises finite resources and capabilities. Intelligence systems and Soldiers trained in specific skills are limited. Once lost to action or accident, these systems and Soldiers are not easily replaceable; in some cases, it may not be possible to replace them during the course of the current operation. The loss of Soldiers, equipment and also interpreters can result in the inability to detect or analyze threat actions.
- In order to provide effective intelligence, the intelligence warfighting function must have adequate communications and network-enabled capabilities.

Commanders must employ organic collection assets as well as plan, coordinate, and articulate requirements to leverage the intelligence enterprise. Commanders and staffs cannot expect higher echelons to automatically provide all of the information and intelligence they need. While intelligence reach is a valuable tool, the push of intelligence products from higher echelons does not relieve subordinate staffs from developing specific and detailed requirements. Commanders and staffs must focus requests for intelligence support by clearly articulating requirements. Intelligence activities are enabled by and subject to laws, regulations, and policies to ensure proper conduct of intelligence operations. While there are too many to list, legal authorities include USC, executive orders (EOs), National Security Council and DOD directives, Army regulations, U.S. SIGINT directives, status-of-forces agreements (SOFAs), rules of engagement (ROE), and other relevant international laws.

The staff focuses information collection plans on answering CCIRs and other requirements, and enables the quick retasking of units and assets as the situation changes. Planning requirements and assessing collection includes continually identifying intelligence gaps. This ensures the developing threat situation and civil considerations—not only the operation order—drive information collection. Specifically, G-2s/S-2s—

- Evaluate information collection assets for suitability (availability, capability, vulnerability, and performance history) to execute information collection tasks and make appropriate recommendations on asset tasking to the G-3/S-3.
- Assess information collection against CCIRs and other requirements to determine the effectiveness of the information collection plan. They maintain awareness to identify gaps in coverage and identify the need to cue or recommend redirecting information collection assets to the G-3/S-3.
- Update the planning requirements tools as requirements are satisfied, added, modified, or deleted. They remove satisfied requirements and recommend new requirements as necessary.

A. Plan and Direct

Each staff element must conduct analysis before operational planning can begin. Planning consists of two separate but closely related components—conceptual and detailed planning. Conceptual planning involves understanding the operational environment and the problem, determining the operation's end state, and visualizing an operational approach. Detailed planning translates the broad operational approach into a complete and practical plan.

For more information on conceptual and detailed planning, refer to ADRP 5-0.

The initial generation of intelligence knowledge about the operational environment occurs far in advance of detailed planning and orders production. This intelligence helps focus information collection once a mission is received. Intelligence planning is also an inherent part of the Army design methodology and the military decisionmaking process (MDMP). Intelligence analysts must prepare detailed planning products for the commander and staff for orders production and the conduct of operations. Through thorough and accurate planning, the staff allows the commander to focus the unit's combat power to achieve mission success.

The plan and direct step also includes activities that identify key information requirements and develops the means for satisfying those requirements. The G-2/S-2 collaborates with the G-3/S-3 to produce a synchronized and integrated information collection plan focused on answering CCIRs and other requirements. CCIRs and other requirements drive the information collection effort. Intelligence planning and directing comprises a broad range of detailed tasks, to include—

- Conducting activities, such as research, intelligence reach, and analysis. These activities produce the initial intelligence knowledge about the operational environment
- Generating intelligence knowledge
- Preparing IPB products and overlays
- Developing the initial intelligence estimate or briefings (usually as part of the mission analysis briefing)
- Establishing the intelligence architecture and testing access to the intelligence enterprise
- Establishing effective analytic collaboration
- Establishing liaisons
- Establishing intelligence team cohesiveness
- Establishing reporting procedures
- Establishing formats and standards for products
- Planning refinements, backbriefs, SOP reviews, and rehearsals, and coordinating with various elements and organizations
- Establishing other troop leading procedures or coordination, as necessary, in accordance with the mission variables (METT-TC)
- Planning requirements and assessing collection
- Assisting the G-3/S-3 with updating the information collection plan
- Assessing continuously
- Providing intelligence portions of the order

Requirements Management

Ref: ADRP 2-0, Intelligence (Aug '12), p. 3-5.

For requirements management, there are three types of requirements resulting from planning requirements and assessing collection. The following three types of validated information requirements are prioritized for purposes of assigning information collection tasks: priority intelligence requirements (PIRs), intelligence requirements, and information requirements.

Refer to FM 3-55 and ATTP 2-01 for more details on requirements and indicators.

The following shows the process of developing requirements and integrating them into the information collection process.

Requirements Development

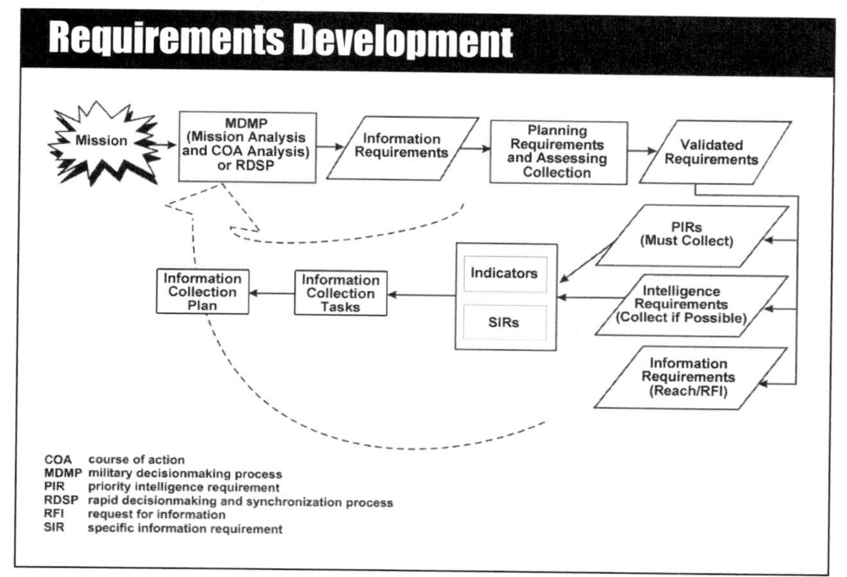

Ref: ADRP 2-0, Intelligence, fig. 3-2, p. 3-5.

Priority Intelligence Requirement (PIR)

An intelligence requirement, stated as a priority for intelligence support, that the commander and staff need to understand the adversary or the operational environment. Also called PIR. See FM 2-01.3.

Intelligence Requirements

A type of information requirement developed by subordinate commanders and the staff (including subordinate staffs) that requires dedicated information collection for the elements of threat, terrain and weather, and civil considerations. (ADRP 2-0)

Information Requirements

In intelligence usage, those items of information regarding the adversary and other relevant aspects of the operational environment that need to be collected and processed in order to meet the intelligence requirements of a commander. See ADRP 2-0. (Army) Any information elements the commander and staff require to successfully conduct operations. (ADRP 6-0)

B. Collect

Collection is synchronized to provide critical information at key times throughout the phases of an operation and during the transition from one operation to another operation. A successful information collection effort results in the timely collection and reporting of relevant and accurate information, which supports the production of intelligence. Collection consists of collecting, processing, and reporting information in response to information collection tasks. Different units and systems collect information and data about threats, terrain and weather, and civil considerations. Collected information is used in intelligence databases, intelligence production, and the G-2's/S-2's awareness—and ultimately supports the commander's situational understanding. Information collection activities transition as requirements change, the unit mission changes, the unit proceeds through the phases of an operation, and the unit prepares for future operations.

It is critical for the staff to plan for and use well-developed procedures and flexible planning to track emerging targets, adapt to changing operational requirements, and meet the requirement for combat assessment. Once the information is collected, it is processed into a form that enables analysts to extract essential information and produce intelligence and targeting products. Processing involves converting, evaluating, analyzing, interpreting, and synthesizing raw collected data and information. Processing examples include—

- Preparing imagery for exploitation
- Translating a document from a foreign language
- Converting electronic data into a standardized reporting format (including database formats) that can be analyzed by a system operator
- Correlating information that groups data into a form all analysts can use

Collected and processed information must then be reported to the appropriate units, organizations, or agencies for analysis or action. The G-2/S-2 coordinates with the unit staff, subordinate and lateral commands, and higher echelon units to ensure specific reporting assets, personnel, equipment (especially communications), and procedures are in place. The G-2/S-2 staff evaluates the reported information for its responsiveness to information collection tasks.

The timely and accurate reporting of combat information and intelligence is critical to successful operations. The most critical information collected may be of little value if not reported in a timely manner. Unit SOPs must clearly state the transmission means of different types of reports (for example, sent by satellite communications, tactical radios, or by automated means). Generally, the transmission of reports for threat contact and actions, CCIRs, combat information, and CBRN is by voice, followed up with automated reports.

Intelligence and time-sensitive combat information that affect the current operation are disseminated immediately upon recognition. Combat information is unevaluated data, gathered by or provided directly to the tactical commander which, due to its highly perishable nature or the criticality of the situation, cannot be processed into tactical intelligence in time to satisfy the user's tactical intelligence requirements (JP 2-01). The routing of combat information proceeds immediately in two directions—directly to the commander and through routine reporting channels for use by intelligence analysis and production elements.

C. Produce

Production is the development of intelligence through the analysis of collected information and existing intelligence. Analysts create intelligence products, conclusions, or projections regarding threats and relevant aspects of the operational environment to answer known or anticipated requirements in an effective format. The G-2/S-2 staff processes and analyzes information from single or multiple sources, disciplines,

Intelligence Reach

Ref: ADRP 2-0, Intelligence (Aug '12), pp. 3-5 to 3-6.

Information can be acquired through the push and pull of information, databases, homepages, collaborative tools, and broadcast services. Intelligence reach also supports distributed analysis. Three important aspects of intelligence reach are searches and queries, data mining, and collaboration.

Searches and Queries

The ability to search networks and query databases is an essential skill for intelligence professionals. The basic search techniques used on the unclassified Internet also apply to the SECRET Internet Protocol Router Network (SIPRNET) and Joint Worldwide Intelligence Communications System (JWICS). In order to conduct a search or query, intelligence professionals must plan the search, conduct the search, refine the search, and record the results. Intelligence professionals use their understanding of the supported unit's mission, specific information requirements, and available sources of information and intelligence to plan and execute their search. Because classified intelligence networks and systems are very compartmentalized, intelligence analysts should first determine which networks or databases are most likely to have the required information. Classified networks usually have specific access requirements, and intelligence professionals must coordinate for access to the classified networks.

Data Mining

Data mining is finding key pieces of intelligence that may be buried in the mass of data available. Data mining uses automated statistical analysis techniques to search for the specific data parameters that intelligence professionals predetermine will answer their information requirements. Data mining can help organize the mass of collected data.

Collaboration

Intelligence professionals work in an environment that is enhanced by collaboration. Collaboration facilitates parallel planning and enhances all aspects of the intelligence process by enriching analysis, incorporating different points of view, and broadening situational understanding. Intelligence professionals develop the ability to work effectively with others on a common task, respect the contributions of others, and contribute to consensus when warranted.

Request for Information

After having exhausted intelligence reach sources, a unit may decide to submit an RFI to higher headquarters, lateral units, or other organizations. Users enter RFIs into an RFI management system where every other system user can see and potentially answer them.

Liaison

In order to coordinate, synchronize, and integrate operations; exchange information and intelligence; move through certain areas; and ensure protection, it may be necessary to establish liaison with many different elements, organizations, and institutions of the host nation. These include the local police, town officials, foreign military forces, and political and other key figures within the AO. Operations may also necessitate coordination, synchronization, and integration with other U.S. and multinational forces. The most effective liaisons are physically co-located with the element, which enables them to build rapport and establish a close relationship.

Refer to The Battle Staff SMARTbook for more information on liaison duties.

and complementary intelligence capabilities, and integrates the information with existing intelligence to create finished intelligence products.

Intelligence products must be timely, relevant, accurate, predictive, and tailored to facilitate situational understanding and support decisionmaking. The accuracy and detail of intelligence products have a direct effect on operational success. Due to time constraints, analysts sometimes develop intelligence products that are not as detailed as they prefer. However, a timely, accurate answer that meets the commander's requirements is better than a more detailed answer that is late.

The G-2/S-2 staff prioritizes and synchronizes the unit's information processing and intelligence production efforts. The G-2/S-2 staff addresses numerous and varied production requirements based on PIRs and other requirements; diverse missions, environments, and situations; and user-format requirements. Through analysis, collaboration, and intelligence reach, the G-2/S-2 and staff use the intelligence capability of higher, lateral, and subordinate echelons to meet processing and production requirements.

Processing is often an important production activity. The G-2/S-2 staff processes information collected by the unit's assets as well as information received from higher, subordinate, and lateral echelons and other organizations. Processing includes sorting through large amounts of collected information and intelligence and converting relevant information into a form suitable for analysis, production, or immediate use.

Analysis occurs to ensure the information is relevant, to isolate significant elements of information, and to integrate the information into an intelligence product. Additionally, analysis of information and intelligence is important to ensure the focus, prioritization, and synchronization of the unit's intelligence production effort is in accordance with the PIRs and other requirements.

D. Disseminate

Commanders must receive combat information and intelligence products in time and in an appropriate format to facilitate situational understanding and support decisionmaking. Timely dissemination of intelligence is critical to the success of operations. Dissemination is deliberate and ensures consumers receive intelligence to support operations.

This step does not include the normal reporting and technical channels otherwise conducted by intelligence warfighting function organizations and units during the intelligence process. Each echelon with access to information may perform analysis on that information. Then each echelon ensures that resulting intelligence products are properly disseminated. Determining the product format and selecting the means to deliver it are key aspects of dissemination.

The commander and staff must establish and support a seamless intelligence architecture including an effective dissemination plan. A dissemination plan can be a separate product or integrated into existing products, such as the planning requirements tools.

Intelligence and communications systems continue to evolve in their sophistication, application of technology, and accessibility to the commander. Their increasing capabilities also create an unprecedented volume of information available to commanders at all echelons. The commander and staff must have a basic understanding of intelligence dissemination systems and their contribution to the intelligence warfighting function.

Dissemination Methods and Techniques

Ref: ADRP 2-0, Intelligence (Aug '12), pp. 3-8 to 3-9.

There are numerous methods and techniques for disseminating information and intelligence. The appropriate technique in any particular situation depends on many factors, such as capabilities and mission requirements. Information presentation may be in a verbal, written, interactive, or graphic format. The type of information, time allocated, and commander's preferences all influence the information format. Answers to PIRs require direct dissemination to the commander, subordinate commanders, and staff. Direct dissemination is conducted person-to-person, by voice communications, or electronic means. Other dissemination methods and techniques include—

- Direct electronic dissemination (a messaging program)
- Instant messaging
- Web posting (with notification procedures for users)
- Printing or putting the information on a compact disk and sending it

Disseminating intelligence simultaneously to multiple recipients is one of the most effective, efficient, and timely methods, and can be accomplished through various means—for example, push or broadcast. G-2s/S-2s must plan methods and techniques to disseminate information and intelligence when normal methods and techniques are unavailable. For example, information and intelligence can be disseminated using liaisons or regularly scheduled logistic packages as long as any classified information is properly protected and individuals are issued courier orders.

Dissemination Channels

Intelligence leaders at all levels assess the dissemination of intelligence and intelligence products. Reports and other intelligence products move along specific channels within the intelligence architecture. The staff helps streamline information distribution within these channels by ensuring dissemination of the right information in a timely manner to the right person or element. There are three channels through which commanders and their staffs communicate:

- **Command Channels.** Command channels are direct chain-of-command links used by commanders or authorized staff officers for command-related activities. Command channels include command radio nets, video teleconferences, and mission command systems.
- **Staff Channels.** Staff channels are staff-to-staff links within and between headquarters. The staff uses staff channels for control-related activities. Through these channels, the staff coordinates and transmits intelligence, controlling instructions, planning information, early warning information, and other information to support mission command. Examples of staff channels include the operations and intelligence radio net, voice-over-Internet phone (VOIP), the staff huddle, and video teleconferences, which provide information and intelligence to the rest of the intelligence architecture.
- **Technical Channels.** Technical channels are the transmission paths between two technically similar units or offices that perform a technical function requiring special expertise. These channels are used to control the performance of technical functions. Technical channels are used only when that control is authorized by an operation order or for those authorities granted specifically in Army regulations or unit SOPs. Staffs typically use technical channels to control specific functions. These functions include fire direction and the technical reporting channels for intelligence operations, reconnaissance, and surveillance.

II. Intelligence Process Continuing Activities

Analyze and assess are two continuing activities that shape the intelligence process. They occur continually throughout the intelligence process.

E. Analyze

Analysis assists commanders, staffs, and intelligence leaders in framing the problem, stating the problem, and solving it. Leaders at all levels conduct analysis to assist in making many types of decisions. Analysis occurs at various stages throughout the intelligence process and is inherent throughout intelligence support to situational understanding and decisionmaking. Collectors perform initial analysis before reporting. For example, a HUMINT collector analyzes an intelligence requirement to determine the best possible collection strategy to use against a specific source.

Analysis in requirements management is critical to ensuring information requirements receive the appropriate priority for collection. The G-2/S-2 staff analyzes each requirement to determine—

- The requirement's feasibility and whether it supports the commander's guidance
- The best method of satisfying the requirement (for example, what unit or capability and where to position that capability)
- If the collected information satisfies the requirement

Analysis is used in situation development to determine the significance of collected information and its significance relative to predicted threat COAs and PIRs and other requirements. Through predictive analysis, the staff attempts to identify threat activity or trends that present opportunities or risks to the friendly force. They often use indicators developed for each threat COA as the basis for their analysis and conclusions.

F. Assess

Assess is part of the overall assessment activity of the operations process. For intelligence purposes, assessment is the continuous monitoring and evaluation of the current situation, particularly significant threat activities and changes in the operational environment. Assessing the situation begins upon receipt of the mission and continues throughout the intelligence process. This assessment allows commanders, staffs, and intelligence leaders to ensure intelligence synchronization. Friendly actions, threat actions, civil considerations, and events in the area of interest interact to form a dynamic operational environment. Continuous assessment of the effects of each element on the others, especially the overall effect of threat actions on friendly operations, is essential to situational understanding.

The G-2/S-2 staff continuously produces assessments based on operations, the information collection effort, the threat situation, and the status of relevant aspects of the operational environment. These assessments are critical to—

- Ensure PIRs are answered
- Ensure intelligence requirements are met
- Redirect collection assets to support changing requirements
- Ensure operations run effectively and efficiently
- Ensure proper use of information and intelligence
- Identify threat efforts at deception and denial

The G-2/S-2 staff continuously assesses the effectiveness of the information collection effort. This type of assessment requires sound judgment and a thorough knowledge of friendly military operations, characteristics of the area of interest, and the threat situation, doctrine, patterns, and projected COAs.

II. Army Intelligence Capabilities

Ref: ADRP 2-0, Intelligence (Aug '12), chap. 4.

The intelligence warfighting function executes the intelligence process by employing intelligence capabilities. All-source intelligence and single-source intelligence are the building blocks by which the intelligence warfighting function facilitates situational understanding and supports decisionmaking. The intelligence warfighting function receives information from a broad variety of sources. Some of these sources are commonly referred to as single-source capabilities. Single-source capabilities are employed through intelligence operations with the other means of information collection (reconnaissance, surveillance, and security operations). The intelligence produced based on all of those sources is called all-source intelligence.

I. All-Source Intelligence

Army forces conduct operations based on all-source intelligence assessments and products developed by the G-2/S-2 staff. All-source intelligence is the integration of intelligence and information from all relevant sources in order to analyze situations or conditions that impact operations. All-source intelligence is used to develop the intelligence products necessary to aid situational understanding, support the development of plans and orders, and answer information requirements. Although all-source intelligence normally takes longer to produce, it is more reliable and less susceptible to deception than single-source intelligence.

In joint doctrine, all-source intelligence also refers to intelligence products and/or organizations and activities that incorporate all sources of information, most frequently including human resources intelligence, imagery intelligence, measurement and signature intelligence (MASINT), SIGINT, and open-source data in the production of finished intelligence.

Refer to JP 2-0.

Fusion facilitates all-source production. For Army purposes, fusion is consolidating, combining, and correlating information together. Fusion occurs as an iterative activity to refine information as an integral part of all-source analysis.

All-source intelligence production is continuous and occurs throughout the intelligence and operations processes. Most of the products resulting from all-source intelligence are initially developed during planning and updated as needed throughout preparation and execution based on information gained from continuous assessment.

The fundamentals of all-source intelligence analysis are intelligence analysis, all-source production, situation development, generating intelligence knowledge, support to IPB, support to targeting, and support to information collection.

Through the receipt and processing of incoming reports and messages, the G-2/S-2 staff determines the significance and reliability of incoming information, integrates incoming information with current intelligence holdings, and through analysis and evaluation determines changes in threat capabilities, vulnerabilities, and probable COAs. The G-2/S-2 staff supports the integrating processes (IPB, targeting, and risk management) and continuing activities (information collection) by providing all-source analysis of threats, terrain and weather, and civil considerations.

II. Single-Source Intelligence

Single-source intelligence includes the joint intelligence disciplines and complementary intelligence capabilities. One important aspect within single-source intelligence is processing, exploitation, and dissemination (PED) activities.

A. The Intelligence Disciplines

In joint operations, the intelligence enterprise is commonly organized around the intelligence disciplines. The intelligence disciplines are—

- CI
- Geospatial intelligence (GEOINT)
- HUMINT
- MASINT
- Open-source intelligence (OSINT)
- SIGINT
- Technical intelligence (TECHINT)

The intelligence disciplines are integrated to ensure a multidiscipline approach to intelligence analysis, and ultimately all-source intelligence facilitates situational understanding and supports decisionmaking. Each discipline applies unique aspects of support and guidance through technical channels.

See following pages (pp. 4-26 to 4-27) for further discussion.

Intelligence (ADRP 2-0)

B. Complementary Intelligence Capabilities

Complementary intelligence capabilities contribute valuable information for all-source intelligence to facilitate the conduct of operations. The complementary intelligence capabilities are specific to the unit and circumstances at each echelon and can vary across the intelligence enterprise. These capabilities include but are not limited to—

- Biometrics-enabled intelligence (BEI)
- Cyber-enabled intelligence
- Document and media exploitation (DOMEX)
- Forensic-enabled intelligence (FEI)

See following pages (pp. 4-28 to 4-29) for further discussion.

C. Processing, Exploitation, and Dissemination (PED)

Processing and exploitation in intelligence usage, is the conversion of collected information into forms suitable to the production of intelligence (JP 2-01). Dissemination and integration, in intelligence usage, is the delivery of intelligence to users in a suitable form and the application of the intelligence to appropriate missions, tasks, and functions (JP 2-01). These two definitions are routinely combined into the acronym PED. PED is exclusive to single-source intelligence and fits within the larger intelligence process.

In joint doctrine, PED is a general concept that facilitates the allocation of assets to support intelligence operations. Under the PED concept, planners examine all collection assets and then determine if allocation of additional personnel and systems is required to support the exploitation of the collected information. Accounting for PED facilitates processing collected information into usable and relevant information for subsequent all-source production in a timely manner. Beyond doctrine, PED plays an important role within larger DOD intelligence programmatics.

Processing, Exploitation, & Dissemination (PED) Activities within Intelligence Operations

Ref: ADRP 2-0, Intelligence (Aug '12), p. 4-13.

Every intelligence discipline and complementary intelligence capability is different, but each conducts PED activities to support timely and effective intelligence operations. Effective intelligence operations allow flexibility and responsiveness to changing situations and adaptive threats. In general, PED activities are part of the single-source information flow into all-source intelligence, allow for single-source intelligence to answer intelligence requirements, and are inextricably linked to the intelligence architecture. PED activities facilitate timely, relevant, usable, and tailored intelligence.

Some PED enablers are organic to the intelligence unit while other enablers are task-organized or distributed through the network. PED activities are key components of the intelligence communications networks, data/information repositories, and the organizational backbone (sometimes referred to as the foundation layer of the intelligence enterprise). These capabilities are also an important part of the Army's contribution to the intelligence enterprise.

PED enablers often enhance a unit's ability to—

- **Task:** Provide input to the tasking of intelligence collection systems and conduct dynamic retasking. PED activities also improve the flow of information and guidance through technical channels.
- **Collect:** Receive collection from systems that would otherwise be inaccessible.
- **Process:** Transform a larger volume of data and convert that data into a usable format. Augmenting the intelligence unit's organic ability to process massive amounts of data is often valuable and improves the unit's operational effectiveness.
- **Exploit:** Quickly use the processed information to refine guidance (using technical channels) and identify specific impacts on the mission. The multifunctional team is an example of a unit specifically designed to accomplish these goals to support time-sensitive requirements.
- **Disseminate:** Report collected information to other intelligence and operational elements and to the commander to support decisionmaking. This reporting facilitates all-source intelligence, targeting, and cueing other collectors.

System developers often combine PED capabilities into intelligence systems, but that is not possible in all cases. In the past, many have viewed PED enablers as systems. Trained operators, analysts, and maintainers are necessary to conduct and sustain some PED activities. PED enablers include a broad list of systems and supporting personnel across the intelligence enterprise.

The effective employment of PED activities within the intelligence architecture allows for the dynamic execution of intelligence operations. The intelligence process moves at significantly different speeds depending on the mission, situation, and other factors. To adapt to the changing pace of the intelligence process, intelligence leaders adjust the pace of intelligence operations. Successfully supporting time-sensitive requirements requires the intelligence process steps (plan and direct, collect, produce, and disseminate) and continuity activities (analyze and assess) to occur almost simultaneously. This results in a rapid or compressed intelligence process and is supported by dynamic intelligence operations. The execution of dynamic intelligence operations depends on a mature intelligence architecture, thorough planning, sufficient PED enablers, and all-source driven situational understanding.

A. The Intelligence Disciplines

Ref: ADRP 2-0, Intelligence (Aug '12), pp. 4-2 to 4-9.

In joint operations, the intelligence enterprise is commonly organized around the intelligence disciplines. The intelligence disciplines are—

Intelligence Disciplines

- **Counterintelligence (CI)**
- **Geospatial Intelligence**
- **Human Intelligence (HUMINT)**
- **Measurement and Signature Intelligence (MASINT)**
- **Open-Source Intelligence (OSINT)**
- **Signals Intelligence (SIGINT)**
- **Technical Intelligence (TECHINT)**

The intelligence disciplines are integrated to ensure a multidiscipline approach to intelligence analysis, and ultimately all-source intelligence facilitates situational understanding and supports decisionmaking. Each discipline applies unique aspects of support and guidance through technical channels.

Refer to JP 2-0.1.

Counterintelligence (CI)

CI counters or neutralizes intelligence collection efforts through collection, CI investigations, operations, analysis, production, and technical services and support. CI includes all actions taken to detect, identify, track, exploit, and neutralize multidiscipline intelligence activities of foreign intelligence and security services (FISS), international terrorist organizations, and adversaries, and is the key intelligence community contributor to protect U.S. interests and equities.

The mission of Army CI is to conduct aggressive, comprehensive, and coordinated investigations, operations, collection, analysis and production, and technical services. These functions are conducted worldwide to detect, identify, assess, counter, exploit, or neutralize the FISS, international terrorist organization, and adversary collection threat.

Refer to FM 2-22.2.

Geospatial Intelligence

Geospatial intelligence is the exploitation and analysis of imagery and geospatial information to describe, assess, and visually depict physical features and geographically referenced activities on the Earth. Geospatial intelligence consists of imagery, imagery intelligence, and geospatial information (JP 2-03). (Section 467, Title 10, USC [10 USC 467], establishes GEOINT.) Note. GEOINT consists of any one or any combination of the following components: imagery, IMINT, and geospatial information and services.

For more information on GEOINT, refer to TC 2-22.7.

Human Intelligence (HUMINT)

Human intelligence is the collection by a trained human intelligence collector of foreign information from people and multimedia to identify elements, intentions, composition, strength, dispositions, tactics, equipment, and capabilities (FM 2-0).

A HUMINT source is a person from whom foreign information is collected for the purpose of producing intelligence. HUMINT sources can include friendly, neutral, or hostile personnel. The source may either possess first- or second-hand knowledge normally obtained through sight or hearing. Categories of HUMINT sources include but are not limited to detainees, enemy prisoners of war, refugees, displaced persons, local inhabitants, friendly forces, and members of foreign governmental and nongovernmental organizations.

For more information on HUMINT, refer to FM 2-22.3.

Measurement and Signature Intelligence (MASINT)

Measurement and signature intelligence is intelligence obtained by quantitative and qualitative analysis of data (metric, angle, spatial, wavelength, time dependence, modulation, plasma, and hydromagnetic) derived from specific technical sensors for the purpose of identifying any distinctive features associated with the emitter or sender, and to facilitate subsequent identification and/or measurement of the same. The detected feature may be either reflected or emitted (JP 2-0).

For more information on MASINT, refer to JP 2-0.

Open-Source Intelligence (OSINT)

Open-source intelligence is information of potential intelligence value that is available to the general public (JP 2-0). For the Army, OSINT is the discipline that pertains to intelligence produced from publicly available information that is collected, exploited, and disseminated in a timely manner to an appropriate audience for the purpose of addressing a specific intelligence requirement. OSINT operations are integral to Army intelligence operations.

For more information on OSINT, refer to ATTP 2-22.9.

Signals Intelligence (SIGINT)

Signals intelligence is intelligence derived from communications, electronic, and foreign instrumentation signals (JP 2-0). SIGINT provides unique intelligence information, complements intelligence derived from other sources, and is often used for cueing other sensors to potential targets of interest. For example, SIGINT, which identifies activities of interest, may be used to cue GEOINT to confirm that activity. Conversely, changes detected by GEOINT can cue SIGINT collection against new targets. The discipline is subdivided into three subcategories:

- Communications intelligence (COMINT)
- Electronic intelligence (ELINT)
- Foreign instrumentation signals intelligence (FISINT)

Technical Intelligence (TECHINT)

Technical intelligence is intelligence derived from the collection, processing, analysis, and exploitation of data and information pertaining to foreign equipment and materiel for the purposes of preventing technological surprise, assessing foreign scientific and technical capabilities, and developing countermeasures designed to neutralize an adversary's technological advantages (JP 2-0). The role of TECHINT is to ensure Soldiers understand the threat's full technological capabilities. With this understanding, U.S. forces can adopt appropriate countermeasures, operations, and tactics, techniques, and procedures.

Every TECHINT mission supports tactical through strategic requirements by the timely collection and processing of materiel and information, follow-on analysis and resulting production of intelligence, and dissemination to a wide range of consumers. Commanders rely on TECHINT to provide them with tactical and technological advantages to successfully synchronize and execute operations. TECHINT combines information to identify specific individuals, groups, and nation states, matching them to events, places, devices, weapons, equipment, or contraband that associates their involvement in hostile or criminal activity.

For more information on TECHINT, refer to TC 2-22.4.

B. Complementary Intelligence Capabilities

Ref: ADRP 2-0, Intelligence (Aug '12), pp. 4-9 to 4-12.

Complementary intelligence capabilities contribute valuable information for all-source intelligence to facilitate the conduct of operations. The complementary intelligence capabilities are specific to the unit and circumstances at each echelon and can vary across the intelligence enterprise. These capabilities include but are not limited to—

Complimentary Intelligence Capabilities

- **Biometrics-Enabled Intelligence (BEI)**
- **Cyber-Enabled Intelligence (CEI)**
- **Document and Media Exploitation (DOMEX)**
- **Forensic-Enabled Intelligence (FEI)**

Biometrics-Enabled Intelligence (BEI)

Joint doctrine defines biometric as a measurable physical characteristic or personal behavioral trait used to recognize the identity or verify the claimed identity of an individual (JP 2-0). Biometrics as a process of confirming identity is not exclusive to the intelligence warfighting function. This enabler supports multiple activities and tasks of other warfighting functions. Biometrics-enabled intelligence is the information associated with and/or derived from biometric signatures and the associated contextual information that positively identifies a specific person and/or matches an unknown identity to a place, activity, device, component, or weapon.

Commanders are employing biometrics with increasing intensity during operations to identify insurgents, verify local and third-country nationals accessing U.S. bases and facilities, and link people to events. Biometric systems are employed to disrupt threat forces freedom of movement within the populace and to positively identify known threat forces and personnel. These systems collect biometric data and combine them with contextual data to produce an electronic biometric dossier on the individual. Affixing an individual's identification using his or her unique physical features and linking this identity to the individual's past activities and previously used identities provide more accurate information about the individual. For example, during counterinsurgency operations, biometric collections and forensic exploitation of improvised explosive devices, cache sites, safe houses, and vehicles provide commanders additional tools to separate insurgents and criminals from the populace.

Biometric collection devices used by U.S. and multinational forces typically collect fingerprints, iris scans, and facial images. Biometrics positively identifies an encountered person and unveils terrorist or criminal activities regardless of paper documents, disguises, or aliases. This data is combined with local and national databases. The intelligence data, coupled with verifiable biometrics, enables the commander to perform more precise and effective targeting missions.

For more information on BEI, refer to TC 2-22.82.

Cyber-Enabled Intelligence (CEI)

The cyber domain provides another means to collect intelligence. Cyber-enabled intelligence is a complementary intelligence capability providing the ability to collect information and produce unique intelligence. All-source intelligence, the intelligence disciplines, and the other complementary intelligence capabilities are facilitated by using computers, technology, and networks. However, their use of computers, technology, and networks does not mean these are cyber operations. The guiding methods and regulations for the conduct of each intelligence discipline or complementary intelligence capability are governed under the appropriate title authority for each specific discipline or capability. Hence, the mission, authority, and oversight of an activity determine whether an activity is cyber-enabled intelligence or cyber-controlled.

Cyber-enabled intelligence is produced through the combination of intelligence analysis and the collaboration of information concerning activity in cyberspace and the electromagnetic spectrum. This intelligence supports cyber situational understanding. Unlike cyber operations, cyber-enabled intelligence is intelligence-centric based on collection within cyberspace and does not include operations and dominance within the electromagnetic spectrum. The results of cyber electromagnetic activities can provide intelligence professionals with a significant amount of information concerning both the physical and information domains.

Cyber-enabled intelligence facilitates decisionmaking at all levels through the analysis and production of relevant and tailored intelligence. Additionally, this complementary intelligence capability includes the integration of intelligence products into staff processes, such as IPB and targeting. The intelligence can range from broadly disseminated products targeted to general users to very specific and narrowly focused analysis and reports distributed via classified channels. The use of cyber-enabled intelligence facilitates an understanding of the threat's capabilities, intentions, potential actions, vulnerabilities, and impact on the environment.

Document and Media Exploitation (DOMEX)

Document and media exploitation is the processing, translation, analysis, and dissemination of collected hardcopy documents and electronic media that are under the U.S. Government's physical control and are not publicly available (TC 2-91.8). Threat intent, capabilities, and limitations may be derived through the exploitation of captured materials. Captured materials are divided into captured enemy documents and captured enemy materiel.

DOMEX products become a force multiplier only when captured materials are rapidly exploited at the lowest echelon possible. DOMEX assets pushed down to the tactical level provide timely and accurate intelligence support. This practice not only enables rapid exploitation and evacuation of captured materials but also hastens the feedback commanders receive from the higher echelon analysis.

For more information on DOMEX, refer to TC 2-91.8.

Forensic-Enabled Intelligence (FEI)

Forensics involves the application of a broad spectrum of scientific processes and techniques to establish facts. Battlefield or expeditionary forensics refers to the use of forensic techniques to provide timely and accurate information that facilitates situational understanding and supports decisionmaking. This includes collecting, identifying, and labeling portable items for future exploitation, and the collection of fingerprints, deoxyribonucleic acid (DNA), and other biometric data, which can aid in personnel recovery, from nontransportable items at a scene. Intelligence personnel can use information from forensic analysis and send it as combat information or incorporate it in the intelligence analysis effort.

FEI helps accurately identify networked and complex threats and attributes them to specific incidents and activities. The effort is often critical in supporting the targeting process. FEI can identify and determine the source of origin of captured materials. Accurate site documentation of incidents or events, material and structural analysis, and supporting data and information from the various forensic processes and techniques provide valuable data and facilitate adjusting friendly tactics and modifying equipment to enhance protection. Additionally, timely trace detection or material analysis of unknown substances can help protect the force from contaminants, toxins, and other hazards. Through toxicology, pathology and other forensic techniques, FEI supports Army medical intelligence. This intelligence includes detailed information on medical conditions of a specific area or of a threat.

There are many enablers that support PED activities. PED enablers are the specialized intelligence and communications systems, advanced technologies, and the associated personnel that conduct intelligence processing as well as single-source analysis within intelligence units. These enablers are distinct from intelligence collection systems and all-source analysis capabilities. PED activities are prioritized and focused on intelligence processing, analysis, and assessment to quickly support specific intelligence collection requirements and facilitate improved intelligence operations. PED began as a processing and analytical support structure for unique systems and capabilities like full motion video from unmanned aircraft systems. Unlike previous GEOINT collection capabilities, full motion video did not have supporting personnel and automated capability to process raw data into a usable format and conduct initial exploitation. PED enablers receive collection from many different intelligence sensors across the terrestrial, aerial, and space layers of the intelligence enterprise.

PED Enablers within the Intelligence Architecture

There are not enough PED enablers to support all intelligence operations, and support often depends on allocation from the joint force. Like other aspects of intelligence, commanders prioritize and resource PED enablers to intelligence units within the intelligence architecture based on thorough planning. Due to the complexity of this task, the G-2/S-2 plans the PED portion of the intelligence architecture and then advises the commander on prioritizing and resourcing PED activities. A thorough assessment of PED activities requires an understanding of the capabilities and requirements for many different types of systems and personnel from across the intelligence enterprise.

When requesting PED enablers, the gaining G-2/S-2 and intelligence unit commander are responsible for coordinating and planning for PED activities. However, the allocating echelon is also responsible for ensuring adequate planning, coordination, and use of PED enablers. Some PED enabler employment considerations for the gaining G-2/S-2 and intelligence unit commander include—

- **Intelligence architecture.** Employment of PED enablers depends on how the collector and supporting PED activity fits in the intelligence architecture. The employment is also specific to the intelligence discipline or complementary intelligence capability and supported echelon. MI units should capture their functional requirements during planning to ensure they request adequate PED capabilities.
- **Communications.** All intelligence operations depend on the various communications systems, networks, and information services that enable intelligence. It is important to consider and understand hardware and software requirements, compatibility issues, bandwidth priority and capacity, and maintenance requirements.
- **Reporting.** Operating effectively within the intelligence architecture requires system operators to understand reporting procedures, requirements, and timelines for operations and intelligence channels as well as for technical channels.
- **Targeting criteria.** Supporting lethal and nonlethal effects requires system operators to be thoroughly knowledgeable with the different criteria, including minimum accuracy and timeliness standards for each specific mission.
- **Technical channels.** System operators must understand how technical channels operate and how to use technical guidance to enhance collection. Additionally, PED activities facilitate the refinement of technical guidance.
- **Training.** Intelligence leaders inform the commander and staff on PED activity capabilities and limitations. Facilitating the integration of PED enablers requires intelligence leaders to conduct training with the intelligence unit and PED system operators, analysts, and maintainers.
- **Sustainment.** PED systems can provide a significant maintenance and logistic challenge to the intelligence unit. Reducing these challenges requires the intelligence unit to conduct thorough planning and coordination.

Chap 5

Fires Warfighting Function

Ref: ADP/ADRP 3-09, Fires (Aug '12), chap. 1 and ADRP 3-0, Operations (Nov '16), p. 5-5.

The fires warfighting function is the related tasks and systems that provide collective and coordinated use of Army indirect fires, air and missile defense, and joint fires through the targeting process. Army fires systems deliver fires in support of offensive and defensive tasks to create specific lethal and nonlethal effects on a target. The fires warfighting function includes the following tasks:

- Deliver fires.
- Integrate all forms of Army, joint, and multinational fires.
- Conduct targeting.

In addition to the characteristics for effective fires in ADRP 3-09, commanders consider:

- The desired effect, available capabilities, and time and resources are required to deliver the appropriate capability.
- Successful integration of information operations into the targeting process is important to mission accomplishment in many operations.

The fires warfighting function includes tasks associated with, integrating, and synchronizing the effects of Army indirect fires, AMD, and joint fires with the effects of other warfighting functions. It includes planning for targeting; providing fire support; countering air, ballistic missile, cruise missile, rocket, artillery, mortars, and unmanned aircraft systems threats; and integrating joint and multinational fires. This represents the tasks the fires function must accomplish to complement and reinforce the other warfighting functions.

Fires organizations require deliberate and dynamic targeting to achieve lethal and nonlethal effects against ground and aerial targets. For ground threats, fires leaders use the Army's targeting methodology to plan, prepare, execute, and assess effects on the ground. For aerial threats, fires leaders use air defense planning to determine air defense priorities and the tailoring of air defense artillery capabilities to defeat aerial threats.

As a warfighting function, fires address requirements associated with offensive and defensive tasks supporting the concept of operations and integrated into the scheme of maneuver.

A. Deliver Fires

Today's operational environments require the integration of Army indirect fires in support of offensive, defensive, and stability tasks. Fires combine the use of air and ground artillery with the capabilities of other Army warfighting functions, special operations forces (SOF), joint forces, and unified action partners to enable the supported commander to seize the initiative. Army forces plan for, integrate, coordinate and synchronize the fires capabilities (sensors and weapon systems) of unified action partners into the concept of operations to achieve synergy, develop a common operational picture (COP), and enable joint interdependencies from the tactical to strategic levels. Additionally, complementary and reinforcing joint and multinational capabilities provide redundancy to mitigate environmental and operational restrictions, resource shortfalls, as well as gaps in coverage from a particular asset.

B. Integrate All Forms of Army, Joint and Multinational Fires

Fires must be integrated with the capabilities of other Army warfighting functions, special operation forces, joint forces and multinational forces. Integration of fires creates an optimal environment that mitigates risks, resource shortfalls and covers gaps within the areas of operations (AO). Ground and air fires must be integrated

II. Fires Overview

Ref: ADP 3-09, Fires (Aug '12).

The role of fires is to enable Army forces to seize and retain the initiative, prevent and deter conflict, defeat adaptive threats and succeed in a wide range of contingencies. Fires in decisive action create effects and set conditions to enable commanders to prevail in unified land operations. Fires are surface-to-surface, surface-to-air, and joint fires including electronic attack.

Fires combine the core competencies of air defense artillery and field artillery by planning, synchronizing, and executing their critical capabilities, using the principles and characteristics of fires to support unified land operations. The core competencies, critical capabilities, principles, and characteristics of fires are listed below.

Fires in Support of Unified Land Operations

Unified land operations recognizes the three-dimensional nature of modern warfare and the need to conduct a fluid mix of Fires in support of offensive, defensive, and stability tasks simultaneously. Integrating fires into unified land operations requires the development and full understanding of and strict adherence to common maneuver coordination measures, airspace coordinating measures and fire support coordination measures, ROE and other constraints/restraints.

The operational environment requires the integration of Army offensive and defensive surface-to-surface and surface-to-air fires capabilities with the capabilities of other Army warfighting functions, special operations forces, joint services, interagency, and multinational partners. Fires personnel integrate the fires capabilities (sensors, weapons, effects) of joint, interagency, and multinational partners into the concept of operations to achieve synergy, develop a common operational picture, and help facilitate joint interdependency. Additionally, complementary and reinforcing joint and multinational capabilities will provide redundancy to mitigate environmental and operational restrictions, resource shortfalls, and gaps in coverage for a supported unit or defended asset.

Fires in Support of Offensive, Defensive and Stability Tasks

Fires align with the tenets of unified land operations in ADP 3-0 which are flexibility, integration, lethality, adaptability, depth, and synchronization. Fires execution supports these tenets and the supported force in seizing and maintaining the initiative.

Fires in support of offensive, defensive, and stability tasks apply from tactical to strategic levels, and are employed in decisive, shaping, and sustaining operations. In Joint and Army doctrine, the effects of fires on a target may be lethal or nonlethal. Fires should be preplanned whenever possible.

ADRP 3-09, Fires (Aug '12)

Army Doctrine Reference Publication (ADRP) 3-09, Fires, is one of the ADRPs released under Doctrine 2015. ADRP 3-09 expands on the foundations and principles found in ADP 3-09. This Army doctrine for fires builds on the collective knowledge and experience gained through recent operations and numerous exercises. It is rooted in time-tested principles and fundamentals, while accommodating new technologies.

ADRP 3-09 makes numerous changes from Field Manuals (FM) 3-01 and 3-09. The most significant change is that air and missile defense (AMD) is moved from the protection warfighting function into the fires warfighting function. AMD contributes to the area air defense plan (AADP) by assisting the protection cell with the planning and development of the defended assets list (DAL). Execution of tasks related to fires which are part of the DAL will be performed by the fires warfighting function.

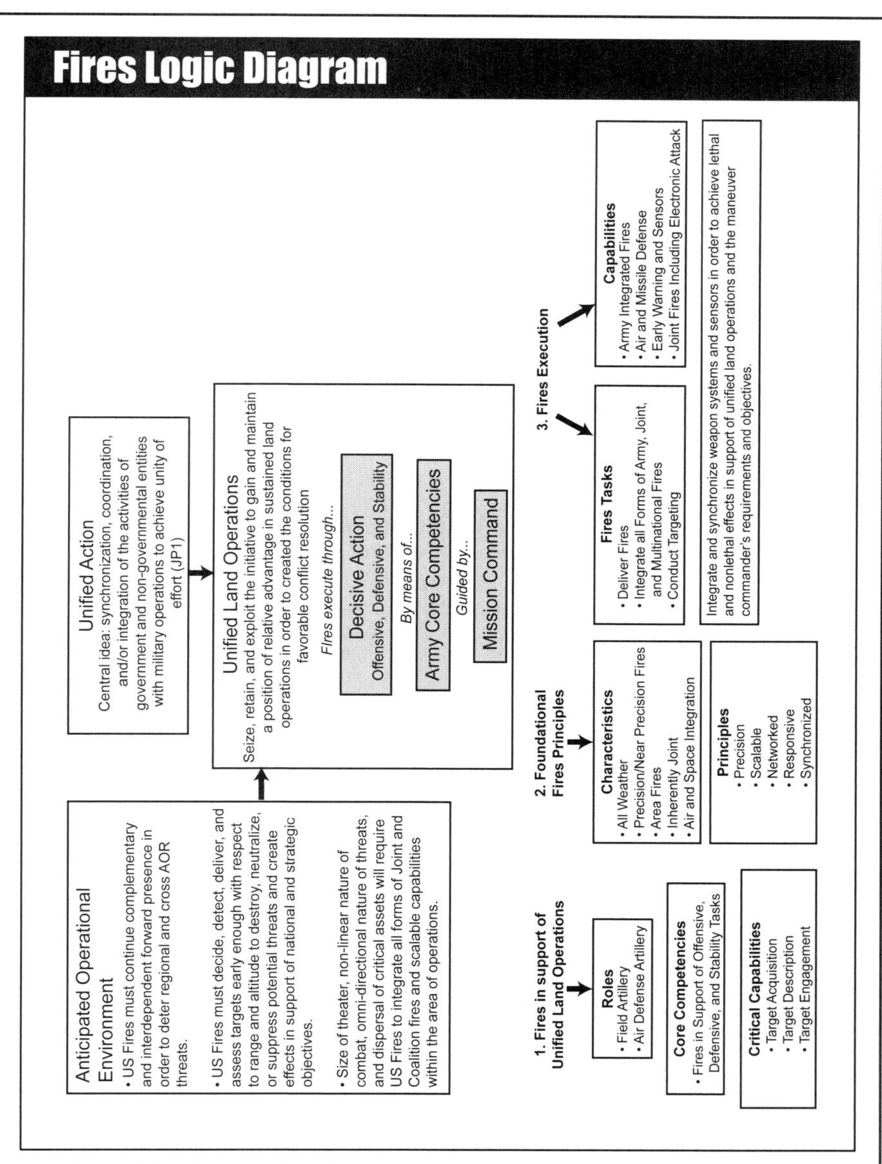

Ref: ADP 3-09, Fires, fig. 1, p. iii.

Additional changes in ADRP 3-09, from FM 3-01 and FM 3-09, include the field artillery (FA) mission statement is updated to read: The mission of the Field Artillery is to destroy, defeat, or disrupt the enemy with integrated fires to enable maneuver commanders to dominate in unified land operations.

ADRP 3-09 remains generally consistent with FM 3-01 and FM 3-09, on key topics while adopting updated terminology and concepts as necessary. These topics include the discussion of fires in support of unified land operations, decisive action and the operational framework.

with decisive action and unified land operations. Fires in unified land operations from air-to-surface, surface-to-air and surface-to-surface assets must be coordinated and cleared on the ground and through the airspace to enable the rapid and timely delivery of fires and to prevent fratricide. Network-enabled mission command systems facilitate access, integration, coordination, and clearances maintaining a relevant common operational picture. Interoperability will also be key, as well as the ability to attack targets and threats identified and located by organic and nonorganic sensors.

Air Defense Artillery (ADA) commanders plan their operations to support the accomplishment of the supported commander's strategic, operational or tactical objectives by providing fires, early warning, and situational understanding to protect critical assets and forces from air and missile attack and aerial surveillance.

ADA commanders, supported by air defense artillery fire control officer elements, tactical directors, fire control officers, and air defense airspace management cells and airspace control cells at the appropriate Army echelon, will integrate joint, unified action partners and Army AMD for the supported commanders and the area air defense commander (AADC). These leaders will perform the integration function for the supported commanders, providing AMD access and contributing to situational understanding, and airspace management. AMD fires, while under the control of the AADC will be allocated accordingly to the defended asset list (DAL) priority and will be integrated with the joint force commander's and supported commander's concept of operations. AMD fires will be coordinated and cleared enabling rapid and timely engagement of threats while mitigating fratricide.

FA commanders, assisted by fire support personnel and organizations at all echelons integrate joint, Army, interagency and multinational fires capabilities during the operations and targeting processes. These leaders will perform the integration function for the maneuver commanders, including electronic attack and SOF, providing access to joint, Army, interagency, and multinational capabilities. Synchronized fire support requires the coordinated interaction of all of the elements of the fire support system, thorough continuous planning, aggressive coordination, and vigorous execution. The fire support system includes the target acquisition, mission command, and attack/delivery systems that must function collectively to ensure effective fires are delivered where and when the commander requires them.

C. Conduct Targeting

Targeting is the process of selecting and prioritizing targets and matching the appropriate response to them, considering operational requirements and capabilities (JP 3-0). Targeting is continuously refined and adjusted between the commander and staff as the operation unfolds. A Target is an entity or object considered for possible engagement or other action (JP 3-60). Targets also include the wide array of mobile and stationary forces, equipment, capabilities, and functions that an enemy commander can use to conduct operations. The identification and subsequent development of targets, the attack of the targets, and combat assessment provide the commander with vital feedback on the progress toward reaching the desired end state. Combat assessment can provide crucial and timely information to allow analysis of the success of the plan or to initiate revision of the plan.

Targeting is a fundamental task of the fires function that encompasses many disciplines and requires participation from many joint force staff elements and components along with numerous nonmilitary agencies. The purpose of targeting is to integrate and synchronize fires into unified land operations. Army targeting uses the functions decide, detect, deliver, and assess (D3A) as its methodology. Its functions complement the development, planning, execution, and assessment of the effectiveness of targeting and weapons employment.

The fires warfighting function must maintain three critical capabilities in order to deliver and integrate fires: target acquisition, target discrimination, and target engagement.

See pp. 5-26 to 5-27 for discussion of the targeting within the operations process.

Chap 5

I. Fires in Support of Unified Land Operations

Ref: ADRP 3-09, Fires (Aug '12), chap. 1.

I. Roles

The role of fires is to enable Army forces to seize and retain the initiative, prevent and deter conflict, defeat adaptive threats and succeed in a wide range of contingencies. Fires in decisive action create effects and set conditions to enable commanders to prevail in unified land operations. Fires are surface-to-surface, surface-to-air, and joint fires including electronic attack.

II. Core Competencies

A. Air Defense Artillery

Army ADA forces, fight interdependently with other elements of unified action partners at strategic, operational, and tactical levels. Army ADA provides AMD and contributes to the situational understanding, airspace management, early warning, and operational force protection. Army ADA forces deter or defeat enemy aerial threats, protect the force and high value assets. This mission is normally executed within a joint theater and requires integration and close coordination between Army ADA forces and other counterair forces. The mission of Air Defense Artillery is to provide fires to protect the force and selected geopolitical assets from aerial attack, missile attack, and surveillance.

ADA forces accomplish their mission by developing procedures and deploying air and missile defense systems to protect forward deployed elements of the U.S. armed forces, and multinational partners. This mission employs multi-tier (lower-tier and upper-tier), that are interoperable, and provide a layered and defense in-depth capability against air breathing threats, ballistic and cruise missiles, and unmanned aircraft systems. The distinction between the upper- and lower-tier systems and capabilities depend on the ranges and altitude of the threat.

Lower-tier systems defeat air breathing threats, short range ballistic missiles, cruise missiles, unmanned aircraft systems, and enemy indirect fire. For example, the lower-tier systems provide air defense of ground combat forces and high value assets against high performance air-breathing threats fixed-wing, rotary-wing and unmanned aircraft, cruise missiles, and ballistic missile threats.

Upper-tier systems defend larger areas and defeat medium and intermediate range ballistic missiles, and increase the theater commanders' effectiveness against weapons of mass destruction.

B. Field Artillery

FA operations are actually two distinct functions; FA, and fire support. The FA provides the nucleus for effective fires coordination through staff personnel, fires agencies, and attack resources. The integration of fires is a critical factor in the success of operations. The commander is responsible for the integration of fires within their operational area. The chief of fires (COF), the fire support coordinator (FSCOORD), and brigade fire support officer (FSO) advise the commander on the allocation and use of available fires resources. The mission of the Field Artillery is to destroy, defeat, or disrupt the enemy with integrated fires to enable maneuver commanders to dominate in unified land operations.

FA cannon, rocket, missile, and sensor systems provide continuously available fires under all weather conditions and in all types of terrain. FA can shift and mass fires rapidly without having to displace. Should a maneuver or other supported force displace, FA units should be as mobile as the units they support. FA forces man the fires cells, act as forward observers, and are employed as fire support teams (FIST) and combat observation and lasing teams (COLT) to integrate all means of fire support for the commander and synchronize fire support with the concept of operations.

FA destroys, disrupts, denies, degrades, neutralizes, interdicts, or suppresses enemy forces, and protects and enables friendly forces in support of the maneuver commander requirements and objectives. A variety of FA munitions provide the commander with tremendous flexibility when attacking targets with fires.

FA forces synchronize and integrate Army, joint and multinational fires assets for use at the designated place and time. Fires are critical to accomplishing offensive and defensive operations. Accomplishing the mission by achieving an appropriate mix of lethal and nonlethal effects remains an important consideration for every commander.

III. Fires and Joint Principles

The nine principles of joint operations that provide guidelines for combining the elements of combat power and for employing fires are listed below.

Objective

The purpose of specifying the objective is to direct every military operation toward a clearly defined, decisive, and achievable goal (JP 3-0). Objective means ensuring all fires actions contribute to the supported commander's mission. The fire support plan and AADP must have clearly defined objectives that support the commander's intent. Objectives allow commanders to focus combat power on the most important tasks and to protect critical assets in their area of operations.

Offensive

The purpose of an offensive action is to seize, retain, and exploit the initiative (JP 3-0). Fires must always be conducted in the spirit of the offense. Effective fires must maintain responsiveness and fire superiority to allow the supported force to seize and retain the initiative.

Regardless of whether the force is engaged in the offense or is in a defensive posture, fires are used offensively to strike HPT and in offensive counterair operations. Optimally, fires are preemptive, with the ability for rapid reaction to unforeseen requirements. The aggressive application of fires can keep an enemy off balance and in a reactive state. Disrupting his operations throughout the AO with synchronized fires can prevent the enemy from establishing his desired tempo of operations and concentration of forces. Additionally AMD fires allow a commander to maintain momentum and operational tempo of the offense. The use of clearly stated essential tasks for fires, concise fire support plans, area air defense plans (AADP) and decentralized control of fires assets are ways to facilitate increased initiative.

Mass

The purpose of mass is to concentrate the effects of combat power at the most advantageous place and time to produce decisive results (JP 3-0). Fires weapons and units are normally not physically massed, but they must be able to provide maximum massed fires when and where they are required. The actual methods of achieving massed fires vary with each attack resource. Commanders select the method that best fits the circumstances. Army forces can mass fires quickly and across large distances. Commanders can use fires to achieve mass by—

- Allocating fires assets to add weight to the main effort
- Assigning priorities of fires and quickfire channels
- Focusing target acquisition, sensors, and information collection assets
- Concentrating fires assets on one aspect of fires such as fires in support of close combat

IV. Fires - Principles and Characteristics

Ref: ADRP 3-09, Fires (Aug '12), p. 1-8.

Principles

Precision. Providing a coordinated effect on a specific target characterized by having a high degree of accuracy using guidance control and correctable ballistics.

Scalable. Fires capabilities that are adaptable, versatile, and capable to a degree that allows intended effects to be achieved through nonlethal to lethal capabilities.

Synchronized. Fires arranged in time, space and purpose in order to produce the desired effect at a decisive place and time; in fires context, the application of sources and methods in concert with the operation plan to ensure lethal and non-lethal effects are executed in time to support the commander's objectives.

Responsive. Employment of fires capabilities in an expedient manner meeting the needs of the supported forces.

Networked. Interconnected weapon systems and sensors that enable mission command and provide rapid target acquisition, target discrimination, and target engagement in accordance with the commander's intent.

Characteristics

All Weather. Fires capabilities are not weather restrictive, and can be sustained and maintained in any or all weather conditions.

Precision/Near Precision Fires. Precision capabilities have a circular error probable of less than 10 meters. Near Precision fires typically have a circular error probable of between 10 and 50 meters. Air Defense fires have a probability of kill percentage associated with their effects.

Mass Area Fires. Fires retain the responsibility and capability of providing 360 degree coverage and early warning to preserve friendly forces and their assets. Fires mass in space and time on a single or multiple targets. Battalion sized firing units down to two weapon systems deliver area fires and their effects. The ability for smaller firing elements to mass is enabled by system range capabilities, weapons platform capabilities, extended range communication and the mission command network.

Air and Space Integration. Fires leaders have the responsibility to integrate air and space control measures to ensure all commanders have the maximum freedom to achieve their objectives and have maximum flexibility to use assets (organic, supporting and joint) within that airspace. Fires requires responsive integrated network connectivity meshed with joint air and space management systems in order to synchronize and deliver timely air and ground fires and early warning in support of unified land operations. Fires enable all users of airspace to synchronize, plan, and execute a cohesive air deconfliction resolution. Fires personnel coordinate airspace integration to ensure that conflicts between ground fires and air operations are minimized using FSCMs and ACMs. This unified action mitigates the possibility of fratricide and duplication of effort.

Inherently Joint. All forms of fires are joint by nature through their association with the development, coordination, and integration of the joint integrated prioritized target list and the critical asset list (CAL) into the DAL development. The joint fires element develops joint targeting guidance, objectives, and priorities. Fires organizations such as the Army air and missile defense command (AAMDC) and the battlefield coordination detachment (BCD) routinely perform joint coordination functions with the joint force commander (JFC) and the joint force air component commander (JFACC).

Maneuver

The purpose of maneuver is to place the enemy in a position of disadvantage through the flexible application of combat power (JP 3-0). Fire support plans and air defense plan (ADP) must have the flexibility to include altered missions, command and support relationships, and priorities. Fires units must also displace rapidly, keep pace with the supported force in the current operation, and position as needed to support future operations. Combating a hybrid threat may demand different fire unit positioning considerations within the same AO. Mission variables may require some fires units to be employed as widely separated elements to achieve the necessary fires range to enable decisive action in one portion of the AO.

Economy of Force

The purpose of economy of force is to expend minimum essential combat power on secondary efforts in order to allocate the maximum possible combat power on primary efforts (JP 3-0). A unit might be required to conduct operations with minimum essential fires and accept risk. Economy of force also implies that the effort allocated to a given unit shall not exceed the effort necessary to produce the commander's desired effects.

Unity of Command

The purpose of unity of command is to ensure unity of effort under one responsible commander for every objective (JP 3-0). Fires must be synchronized with the supported commander's concept of operation based on his intent and guidance for fires. For Air Defense, unity of command is exercised through the tactical control of ADA fires utilizing positive and procedural controls in support of the AADC. While at corps and below, the maneuver commander normally delegates to his COF/FSCOORD/ FSO/ air defense airspace management/brigade aviation element (ADAM/BAE) the requisite authority to direct and coordinate all joint and Army fires on his behalf.

Security

The purpose of security is to prevent the enemy from acquiring an unexpected advantage (JP 3-0). There are two aspects of security in relation to fires. The first aspect concerns general security, which fires helps provide for the supported force. The second aspect is the continued survivability of trained fires personnel, fires command networks and control facilities, target acquisition and sensors, and fires weapons systems.

Surprise

The purpose of surprise is to strike at a time or place or in a manner for which the enemy is unprepared (JP 3-0). Fires enable the commander to achieve surprise with the delivery of a high volume of fire or precision munitions on the enemy without warning. Commanders can use fires to achieve surprise by rapidly and discreetly repositioning fires assets and/or shifting and massing fires; using short, intense programs of fires, such as those for suppression of enemy air defenses and counterfire, against key enemy functions at critical times; using TGMs to strike a target; deceiving the enemy as to the types, numbers, locations, and capabilities of friendly fires assets; and conducting offensive counterair to destroy or negate enemy aerial platforms and sensors.

Restraint

The purpose of restraint is to limit collateral damage and prevent the unnecessary use of force (JP 3- 0). Fires restraint typically concerns the munitions employed and the targets engaged to achieve lethal effects. Having the ability to employ a weapon does not mean it should be employed. In addition to collateral damage considerations, the employment of some weapons—bombs, missiles, rockets, artillery, and mortars—could have negative psychological impacts that create or reinforce instability or security concerns. Collateral damage could adversely affect efforts to gain or maintain legitimacy and impede the attainment of both short-term and long-term goals. Restraint increases the legitimacy of the organization that uses it while potentially damaging the legitimacy of an opponent.

V. Fires in Support of Unified Land Operations

Ref: ADRP 3-09, Fires (Aug '12), pp. 1-4 to 1-5.

The Army operational concept is unified land operations. Army fires support this operational concept with fire support operations through the application of FA and ADA. Fires align with the tenets of unified land operations in ADP 3-0 which are flexibility, integration, lethality, adaptability, depth, and synchronization. Fires execution supports these tenets and the supported force in seizing and maintaining the initiative.

Refer to ADP 3-0 and ADRP 3-0.

Flexibility

To achieve tactical, operational, and strategic success, commanders seek to demonstrate flexibility in the employment of fires. Fires utilize a versatile mix of capabilities, formations, and equipment for conducting operations. Flexible fires planning help units adapt quickly to changing circumstances in operations. Decentralizing fires execution to the lowest level possible gives commanders' flexibility during operations.

Integration

Army forces rely on joint inter-dependence as a part of a larger joint, interagency, intergovernmental, and multinational effort. Integration of joint, interagency, intergovernmental, and multinational capabilities improves collectivity, efficiency, and effectiveness. Fires commanders, staffs, and Soldiers leverage and combine these capabilities, sensors and weapons systems to quickly adapt to changing conditions in various operational environments. Fires utilize mission command to facilitate the integration of these capabilities at all levels of war in support of unified land operations and decisive action. Commanders extend their operational reach through the integration of joint fires.

Lethality

The capacity for physical destruction is fundamental to all other military capabilities and the most basic building block for military operations (ADP 3-0). Lethality is a foundation for effective decisive action. Fires contribute to the Army's lethality through the application of scalable capabilities to create lethal effects by destroying, neutralizing, or suppressing the enemy.

Adaptability

Army leaders accept that no prefabricated solutions for fires exist. Commanders and staffs must adapt their way of thinking and the manner in which they employ fires to the specific situation they face. Fires leaders achieve adaptability by understanding the operational environment, the abilities of their Soldiers and the capabilities of fires systems.

Depth

Depth is the extension of operations in space, time, or purpose. Commanders and staffs provide depth through the successful planning and execution of joint fires. Fires provide the commander depth and breadth to the battlefield through long-range acquisition and early engagement of targets. Fires utilize the employment of various long and short range weapons to facilitate multiple engagements as required.

Synchronization

Synchronization is the arrangement of military actions in time, space, and purpose to produce maximum relative combat power at a decisive place and time (JP 2-0). Synchronization of fires, in conjunction with other warfighting functions, goes beyond integration, optimizes the elements of combat power, including Army indirect fire, AMD, joint fires and unified action partner capabilities to fully maximize their complementary effects through the operation and targeting processes.

VI. Fires in Support of Decisive Operations

Ref: ADRP 3-09, Fires (Aug '12), pp. 1-5 to 1-8.

Decisive action is the continuous, simultaneous combination of offensive, defensive, and stability or defense support of civil authorities tasks (ADRP 3-0).

Fires organizations and leaders support decisive action through the purposeful and simultaneous execution of the Army's core tasks to achieve the commander's intent and end state. In unified land operations, fires facilitate commanders' ability to seize, retain, and exploit the initiative while synchronizing their actions to achieve the best effects possible. Decisive action may require simultaneous combinations of offensive, defensive, and stability tasks, or defense support of civil authorities. Fires organizations and commanders support all the primary tasks and the numerous subordinate tasks associated with decisive action. These tasks, when combined with; the who (unit), when (time), where (location), and why (purpose), may become the essential tasks, or mission statement, for fires organizations.

Scalable Capabilities

Scalable capabilities provide a range of nonlethal to lethal actions commensurate with the commander's intent. Scalable capabilities can create desired effects while reducing collateral damage. Scalable capabilities allow the commander to find the right balance between effects and collateral damage. These capabilities assist in protecting joint, Army, and multinational partners and populations residing in the AO. Scalable capabilities can be addressed in the selection of the appropriate weapon system, number and type of munitions fired and the method used to engage a target through mission command nodes and networked sensors. Nonlethal effects typically neutralize or incapacitate a target or modify adversarial behavior without causing permanent injury, death, or gross physical destruction.

Fires in Support of Offensive Tasks

An offensive task is a task conducted to defeat and destroy enemy forces and seize terrain, resources, and population centers (ADRP 3-0). Fires in support of offensive tasks that preempt enemy actions include preparation fire, close support fires, interdiction, electronic attack, early warning, early engagement, and counterfire. These actions protect and enable friendly forces to seize the initiative, support the scheme of maneuver, and follow on operations.

Supporting the concept of operations during the offense involves acquiring, discriminating, and engaging targets throughout the AO with massed and precision fires to include joint and electronic warfare assets. Considerations for supporting the scheme of maneuver during the offense include:

- Weight the main effort
- Consider positioning fires assets to exploit weapons ranges and preclude untimely displacement when fires are needed the most
- Provide counterfire
- Provide early warning and dissemination
- Provide wide area surveillance
- Provide fires to protect forces preparing for and assets critical to offensive actions
- Disrupt enemy counterattacks
- Plan fires to support breaching operations
- Plan fires to deny enemy observation or screen friendly movements
- Allocate responsive fires to support the decisive operation
- Allocate fires for the neutralization of bypassed enemy combat forces
- Plan for target acquisition and sensors to provide coverage of named areas of interest, target areas of interest and critical assets

Fires in Support of Defensive Tasks

A defensive task is a task conducted to defeat an enemy attack, gain time, economize forces, and develop conditions favorable for offensive or stability tasks (ADRP 3-0). Successful defenses are aggressive, and they maximize protection and maneuver to defeat enemy forces and regain the initiative. Fires in support of defensive tasks are fires that protect friendly forces, populations, and critical infrastructure to enable maneuver forces to transition to offensive tasks.

Supporting the concept of operations during the defense involves attacking/engaging targets throughout the AO with massed or precision indirect fires, AMD fires, defensive counterair, air support, and electronic warfare assets. In the defense, general fires support considerations for supporting the concept of operations include:

- Weight the main effort
- Provide 360 degree AMD coverage
- Provide and disseminate early warning
- Contribute targeting information
- Engage critical enemy assets with fires before the attack
- Plan counterfire against enemy indirect fire systems attacking critical friendly elements
- Use both lethal and nonlethal attack means to apply constant pressure to the enemy's command and control structure
- Provide fires in support of defensive counterair to defeat enemy aerial attacks
- Plan the acquisition and attack of high payoff targets (HPT) throughout the AO
- Provide integrated fires in synchronization with maneuver and electronic warfare countermeasures in the conduct of decisive and shaping operations
- Retain maximum feasible centralized control of fires resources in order to concentrate fires at the decisive place and time
- Provide fires to support counterattacks
- Plan fires in support of the barrier and/or obstacle plan
- Provide fires in support of decisive, shaping and sustaining operations
- Plan for target acquisition and sensors to provide coverage of named areas of interest, target areas of interest and critical assets

Fires in Support of Stability Tasks

Stability tasks are tasks and activities conducted outside the United States to maintain or reestablish a safe and secure environment and to provide essential governmental services, emergency infrastructure reconstruction, and humanitarian relief.

Fires considerations in support of stability tasks include the considerations identified for offensive and defensive tasks. Commanders must analyze each mission and adapt to the mission variables that fit the situation. Characteristics of stability tasks include:

- They are often conducted in noncontiguous AOs. This can complicate the use of airspace coordinating measures (ACM), fire support coordination measures (FSCM), the ability to mass and shift fires, and clearance of fires procedures.
- What constitutes key terrain may be based more on political, cultural, and/or social considerations than physical features of the landscape. Fires may be used more frequently to defend key geopolitical sites.
- Rules of engagement (ROE) are often more restrictive when conducting stability tasks.
- Improper application of fires can have a long-term adverse impact on achievement of the desired end state.
- Use of fires to demonstrate capabilities, show of force, or area denial.
- Use of targeting to synchronize non-lethal effects and conduct engagement planning
- Provide force protection through counterfire and counterair operations.

VII. Employment of Fires

Ref: ADRP 3-09, Fires (Aug '12), pp. 1-13 to 1-14.

To employ fires is to use available weapons and other systems to create a specific lethal or nonlethal effect on a target (JP 3-0). Planning and assessment contribute to the execution of fire support and AMD in the successful employment of fires to achieve the commander's intent. This contribution employs principles of planning, coordination, and execution as a guide. In advising the supported commander on the application of fires, the fires planner also reviews fires requirements against several basic fires considerations that guide planning in the development of the fire support plan and AADP. This function encompasses the fires associated with a number of tasks, missions, and processes, including:

A. Air and Missile Defense Employment

Employment Principles. ADA employment principles enable air defense forces to successfully perform combat missions and support overall force objectives. The four principles are:

Mass

Mass is the concentration of air defense combat power. Mass may also be interpreted to include the launching of more than one interceptor against a target.

Mix

Mix is the employment of a combination of weapon and sensor systems to protect the force and assets from the threat. Mix offsets the limitations of one system with the capabilities of another.

Mobility

Mobility is defined as a quality or capability of military forces, which permits them to move from place to place while retaining the ability to fulfill their primary mission.

Integration

Integration is the combination of the forces, systems, functions, processes and information acquisition and distribution required to efficiently and effectively perform the mission. Integration combines separate systems, capabilities, functions, etc. in such a way that those individual elements can operate independently or in concert without adversely affecting other elements.

AMD Employment Guidelines

Planning during defense design and positioning ADA units involves applying six employment guidelines. Optimum protection of the items on the JFC's CAL must be the goal. The six guidelines are:

Mutual Support

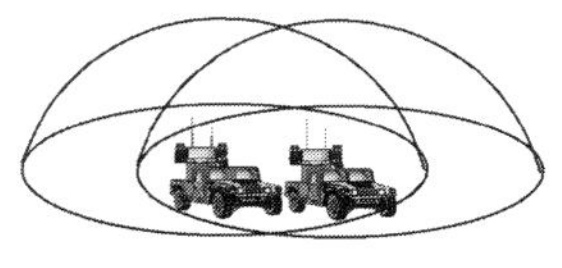

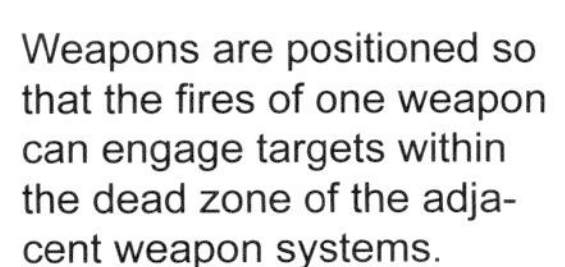

Weapons are positioned so that the fires of one weapon can engage targets within the dead zone of the adjacent weapon systems.

Overlapping Fires

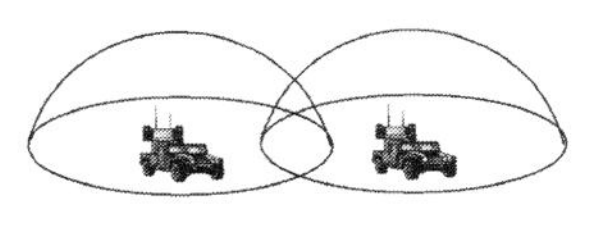

Weapons are positioned so that their engagement envelopes overlap. Defense planners must apply mutual supporting and overlapping fires vertically and horizontally.

Balanced Fires

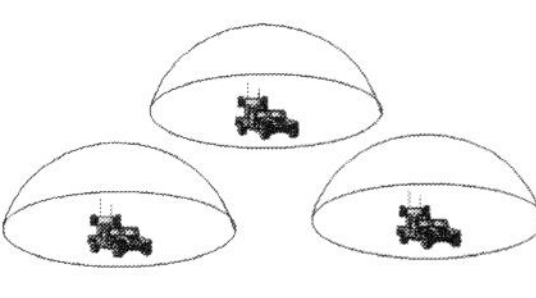

Weapons are positioned to deliver an equal volume of fires in all directions.

Weighted Coverage	Early Engagement	Defense in Depth
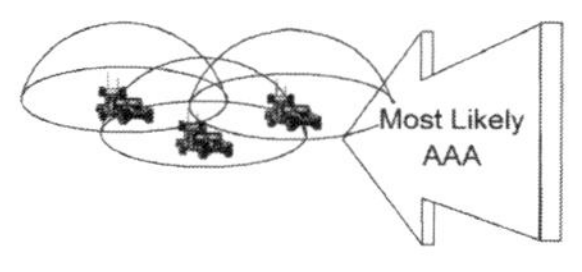	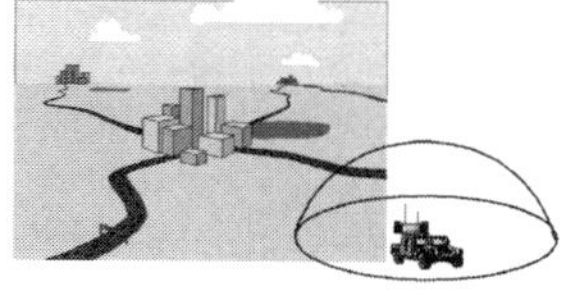	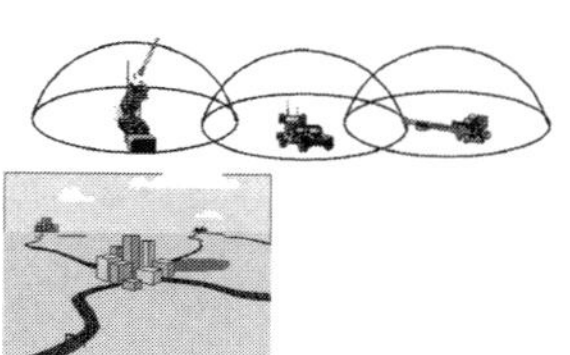
Combining and concentrating weapons coverage toward the most likely enemy air avenues of approach or direction of attack achieve **weighted coverage**.	**Early engagement** is achieved by positioning sensors and weapons so they can engage the enemy before aircraft ordnance release or friendly target acquisition by the enemy.	**Defense in depth** is achieved by positioning weapons and sensors so the enemy is exposed to a continuously increasing volume of fire as it approaches the friendly protected asset or force.

B. Field Artillery Employment

When planning for the integration, synchronization and coordination of FA employment, commanders and staffs should consider the steps listed below:

Adequate Fire Support for the Committed Units

Organic fire units are most responsive to maneuver elements. The minimum adequate support for committed units is considered to be one organic FA battalion for each committed brigade. In no instance can there be more than one fires unit in direct support of a maneuver unit.

Weight the Main Effort

Support relationships of reinforcing or general support reinforcing (GSR) can be assigned to provide additional responsive fires to an organic FA battalion or a FA battalion with a direct support relationship.

Immediate Responsive Fires

The force commander should place some artillery on call with which they can immediately influence unified land operations.

Facilitate Future Operations

This fundamental is essential to ensure success in the face of unforeseen events and to ensure smooth transition from one phase of an operation to another. The fundamental can be implemented through the assignment of a support relationship, positioning of fires elements, and allocation of ammunition. The assignment of an on-order mission facilitates a future mission. Another way to facilitate future operations is to modify the current command or support relationship in accordance with anticipated requirements.

Maximize Feasible Centralized Control

Fires are most effective when control is centralized at the highest level consistent with the fire support capabilities and requirements of the overall mission. Centralized control of fires permits flexibility in their employment and facilitates effective support to each subordinate element of the command and to the force as a whole. Command and support relationships represent varying degrees of centralized control and responsiveness to committed units. The optimum degree of centralized control varies with each tactical situation. Decisive action will require more careful planning because of the limited resources available to attack targets and the need for carefully coordinated employment of acquisition, attack, and assessment means. A high degree of centralized control is desired in a defensive situation. Since the enemy has the initiative, it is difficult to accurately predict where and when he will strike. A lesser degree of centralized control is required in an offensive situation, because the supported force has the initiative.

VIII. Fires and the Operational Framework

Ref: ADRP 3-09, Fires (Aug '12), pp. 1-12 to 1-13.

Fires provide a collective and coordinated use of Army indirect fires, AMD, and joint fires in support of operations no matter how the unit defines and describes the operational framework.

See pp. 1-20 to 1-21 for discussion of the operational framework from ADRP 3-0.

A. Decisive-Shaping-Sustaining Framework

Fires contribute to the overall effect of maneuver in which commanders use them separately in decisive, shaping, and sustaining operations to directly influence the mission objectives.

- **Fires in Support of Decisive Operation.** Decisive operations lead directly to the accomplishment of a commander's purpose (ADP 3-0). Fires supporting decisive operations include preparation fires, close support fires, interdiction, Army support to offensive counterair (OCA)/DCA, AMD, final protective fires, electronic attack, and counterfire. Fires in the decisive operation integrate and synchronize weapon systems and sensors to achieve lethal and nonlethal effects in support of the scheme of maneuver.
- **Fires in Support of Shaping Operations.** Shaping operations create and preserve conditions for the success of the decisive operation (ADP 3-0). Fires in support of shaping operations disrupt or destroy the enemy's attacking echelons and fire support, mission command, and logistic infrastructure. Fires may be used to limit the enemy's ability to shift forces to meet attacking friendly maneuver forces and to sustain the momentum of the attack. Fires in support of shaping operations employ the same types of fires as during the decisive operation.
- **Fires in Support of Sustaining Operations.** Fires in sustaining operations protect and enable friendly forces to retain freedom of action. Fires must be responsive and positioned to attack and disable enemy forces or any potential threat.

B. Deep-Close-Security Framework

The Deep-Close-Security framework has historically been associated with a terrain orientation but can be applied to temporal and organizational orientations as well.

- **Fires in Support of Deep Operations.** Deep operations involve efforts to disrupt uncommitted enemy forces (ADP 3-0). These types of operations frequently tie to events in time or space. Fires in support of deep operations disrupt enemy movement, command and control, sustainment and fires assets. Fires used in deep operations include interdiction, counterair, and electronic attack.
- **Fires in Support of Close Operations.** Close operations involve efforts to have immediate effects with committed friendly forces-potentially in direct contact with enemy forces-to include enemy reserves available for immediate commitment (ADP 3-0). Fires in support of close operations include counterfire, indirect fire protection capabilities, combined arms for air defense, close air support (CAS), and final protective fires. When employing fires in support of close operations, commanders must mitigate risk of fratricide by selecting the most appropriate fires capability and implementing ACMs and FSCMs.
- **Fires in Support of Security.** Security operations involve efforts to provide early and accurate warning of enemy operations, provide the force with time, and maneuver space within which to react to the enemy, protect the force from surprise, and develop the situation so the commander can effectively use the force (ADP 3-0). Fires in support of security operations include AMD, sensor early warning, indirect fires, and CAS.

II. Fires Organizations and Key Personnel

Ref: ADRP 3-09, Fires (Aug '12), chap. 2.

The fires warfighting function uses a diverse group of systems, personnel, and materiel—most of which operate in various ways to provide different capabilities. This chapter discusses the organizations and key personnel of the fires warfighting function from the strategic level to the tactical level. The following table provides an overview of all fires organizations and key personnel down to the battalion and battery level. This chapter will only cover organizations and key personnel down to brigade.

Fires Organization and Key Personnel

Strategic Level Fires Organizations	
Air Defense	**Field Artillery**
• Joint Functional Component Command-Integrated Missile Defense (JFCC-IMD) • Army Space and Missile Defense Command (SMDC)/Army Strategic Command (ARSTRAT)	• Joint Land Component/Joint Task Force Fires Cell • Battlefield Coordination Detachment
Operational Level Fires Organizations	
Air Defense	**Field Artillery**
• Army Air and Missile Defense Command (AAMDC) - Theater Air and Missile Defense Coordinator (TAMCOORD) - Deputy Area Air Defense Coordinator (DAADC—when designated) • Corps Air Defense/Airspace Management Cell	• Corps and Theater Fires Cell • Expeditionary Air Support Operations Group • Ground Liaison Detachment
Tactical Level Fires Organizations	
Air Defense	**Field Artillery**
• Air Defense Artillery Brigade - Air Defense Fire Control Officer (ADAFCO) - Air Defense Artillery Battalion - Patriot Battalion - Air and Missile Defense Battalion - Indirect Fires Protection Capability (IFPC) Battalion - Terminal High-Altitude Area Defense (THAAD) Battery	• Division Fires Cell - Division Chief of Fires • Air Liaison Officer • Fires Brigade - Fire Support Officer - Field Artillery Battalions o Multiple Launch Rocket System (MLRS) o High-Mobility Artillery Rocket System (HIMARS) o M109-series Paladin o M777-series o M119-series
Fires Cell (Division/Brigade Combat Team)	
• Air Defense Element • Air Defense Airspace Management/Brigade Aviation Element (ADAM/BAE)	• Fires Element • Air Support Operations Center/Air Liaison Officer (ASOC/ALO)
Legend : D3A – decide, detect, deliver, and asses	

Ref: ADRP 3-09, Fires, table 2-1, p. 2-2.

For further breakdown of organizations and key personnel refer to FM 3-09 and FM 3-01.

I. Strategic Level (Fires Organizations and Personnel)

Ref: ADRP 3-09, Fires (Aug '12), pp. 2-2 to 2-3.

A. Joint Functional Component Command-Integrated Missile Defense (JFCCIMD)

The JFCC-IMD mission is to support U.S. strategic command's mission by synchronizing sea, land, air and space based assets in support of global missile defense operations and missile defense. U.S. strategic command through the JFCC-IMD is the proponent responsible for identifying and recommending missile defense assets in response to the ground combatant commander (FM 3-27).

B. Army Space and Missile Defense Command (SMDC) and Army Strategic Command (ARSTRAT)

SMDC/ARSTRAT conduct space and missile defense operations and provide planning, integration, control, and coordination of Army forces and capabilities in support of U.S. strategic command missions of deterrence, integration of missile defense and space operations. SMDC is also the proponent for integration and force modernization for space, high altitude, and global missile defense.

C. Joint Force Land Component/Joint Task Force Fires Cell

When designated by the JFC a theater Army, corps, or division commander serves as the commander of the joint force land component or a joint task force (JTF). The corresponding fires cell function at the designated headquarters (HQ) include:

- Advise on the application of fires to achieve the desired effect
- Identify requirements for the supported commander's desired effects from fires provided by other components (air interdiction/naval surface fires)
- Review and comment on the JFACC's apportionment recommendation
- Recommend joint forces land component commander (JFLCC) assets for JFC allocation
- Advise on fires asset distribution (priority) to land forces
- Recommend JFLCC priorities, timing, and effects for air interdiction within the JFLCC AO
- Recommend JFLCC targeting guidance and priorities to include information from space-based assets that will aid in targeting
- Recommend JFLCC command target lists and FSCMs
- Lead the joint targeting coordination working group; participate in the Joint Targeting Coordination Board

D. Battlefield Coordination Detachment (BCD)

The BCD is an Army coordination element that provides selected operational functions between the Army and the air component commander (ACC). The BCD mission is to provide Army forces liaison at the joint air operations center/combined air and space operations center. The BCD provides the critical and continuous coordination between the air and land commanders. The BCD staff clearly articulates the Army forces commander's or Land Component Commander's requests for air operations support for the ground operations to complement the joint forces commander's end state.

For additional information on the BCD refer to ATTP 3-09.13.

II. Operational Level (Fires Organizations and Personnel)

A. Army Air and Missile Defense Command (AAMDC)

The area air defense commander (AADC) is normally the JFACC with the authority to plan, coordinate and integrate overall joint force air operations. The JFC and AADC will determine whether a deputy area air defense commander (DAADC) for air and missile defense (AMD) is designated. If so designated, the AAMDC commander assumes the role of the DAADC (AMD). This designation formalizes the relationship between the land-based air defense artillery (ADA) assets dedicated to theater level missions and the AADC and also ensures fully integrated and synchronized counterair and Army forces AMD operations. The AAMDC coordinates with joint and multinational partners to develop procedures for combined AMD operations, interoperability, and training. The AAMDC plans, coordinates, integrates, and executes AMD for the combatant commander (CCDR)/Army forces or the combined/joint force land component commander. The AAMDC commander is the senior Army ADA commander, and coordinator. This organization is the communications system, and intelligence headquarters for ADA forces.

B. Deputy Area Air Defense Commander (DAADC)

If so designated, the AAMDC commander assumes the role of the DAADC (AMD). This designation formalizes the relationship between the land-based ADA assets dedicated to theater level missions and the AADC and also ensures fully integrated and synchronized counterair and Army forces AMD operations.

C. Theater Air and Missile Defense Coordination Officer

The theater army air and missile defense coordinator (TAAMDCOORD) is the commander of the highest echelon Army ADA command in the theater. When the AAMDC is in theater, the commander will be designated the TAAMDCOORD. The TAAMDCOORD is the Army AMD coordinator for the Army forces commander, and the combined forces land component commander (CFLCC), JFLCC (if designated). The TAAMDCOORD ensures that the Army is an integral part of joint counterair and active missile defense operations and planning at the theater level. The TAAMDCOORD, as a special staff officer to the Army forces commander and JFLCC, participates in the operations directorate of a joint staff (J3)/plans directorate of a joint staff (J5) cells and assists in developing Army OCA and DCA input to the air operations plan. The TAAMDCOORD ensures that corps AMD requirements are integrated into joint counterair and Army AMD planning.

D. Ground Liaison Detachment

The ground liaison detachment advises Air Force commanders primarily on Army organization, operations, tactics, and equipment and assists them by coordinating Army units during joint operations. The detachment supports Air Force tactical fighter, tactical air control and airlift units.

E. Fires Cells-Corps and Theater

The corps and theater fires cell (FC) plans, coordinates, integrates, and synchronizes the employment and assessment of fires in support of current and future theater wide operations. The FC recommends targeting guidance to the commander and develops high-payoff targets and selects targets for attack. The cell coordinates, integrates, and assigns joint, interagency and multinational firepower to targets/ target systems. It synchronizes firepower to include Army, joint, interagency, and multinational component air assets, special operations forces, naval surface fire support, cyber/electromagnetic activities, and Army missiles.

The FC participates in combat assessments (battle damage, munitions effectiveness, reattack recommendations); develops planning guidance; provides target intelligence for planning and execution and coordinates with the Battlefield Coordination Detachment (BCD) collocated with the respective air operations center.

F. Chief of Fires

The chief of fires is the senior fires officer at division and higher headquarters who is responsible for advising the commander on the best use of available fire support resources, providing input to necessary orders, and developing and implementing the fire support plan. These duties and responsibilities should be fully delineated by the commander. The COF may be given authority by the commander to:

- Provide for consolidated and focused fire support specific training, readiness, and authority (personnel management, equipment issue, and training)
- Facilitate establishing standard operating procedures across the force
- Ensure efficiently resourced training packages
- Plan for the allocation of fires assets

The fires personnel organic to the force assist the COF in these duties, particularly in staff functions. The COF should be authorized to conduct fire support specific training and recommend to the commander the certification for the various elements of the force's entire fire support system, preferably after live fire exercise.

The COF plans and coordinates the Army indirect fires, and joint fires portion of the fires warfighting function to include integration of electronic attack. He works closely with the Chief of Staff/Executive Officer and G-3 to ensure mutual understanding of all aspects of planning, preparation, execution, and assessment of fires for operations. The COF's responsibilities include, but are not limited to:

- Planning, preparing, executing, and assessing all fires tasks in support of offensive, defensive and stability tasks and provides inputs to preparation of the operations plan and operations order
- Developing, with the commander and G-3 a scheme of fires to support the operation
- Developing a proposed HPT list, target selection standards, and an attack guidance matrix
- Identifying named and target areas of interest, HPTs, and additional events that may influence the positioning of fires assets
- Coordinating positioning of fires assets
- Providing information on the status of fires systems, target acquisition assets, and munitions
- Recommending FSCMs to support current and future operations and managing changes to them
- Recommending and implementing the commander's counterfire and target engagement priorities
- Recommending to the commander the establishment, responsibilities, authorities, and duties of a force FA headquarters as necessary
- Conducting the tasks associated with integrating and synchronizing joint fires, and multinational fires with the other warfighting functions
- Training fires cell personnel to perform all of their functions
- Advising the commander and staff of available fires capabilities and limitations

- Leading the targeting working group
- Working with the chief of staff/executive officer, and G-3 to integrate all types of fires into the commander's concept of operation
- Accompanying the commander in the command group during execution of tactical operations (when directed)

The COF translates objectives into specific targeting and attack guidance that he/she recommends to the commander. This guidance includes instructions for attacking predetermined HPTs.

G. Corps Air Defense/Airspace Management Cell

The corps HQ has an air defense/airspace management (ADAM) cell that integrates with the FC when deployed. The cell is fielded with the air missile defense planning and control system (AMDPCS) components. ADAM cell functions include:

- Conduct AMD augmentation planning and coordination
- Conduct aviation augmentation planning and coordination
- Conduct composite risk management to minimize the potential for fratricide (air/ground positive/procedural identification) for the brigade combat team (BCT)
- Provide early warning of enemy aerial attack
- Develop, display, and disseminate the COP/single integrated air picture to provide and facilitate situational understanding
- Contribute to airspace control planning and execution
- Contribute to joint/local airspace deconfliction including clearance of fires
- Contribute to operational protection
- Advise and update the commander on adjacent ADA unit location, plans, and intent
- Continuous assessment of AMD augmentation requirements
- Integrate operations using assigned Army Battle Command System equipment with units/organizations
- Request, maintain, and disseminate ACMs or restrictions

H. Fire Support Officer

The fire support officer is a FA officer from the operational to tactical level responsible for advising the supported commander and assisting the senior fires officer of the organization on fires functions and fire support.

I. Corps - United States Air Force (USAF) Elements

The corps air liaison officer (ALO), when designated the Expeditionary Air Support Operations Group Commander, commands all Air Force personnel within the corps and is the air component commander's direct liaison to the corps commander. The corps main command post tactical air control party (TACP), Air Force weather, and the air mobility liaison officer locate in or adjacent to the current operations, future operations, plans, fires and intelligence cells. Air Force personnel at the corps command posts provide planning expertise to integrate and use air, space, and cyberspace. This is in addition to the space integration support provided by Army space forces assigned and attached to the corps.

The air support operations center (ASOC) is the principal air control agency of the theater air control system responsible for the direction and control of air operations directly supporting the ground forces. It processes and coordinates requests for immediate air support and coordinates air, space, and cyberspace missions requiring integration with other supporting arms and ground forces. The ASOC normally collocates with the Army senior tactical headquarters, normally at corps or division level.

III. Tactical Level (Fires Organizations and Personnel)

Ref: ADRP 3-09, Fires (Aug '12), pp. 2-7 to 2-9.

A. Fires Cells-Division and BCT

The FC plans, coordinates, integrates, and synchronizes the employment and assessment of fires in support of current and future operations. The FC recommends targeting guidance to the commander and develops HPTs and selects targets for attack. The cell plans, synchronizes, coordinates, and integrates adaptable fires matched to a wide range of targets/target systems. The FC coordinates target acquisition, target dissemination and target engagement functions for the commander. At the division level, the AMD Cell, the ASOC, and the emerging capability of the joint air ground integration cell (JAGIC) are integrated within the fires cell to ensure coordination, synchronization of fires, airspace integration functions and sense and warn systems. Similarly within the BCT, the ADAM/BAE, JAGIC and ALO are also integrated within the fires cell. Functions include:

- Plan, integrate, coordinate, and synchronize through targeting, Army and joint fires, and when directed by the maneuver commander other nonlethal effects.
- Integrate and synchronize airspace coordination functions with Army and joint air capabilities and provides input to the ATO, airspace control plan (ACP), and ACO which includes FSCMs and ACMs. The JAGIC capability provides these functions by integrating fixed wing, rotary wing, and fire control management cells within the Division COIC.
- Produce and execute the fire support plan.
- Manage target nominations and track the lifecycle of the nomination.
- Interface with all boards/cells.
- Provide input to the collection plan.
- Request and coordinate CAS and air interdiction.
- Conduct fires, assess, and recommend reattack.
- Coordinate position areas for fires units with maneuver and airspace control agencies.
- Recommend FSCMs and ACMs.

Additional augmentation to the fires cell includes:

- **Naval surface fire support (NSFS) Liaison Officer.** The NSFS liaison officer supervises a NSFS team that may be attached to the BCT fires cell to advise the commander and staff on the planning, preparation, execution, and assessment of naval surface fires.
- **Marine Corps Liaison Officer.** A U.S. Marine Corps liaison officer or a liaison team may augment the fires cell based on mission, enemy, terrain and weather, troops and support available, time available, civil considerations to coordinate naval and/or U.S. Marine Corps air support to the BCT. The fires cell processes requests for naval/U.S. Marine Corps air support through this liaison officer and/or team. A firepower control team may be attached to the maneuver battalions and/or reconnaissance squadron to perform terminal control of naval/U.S. Marine Corps air support. In the absence of an observer from the firepower control team, the company/troop FIST, joint fires observer (JFO), or the Air Force JTAC may control naval and/or U.S. Marine Corps air.
- **Army Space Support Teams.** Army space operations personnel provide space-related tactical planning and support, expertise, advice, and liaison regarding available space capabilities.

B. Force Field Artillery Headquarters

The force field artillery headquarters, if designated by the supported maneuver commander, is normally the senior FA headquarters organic, assigned, attached, or placed under the operational control of that command. The supported maneuver commander specifies the commensurate responsibilities of the force FA headquarters and the duration of those responsibilities. These responsibilities are based on the mission variables and may range from simple mentoring and technical oversight to established command relationships with all FA units organic, assigned, attached, or placed operational control (OPCON) of that command.

The force FA commander is the commander of the assigned force FA headquarters. He recommends a command or support relationship for the U.S. Army FA units. U.S. Marine Corps or North Atlantic Treaty Organization (NATO) FA units that are attached or OPCON are given FA tactical missions and responsibilities in accordance with standard NATO agreement (STANAG) 2484, which guides those units. Other multinational FA units that are attached or OPCON are given tactical missions and responsibilities in accordance with their national guidance. A FIB assigned, attached, or placed OPCON to a division, corps, JFLCC, joint task force (JTF) or other command may serve as that command's force FA HQ. The BCT's organic FA battalion, when directed to do so by the BCT commander, may serve as the BCT's force FA HQ of any additional FA assets attached or placed OPCON to the BCT. The force FA HQ functions include:

- Serving as the single point of contact for recommending the fires organization for combat and positioning all units organic to, assigned to and supporting the maneuver force commander
- Executing fires for close support of engaged forces, and in support of counterfire, decisive and shaping operations
- Providing critical centralized mission command and integration for the full complement of Army and joint fires capabilities, provided in support of the command (division, corps, JFLCC)
- Establishing common survey, meteorological, and radar TA plans for the command.
- Coordinating and synchronizing fires in support of all operations
- Training FA units that are assigned, attached, or placed under the OPCON of the command and mentoring of the commanders and leaders of these FA units
- Providing mission command for FA units organic, assigned, attached, or placed under the OPCON or tactical control of the command
- Assisting the fires cell in producing Annex D (Fires) for the operations order
- Advising the supported commander on FA related new equipment fielding and software updates within FA units
- Working with the command's assistant chief of staff, operations (G-3) and fires cell in planning, coordinating, and executing fires tasks assigned to the command by its higher HQ
- Facilitating and participating in the commander's targeting process

Whether a FA battalion or FIB is organic, assigned, attached, or OPCON, it can only be the force FA HQ when the supported commander specifically designates it.

Refer to FM 3-09.22 for a discussion of the force FA HQ for a supported command.

Liaison

Coordination/liaison at division with other organizations essential to effective fires includes the Air- Naval Gunfire Liaison Company (ANGLICO). The division coordinates naval fire support through the division air-naval gunfire section of the ANGLICO. This U.S. Marine Corps organization also collocates with the division airspace control element and the fires cell. The ANGLICO commander serves as the divisional naval gunfire officer. Because of the design of the ANGLICO, the division is normally the highest echelon that establishes liaison with naval fire support assets.

IV. Fires Brigade (FIB) (Fires Organizations and Personnel)

Ref: ADRP 3-09, Fires (Aug '12), pp. 2-10 to 2-11.

A FIB's primary task is conducting strike operations. The FIB has an organic multiple launch rocket system (MLRS)/ high mobility artillery rocket system (HIMARS) battalion, and target acquisition battery, however, the FIB can be task organized with additional fires delivery, sensor systems, and IFPC to support the maneuver commander's mission requirements. The FIB is the only Army FA organization above the BCT and can be directed to execute tasks for any joint, Service, or functional headquarters. The FIB is neither organic to any Army organization or echelon, nor is it focused on any specific region or geographic combatant commander's area of responsibility. A division, corps, joint force land component command, JTF or other force may have a FIB assigned, attached or placed under OPCON; however, the FIB is normally attached to a division HQ. FIBs are task-organized to accomplish assigned tasks. The FIB's higher HQ usually assigns missions in terms of target sets to engage, target priorities, or effects to achieve. The situation may also require the FIB to control joint fires assets.

A. Fire Support Coordinator (FSCOORD)

The fire support coordinator is the BCTs organic FA battalion commander; if a FIB is designated as the division force FA headquarters, the FIB commander is the division's fire support coordinator and is assisted by the chief of fires who then serves as the deputy fire support coordinator during the period the force FA headquarters is in effect. The fire support coordinator is the primary advisor on the planning for and employment of fires. The responsibilities and authority given to the FSCOORD should be fully delineated by the supported commander. The FSCOORD may be given authority by the commander to—

- Provide for consolidated and focused fire support-specific training certification, readiness, and oversight (personnel management, equipment issues, and training)
- Facilitate establishing standard operating procedures across the brigade (to save time and ensure a single standard)
- Ensure efficiently resourced training packages (limit requirements for unit tasking and reduce coordination requirements between units)
- Oversee the professional development of the 13-series career management field Soldiers assigned to the BCT
- Mentor, train, and educate junior fires leaders and maintain a habitual supervisory role for the brigade and battalion FSOs

B. Fire Support Officer (FSO)

The FSO is the senior FA staff officer responsible for all fires planning and execution. The brigade FSO's duties and responsibilities are similar to those of the COF.

C. Joint Fires Observer

A joint fires observer (JFO) is a trained and certified Servicemember who can request, adjust, and control surface-to-surface fires, provide targeting information in support of Type 2 and 3 close air support terminal attack controls, and perform autonomous terminal guidance operations (FM 3-09.32). The JFO is not an additional Soldier in his Army fire support organization, but rather an individual who has received the necessary training and certification to receive the JFO's additional skill identifier. JTACs cannot be in a position to see every target on the battlefield. Trained JFOs, in conjunction with JTACs, will assist maneuver commanders with the timely planning, synchronization, and responsive execution of all joint fires. JFOs provide the capability to exploit opportunities that exist in the AO to efficiently assist air delivered fires and facilitate targeting for the JTAC. The goal is to have a JFO-trained and certified Servicemember with each armor company and each infantry platoon.

C. BCT Combat Observation and Lasing Team (COLT)

A combat observation and lasing team is a FA team controlled at the brigade level that is capable of day and night target acquisition and has both laser range finding and laser-designating capabilities. Each BCT typically has organic COLTs under brigade HQ control. The responsibility for COLT training, certification, and recommendation for employment falls underneath the brigade combat team's organic FA battalion commander. The BCT fires cell supervises the planning and execution of COLT employment and ensures the integration of the COLTs into the BCT reconnaissance and surveillance plan. The BCT often employs COLTs as independent observers to weight the decisive operation or key or vulnerable areas. The COLT's self-location and precise target location capabilities can facilitate first round fire for effect (FFE) and the employment of terminally guided munitions.

D. Air Force Tactical Air Control Party (TACP)

An Air Force TACP is under the direction of the brigade FSO and ALO. The TACP is assigned to the maneuver battalion. The ALO leads the TACP and is the principal advisor to the brigade commander and staff on air support. He leverages the expertise of his TACP with linkages to the division and corps TACPs to plan, prepares, execute, and assess air support for brigade operations. He also maintains situational understanding of the total air support picture.

E. Fire Support Team (FIST)

A fire support team is a field artillery team organic to each maneuver battalion and selected units to plan and coordinate all available company supporting fires, including mortars, FA, naval surface fire support, and close air support integration. The battalion commander can direct that FISTs be task-organized within the battalion and employed according to the observation plan. FISTs employed at company level can provide the maneuver companies and reconnaissance troops with fires coordination, targeting, and assessment capabilities. Each fire support team vehicle possesses a target acquisition/communications suite with the capability to designate for laser-guided munitions. A FIST member may conduct target coordinate mensuration if he is trained and certified, and the target acquisition/communications suite is updated with the necessary equipment and software.

V. ADA Brigade

ADA brigades mainly support theater-level operations through the use of terminal high altitude area defense (THAAD), and Patriot assets, but also can provide forces which include IFPC and SHORAD capabilities. AMD operations and ADA task forces provide support from theater to division levels to enable the scheme of maneuver. ADA forces at the brigade level include both non-divisional and maneuver base systems. These systems are employed to protect operational forces and assets from air and missile attack and provide global missile defense. The ADA brigade commander advises the AAMDC commander on overall counterair and AMD integration, synchronization, and employment.

Air Defense Artillery Fire Control Officer (ADAFCO)

The ADAFCO is responsible to the commander for coordinating air defense of designated facilities and areas, as well as coordinating and monitoring the command, air picture, and fire unit exchange between the control reporting center/airborne warning and control system/Aegis/tactical air operations center and the Patriot information coordination central or the ADA battalion fire control/direction center.

An ADAFCO is required in any regional/sector air defense command in which an Army air/missile defense capability is employed. The ADAFCO has the expertise to advise the RADC/sector air defense commander (SADC) on what course of action Army AD units would likely follow during non-standard situations, especially with degraded communications, what limitations ROE can have on autonomous Army ADA units, what tactics may be more effective, etc. ADAFCO elements should be part of/liaison to any of the Service air/missile defense operations centers that may have control of or support from Army ADA assets. Typically, an ADAFCO element deploys to the appropriate air defense region/sector location and is responsible to the RADC/SADC for integrating Army ADA capabilities into that part of the integrated air defense system.

The ADAFCO must have access to dedicated AD communications links (for example, dedicated AD voice circuit) and with Army AD communications nodes when conducting active air defense operations. Unless very unusual circumstances dictate, an ADAFCO should not be placed on an airborne warning and control/airborne command and control aircraft that are not a full-time SADC directing ground-based AD in conjunction with active air intercepts. Those aircraft normally lack a dedicated seat position and communications for the ADAFCO, and they do not have as reliable situational awareness available as does a RADC/SADC with a tactical data link and a common tactical picture or a COP.

III. Fires in the Operations Process

Ref: ADRP 3-09, Fires (Aug '12), chap. 3.

Fires are an integral part of the operations process—the major mission command activities performed during operations: planning, preparing, executing, and continuously assessing the operation (ADP 5-0). The commander drives the operations process.

I. The Operations Process

Planning, preparing, executing, and continuously assessing the operation serve as a template for coordinating other actions associated with an operation including integrating processes, continuing activities, and actions specific to each operations process activity. Both integrating processes and continuing activities occur throughout an operation.

The following table illustrates how the integrating processes and continuing activities last throughout the operations process. Commanders synchronize them with each other and integrate them into all operations process activities.

Operations and Integrating Processes

Plan	Prepare	Execute
	Assess	

Integrating Processes →

- Intelligence Preparation of the Battlefield
- Targeting
- Risk Management

Continuing Activities →

- Information Collection
- Security Operations
- Protection
- Liaison and Coordination
- Terrain Management
- Airspace Control

Ref: ADRP 3-09, Fires, table 3-1, p. 3-1.

II. Fires and Targeting (D3A)

Army targeting uses the functions decide, detect, deliver, and assess (D3A) as its methodology. Its functions complement the planning, preparing, executing, and assessing stages of the operations process. Army targeting addresses two targeting categories—deliberate and dynamic.

- **Deliberate targeting** prosecutes planned targets
- **Dynamic targeting** prosecutes targets of opportunity and changes to planned targets or objectives

See following pages (pp. 5-26 to 5-27) for an overview and discussion of the operations process and targeting relationship.

Fires in the Operations Process

Ref: ADRP 3-09, Fires (Aug '12), pp. 3-1 to 3-4.

Fires and Targeting (D3A)

Army targeting uses the functions decide, detect, deliver, and assess (D3A) as its methodology. Its functions complement the planning, preparing, executing, and assessing stages of the operations process. Army targeting addresses two targeting categories—deliberate and dynamic. Deliberate targeting prosecutes planned targets. Dynamic targeting prosecutes targets of opportunity and changes to planned targets or objectives.

A. Decide

Decide is the first function in targeting and occurs during the planning portion of the operations process. The "decide" function continues throughout the operation.

B. Detect

Detect is the second function in targeting and occurs primarily during the prepare portion of the operations process. A key resource for fires planning and targeting is the intelligence generated through reconnaissance, surveillance, and intelligence operations to answer the targeting information requirements. Requirements for target detection and action are expressed as PIR and information requirements. Their priority depends on the importance of the target to the friendly course of action and tracking requirements. PIR and information requirements that support detection of HPTs are incorporated into the overall unit information collection plan. Named areas of interest and target areas of interest are focal points particularly for this effort and are integrated into the information collection plan.

C. Deliver

Deliver is the third function in targeting and occurs primarily during the execution stage of the operations process. The main objective is to attack/engage targets in accordance with the commander's guidance. The selection of a weapon system or a combination of weapons systems leads to a technical solution for the selected weapon.

D. Assess

Assess is the fourth function of targeting and occurs throughout the operations process. The commander and staff assess the results of mission execution.

* **Combat Assessment.** Combat assessment is the determination of the effectiveness of force employment during military operations. Combat assessment is composed of three elements:

- Battle damage assessment
- Munitions effectiveness assessment
- Reattack recommendation

* Dynamic Targeting

Dynamic targeting has six distinct steps: find, fix, track, target, engage, and assess. Targets of opportunity have been the traditional focus of dynamic targeting because decisions on whether and how to engage must be made quickly. Planned targets are also covered during dynamic targeting but the steps simply confirm, verify, and validate previous decisions (in some cases requiring changes or cancellation). *Refer to JP 3-60.*

* Find, Fix, Finish, Exploit, Analyze and Disseminate (F3EAD)

Find, fix, finish, exploit, analyze, and disseminate (F3EAD) provides maneuver leaders at all levels with a methodology that enables them to organize resources and array forces. While the targeting aspect of F3EAD is consistent with D3A methodology, F3EAD provides the maneuver commander an additional tool to address certain targeting challenges, particularly those found in a counterinsurgency environment.

Operations Process & Targeting Relationship

Fires are an integral part of the operations process—the major mission command activities performed during operations: planning, preparing, executing, and continuously assessing the operation (ADP 5-0). The commander drives the operations process.

Army targeting uses the functions decide, detect, deliver, and assess (D3A) as its methodology. Its functions complement the planning, preparing, executing, and assessing stages of the operations process. Army targeting addresses two targeting categories—deliberate and dynamic.

<table>
<tr><th colspan="2">Operations Process</th><th>D3A</th><th>Targeting Task</th></tr>
<tr><td rowspan="7">Continuous Assessment</td><td rowspan="4">Planning</td><td rowspan="4">Decide</td><td>• Perform target value analysis to develop fire support, high-value targets, and critical asset list.
• Provide fires running estimates and information/influence to the commander's targeting guidance and desired effects.</td></tr>
<tr><td>• Designate potential high-payoff targets.
• Deconflict and coordinate potential high-payoff targets.
• Develop high-payoff target list/defended asset list.
• Establish target selection standards and identification matrix (air and missile defense).
• Develop attack guidance matrix, fire support, and cyber/electromagnetic activities tasks.
• Develop associated measures of performance and measures of effectiveness.</td></tr>
<tr><td>• Refine high-payoff target list.
• Refine target selection standards.
• Refine attack guidance matrix and surface-to-air-missile tactical order.
• Refine fire support tasks.
• Refine associated measures of performance and measures of effectiveness.
• Develop the target synchronization matrix.
• Draft airspace control means requests.</td></tr>
<tr><td>• Finalize the high-payoff target list.
• Finalize target selection standards.
• Finalize the attack guidance matrix.
• Finalize the targeting synchronization matrix.
• Finalize fire support tasks.
• Finalize associated measures of performance and measures of effectiveness.
• Submit information requirements to staff and subordinate units.</td></tr>
<tr><td>Preparation</td><td>Detect</td><td>• Collect information (surveillance, reconnaissance).
• Report and disseminate information.
• Update information requirements as they are answered.
• Focus sensors, locate, identify, maintain track, and determine time available.
• Update the high-payoff target list, attack guidance matrix, targeting synchronization matrix, identification matrix (air and missile defense) and surface-to-air-missile tactical order as necessary.
• Update fire support tasks.
• Update associated measures of performance and measures of effectiveness.
• Target validated, deconfliction and target area clearance resolved, target execution/ engagement approval.</td></tr>
<tr><td>Execution</td><td>Deliver</td><td>• Order engagement.
• Execute fires in accordance with the attack guidance matrix, the targeting synchronization matrix, identification matrix (air and missile defense), and surface-to-air-missile tactical order.
• Monitor/manage engagement.</td></tr>
<tr><td>Assess</td><td>Assess</td><td>• Assess task accomplishment (as determined by measures of performance).
• Assess effects (as determined by measures of effectiveness).
• Reporting results.
• Reattack/reengagement recommendations.</td></tr>
<tr><td colspan="4">Legend: D3A – decide, detect, deliver, and assess</td></tr>
</table>

Ref: ADRP 3-09, Fires (Aug '12), table 3-2, p. 3-2.

Refer to BSS5: The Battle Staff SMARTbook (Leading, Planning & Conducting Military Operations) for discussion of the fires and targeting (D3A - decide, detect, deliver and assess) as it relates to the operations process. In-depth topics include high-payoff target list, intelligence collection plan, target selection standards, attack guidance matrix, attack of targets, tactical and technical decisions, and combat assessments.

III. Fires Planning

Ref: ADRP 3-09, Fires (Aug '12), pp. 3-5 to 3-7.

The commander's ability to orchestrate and employ all available fires related resources as a system and to integrate and synchronize fires with his concept of operations results from an established process known as fires planning. The objective of fires planning is to optimize combat power. It is performed as part of the operations process. Fires planning coordinates, integrates and synchronizes scalable Army indirect fires, AMD fires, joint fires (including electronic attack), and multinational fires with the other warfighting functions into the commander's concept of operations. Fires planners work closely with other elements of the warfighting functions to achieve lethal and nonlethal effects through targeting. Fires planning typically results in the development of the AADP and the fire support plan.

For more information refer to FM 3-01 and FM 3-09.

Fires planning and coordination is central to the effectiveness of fires. It requires continually coordinating plans and managing the fires assets that are available to a supported force. Formal coordination binds fires resources together in a common effort so that the employment of each fires asset is synchronized with the commander's intent and concept of operations. Effective coordination during both planning and execution is required to ensure that a suitable weapon system(s) adequately attacks/engages the desired targets at the correct time and place. Cooperation among the various organizations is necessary for the effective delivery of fires. The action of preparing an integrated fire support plan and area air defense plan (AADP) coordinates and integrates with the other warfighting functions to maximize the results of each attack in generating desired effects.

Commander's Intent

Fires personnel must thoroughly understand the commander's intent and end state in order to design a plan that best supports the commander's concept of operations and, as circumstances change, to rapidly and effectively make the necessary adjustments to the plan. Understanding the commander's intent also makes it easier for fires personnel to advise the commander and his staff on how to best employ fires to support all phases of the operation and to achieve the desired end state.

Commander's Guidance for Fires

The purpose of commander's guidance is to provide his intent and end state to focus staff activities in planning the operation. The commander's guidance for fires provides the staff, and fires personnel, and subordinate units with the general guidelines and restrictions for the employment of fires and their desired effects. The guidance emphasizes in broad terms when, where, and how the commander intends to synchronize the effects of fires with the other elements of combat power to accomplish the mission. Commander's guidance should include priorities and how he envisions that fires will support his concept of operations.

Priority of fires is the commander's guidance to his staff, subordinate commanders, fires planners, and supporting agencies to employ fires in accordance with the relative importance of a unit's mission.

The JFC may prohibit or restrict attacks/engagements on specific targets or objects without specific approval based on military risk, the law of war, ROE, or other civil-military considerations.

Targeting restrictions are typically identified on two lists:

No-Strike List

A no-strike list is a list of objects or entities characterized as protected from the effects of military operations under international law and/or the rules of engagement. Attacking these may violate the law of armed conflict or interfere with friendly relations with indigenous personnel or governments (JP 3-60). The no-strike list is independent of and in parallel to the candidate target list. It is important to note, however, that entities from the candidate target list may be moved to the no-strike list if, as a result of additional target development, it is determined that attacking them may violate the Law of Armed Conflict. Conversely, targets placed on a no-strike list may be removed from that list and become subject to military action if their status as a protected object or entity has changed. For example, a church that functions as a weapons storage facility or a barracks may lose its protected status and could legally be attacked.

Restricted Target List

A restricted target is a valid target that has specific restrictions placed on the actions authorized against it due to operational considerations (JP 3-60). A restricted target list is a list of restricted targets nominated by elements of the joint force and approved by the joint force commander. This list also includes restricted targets directed by higher authorities (JP 3-60). Actions that exceed specified restrictions are prohibited until coordinated and approved by the establishing headquarters. Attacking restricted targets may interfere with projected friendly operations. Targets may have certain specific restrictions associated with them that should be clearly documented in the restricted target list (for example, do not strike during daytime, strike only with a certain weapon). Some targets may require special precautions (for example, chemical, biological, or nuclear facilities, or proximity to no-strike facilities) due to possible collateral effects of using fires on the target. When targets are restricted from lethal attacks, commanders should consider nonlethal capabilities as a means to achieve or support the commander's desired objectives.

For more on no-strike and restricted targets and legal considerations for targeting refer to JP 3-60.

Field Artillery Fire Support Planning

Fire support planning is accomplished using targeting and the running estimate. Fire support coordination is the planning and executing of fire so that targets are adequately covered by a suitable weapon or group of weapons (JP 3-09). The FA provides the nucleus for effective fire planning and coordination through staff personnel, fires agencies, and attack resources. The commander at all levels is responsible for the effective integration of fires with his whole operation. Initiated during mission analysis and continuing through post-execution assessment, fire support planning includes the end state and the commander's objectives; target development and prioritization; capabilities analysis; commander's decision and force assignment; mission planning and force execution; and assessment. Commanders use a common doctrinal language to visualize and describe their operational approach. The operational approach describes a framework that relates tactical tasks to the desired end state through a unifying purpose to focus all operations. This approach includes the scheme of fires, which enables commanders to shape the operational environment with fires to support the maneuver commander's requirements and objectives.

IV. Air Defense Planning

Ref: ADRP 3-09, Fires (Aug '12), pp. 3-6 to 3-7.

Air Defense Artillery has overall responsibility for planning Army AMD in support of the joint force commander. Air defense planning will integrate AMD capabilities and airspace requirements to include air and missile warning/cueing information, combat identification procedures and engagement authority as required.

AMD planning involves consideration of joint and multinational Army units including the JFC, service, functional component commands, AAMDC, ADA brigades, ADA battalions, and their integrated capabilities such as IFPC. AMD planning is performed concurrently at all echelons of command in a process known as "parallel planning". This planning process begins with the receipt of a mission from higher headquarters and culminates in the issuance of an operations plan that provides subordinate commands planning direction. An operations plan (OPLAN) may be put into effect at a prescribed time, or on signal, and becomes the OPORD.

Based on this planning, the AAMDC task organizes the subordinate ADA brigade(s) and assigns missions to the brigade(s). If the AAMDC is not present within a designated AOR, the responsibility for this planning falls to the designated ADA brigade.

After reviewing the initial defended assets list (DAL), AMD planners must ensure that sufficient resources are available and allocated to provide adequate protection. Subordinate commanders and their staffs may nominate additional assets for inclusion into a re-prioritized DAL which becomes the basis for AMD planning and defense design.

Additional critical planning guidance provided by JFC includes the Air Tasking Order (ATO) and the Airspace Control Order (ACO). The ATO provides the rules of engagement (ROE) for all ADA units. The ATO also provides specific instructions for tasking forces/capabilities/sorties to specific missions and targets. The ACO is developed to support the ATO and implements the airspace control plan and provides the details of the approved request for airspace coordinating measures (ACM).

Defended Asset Development

The JFC provides his guidance and prioritization of key assets he wants defended. Two prioritized lists of assets are developed and approved from his guidance. The two lists are:

- **Critical Asset List (CAL).** A critical asset list is defined as a prioritized list of assets, normally identified by phase of the operation, and approved by the joint force commander that should be defended against air and missile threats (JP 3-01). This list is developed by the J3/J5 with input from the components of the joint forces. They identify candidate assets to protect from attack. The JFC or combatant commander approves the critical asset list (CAL) listing and it is included in the OPLAN and AADP. The CAL development process evaluates critical assets based on criticality, vulnerability and threat, AMD resources, possible defense designs and the element of risk to develop the DAL. This process originates in the Protection cell.
- **Defended Asset List (DAL).** A defended asset list, in defensive counterair operations, is a listing of those assets from the critical asset list prioritized by the joint force commander to be defended with the resources available (JP 3-01). This list is included in the OPLAN and the AADP. The DAL specifies required levels of protection for each asset. The DAL may be developed by the AADC with component input but normally the AADC delegate the DAL process to the DAADC. The DAADC DAL recommendation (with component commanders input) is submitted through the AADC to the JFC for approval. This process originates in the Protection cell.

For more information about the CAL and DAL refer to ADRP 3-37 or FM 3-01.

Chap 6

The Sustainment Warfighting Functio..

Ref: ADP 4-0, Sustainment (Jul '12) and ADRP 3-0, Operations (Nov '16), pp. 5-5 to 5-6.

The sustainment warfighting function is the related tasks and systems that provide support and services to ensure freedom of action, extend operational reach, and prolong endurance. The endurance of Army forces is primarily a function of their sustainment. Sustainment determines the depth and duration of Army operations. It is essential to retaining and exploiting the initiative. Sustainment provides the support necessary to maintain operations until mission accomplishment. The sustainment warfighting function includes the following tasks:

- Conduct logistics.
- Provide personnel services.
- Provide health service support.

A. Logistics

Logistics is planning and executing the movement and support of forces. It includes those aspects of military operations that—

- Design, develop, acquire, store, move, distribute, maintain, evacuate, and dispose of materiel.
- Acquire or build, maintain, operate, and dispose of facilities.
- Acquire or furnish services.

Although joint doctrine defines logistics as a science, logistics involves both military art and science. Knowing when and how to accept risk, prioritizing a myriad of requirements, and balancing limited resources all require military art. Logistics integrates strategic, operational, and tactical support of deployed forces while scheduling the mobilization and deployment of additional forces and materiel. Logistics includes—

- Maintenance.
- Transportation.
- Supply.
- Field services.
- Distribution.
- Operational contract support.
- General engineering support.

Refer to SMFLS4: Sustainment & Multifunctional Logistics SMARTbook (Warfighter's Guide to Logistics, Personnel Services, & Health Services Support) -- updated with the latest doctrinal references (ADRP 4-0 Sustainment, ATP 4-93 Sustainment Brigade, JP 4-0 Joint Logistics, and more than 20 other joint and service publications) -- for complete discussion of strategic, operational, and tactical logistics.of force projection, deployment and redeployment, and RSO&I operations.

II. Sustainment Overview

Ref: ADP 4-0, Sustainment (Jul '12).

For the Army, sustainment is the provision of logistics, personnel services, and health service support necessary to maintain operations until successful mission completion.

Sustainment of Unified Land Operations

Army forces are employed within a strategic environment. Army forces operate as part of a larger national effort characterized as unified action. Unified action is the synchronization, coordination, and/or integration of the activities of governmental and nongovernmental entities with military operations to achieve a unity of effort (JP 1). Unified land operations acknowledge that strategic success requires fully integrated U.S. military operations to include the efforts of unified action partners.

Joint Interdependence

Joint interdependence is the purposeful reliance by one Service's forces on another Service's capabilities to maximize the complementary and reinforcing effects of both (JP 3-0). Army forces operate as part of an interdependent joint force. For example:

The United States Air Force through the Air Mobility Command, provides worldwide cargo and passenger airlift, air refueling, and aeromedical evacuation. Air Mobility command also provides Contingency Response Elements that provide enroute ground support for airlift operations.

Joint logistics over-the-shore operations occur when Navy and Army forces conduct logistics over-the-shore operations together under a joint force commander. The Navy's cargo off-load and discharge system is comprised of the container off-loading and transfer system and the offshore bulk fuel system. Army provides lighterage, roll-on/rolloff discharge facilities, causeway systems, and shore-based water storage systems.

The Army plays a critical role in setting the theater and is the primary Service with a sustainment capability to conduct this mission on a large and long term scale. The Army is responsible for theater opening; port and terminal operations; conducting reception, staging, onward movement, and integration; force modernization and theater-specific training; and common-user logistics to joint and multinational forces.

Army Sustainment Responsibilities

Title 10, U.S. Code, specifies that individual Services retain sustainment responsibility. As such, each Service retains responsibility for the sustainment of forces it allocates to a joint force. The Secretary of the Army exercises this responsibility through the Chief of Staff of the Army and the Theater Army assigned to each combatant command.

The Theater Army is responsible for the preparation and administrative support of Army forces assigned or attached to the combatant command. However, the purposeful combination of service capabilities to create joint interdependent forces is often the most effective and efficient means by which to sustain a joint force. The options for executing sustainment of a joint force may include any combination of Directive Authority for Logistics, Executive Agency, lead service and/or establishing a joint command for logistics. In order for the joint command for logistics to succeed, the CCDR must augment it with the capabilities needed to integrate and control the delivery of theater support to meet the joint force requirements. If the Army is designated for establishing a joint command for logistics, the Army Theater Sustainment Command will fulfill that role.

The Secretary of Defense may designate the head of a DOD component (such as Chief of a Service, CCDR, or director of a Combat Support Agency) as an Executive Agent for specific responsibilities, functions, and authorities. When designated as an Executive Agent, the Army is specifically tasked by the Secretary of Defense for certain responsibilities sometimes limited by geography, sometimes for a particular operation, and sometimes for the entire DOD on a continuing basis.

Sustainment Underlying Logic

ADP 4-0 Underlying Logic

Anticipated Operational Environment
- US must project power into region, opposed
- US must seize at least one base of operations (maybe more)
- Threat of WMD will require dispersion of US forces and decentralized operations
- Size of theater (space and population) will exceed US ability to control

Sustainment in Joint Operations

Sustainment is the provision of logistics and personnel services necessary to maintain and prolong operations until successful mission completion. Sustainment in joint operations provides the JFC flexibility, endurance, and the ability to extend operational reach. (JP 4-0)

Unified Action

Central idea: synchronization, coordination, and/or integration of the activities of governmental and non-governmental entities with military operations to achieve unity of effort (JP 1)

Sustainment of Unified Action

Joint Interdependence: The purposeful reliance by Service forces on another Service's capabilities

Unified Land Operations

Seize, retain, and exploit the initiative to gain and maintain a position of relative advantage in sustained land operations through simultaneous offensive, defensive, and stability operations in order to prevent or deter conflict, prevail in war, and create the conditions for favorable conflict resolution.

ADP 4-0, Sustainment

Principles of Sustainment
- Integration
- Anticipation
- Responsiveness
- Simplicity
- Economy
- Survivability
- Continuity
- Improvisation

Joint Doctrine

ADP 4-0 Sustainment

Decisive Operations

Enabling CCDR and ARFOR to conduct...

Operational Reach

Freedom of Action

Endurance

Strategic Base leverages National capability to generate Theater capabilities

Intrinsically Linked

Sustainment HQ cognitively link strategic capability with tactical success

TSC ESC SB AFSB

FMC MEDCOM (DS) HRSC

Synchronizing Strategic and Operational Support Occurs through Mission Command

Joint Interdependence
- Joint Deployment and Distribution Enterprise (JDDE)
- Common User Logistics (CUL)
- Army Support to Other Services (ASOS)

Logistics
- Maintenance
- Transportation
- Supply
- Field Services
- Distribution
- Operational Contracting
- General Engineering

Personnel Services
- Human Resources Support
- Financial Management Operations
- Legal Support
- Religious Support
- Band

Health Service Support
- Casualty Care
  - Organic and Area Medical Support
  - Hospitalization
  - Dental Care
  - Behavioral Health/Neuropsychiatric Treatment
  - Clinical Laboratory Services
  - Treatment of CBRN Patients
- Medical Evacuation
- Medical Logistics

Sustainment Capabilities

Ref: ADP 4-0, Sustainment, fig. 1, p. iv.

Refer to SMFLS4: Sustainment & Multifunctional Logistics SMARTbook (Guide to Logistics, Personnel Services, & Health Services Support). Includes ATP 4-94 Theater Sustainment Command, ATP 4-93 Sustainment Brigade, ATP 4-90 Brigade Support Battalion, Sustainment Planning, JP 4-0 Joint Logistics, ATP 3-35 Army Deployment and Redeployment, and more than a dozen new/updated Army sustainment references.

B. Personnel Services

Personnel services are those sustainment functions related to Soldiers' welfare, readiness, and quality of life. Personnel services complement logistics by planning for and coordinating efforts that provide and sustain personnel. Personnel services include—

- Human resources support.
- Financial management.
- Legal support.
- Religious support.
- Army music support.

C. Health Service Support

The Army Health System is a component of the military health system that oversees operational management of the health service support and force health protection missions. The Army Health System includes all mission support services performed, provided, and arranged by the Army Medical Department to support health service support, and it includes force health protection mission requirements for the Army. Health service support is part of the sustainment warfighting function, while force health protection is a part of the protection warfighting function.

The health service support mission promotes, improves, conserves, or restores the mental and physical well-being of Soldiers and, as directed, other personnel. This mission consists of casualty care, medical evacuation, and medical logistics. Casualty care encompasses the treatment aspects of a number of Army Medical Department functions including—

- Organic and area medical support.
- Hospitalization (including treatment of chemical, biological, radiological, and nuclear patients).
- Dental treatment.
- Behavioral health and neuropsychiatric treatment.
- Clinical laboratory services.
- Medical evacuation (including en route care and medical regulating).
- Medical logistics (including blood and blood products).

Health service support closely relates to force health protection: the measures to promote, improve, or conserve the mental and physical well-being of Soldiers. These measures enable a healthy and fit force, prevent injury and illness, and protect the force from health hazards.

In addition to the principles of sustainment in ADRP 4-0, commanders consider the following when performing sustainment warfighting function tasks:

- Commanders need to plan for the early acquisition of locations and facilities for force and logistic bases where temporary occupancy is planned or when the host nation fails to provide, or provides inadequate, locations and facilities.
- Sustainment forces, like all other forces, must be capable of self-defense, particularly if they deploy alone or in advance of other military forces.

Chap 6

I. Sustainment of Unified Land Operations

Ref: ADRP 4-0, Sustainment (Jul '12). chap. 2.

Unified action is the synchronization, coordination, and/or integration of the activities of governmental and nongovernmental entities with military operations to achieve a unity of effort (JP 1). Unified land operations acknowledges that strategic success requires fully integrating U.S. military operations with the efforts of interagency and multinational partners. The sustainment of unified land operations requires a continuous link between the strategic, operational, and tactical levels. It also requires close coordination and collaboration with other Services, allies, host nation, and other governmental organizations. This chapter demonstrates the important roles that the U.S. military and intergovernmental partners play during the sustainment of Army forces. It also builds the doctrinal bridge between our strategic and inter-organizational partners and sustainment of Army forces conducting operations.

I. Strategic Context

In the U.S., sustainment originates at the strategic base. The strategic base consists of the Department of Defense and industrial base. The DOD acquisition(s) sustainment resources and capabilities and then provide(s) them for use in support of national strategic objectives. The industrial bases, privately and government-owned capabilities, manufactures, maintains, modifies, and repairs resources required by U.S. forces. The strategic base generates Army capabilities which are employed across the strategic environment. Army forces through joint interdependence rely upon joint capabilities, air and maritime, to deliver sustainment to a theater of operations. Through coordination and collaboration between strategic and operational partners, a continuous and accountable flow of sustainment is provided to achieve national military objectives. Also through coordination, collaboration, and agreements with host nation, allies and intergovernmental organizations certain sustainment efficiencies are achieved to facilitate a unity of effort.

II. Joint Interdependence

Joint interdependence is the purposeful reliance by one Service's forces on another Service's capabilities to maximize the complementary and reinforcing effects of both. Army forces operate as part of an interdependent joint force.

There are many services that joint forces provide each other. The U.S. Air Force (USAF) provides lift capabilities to quickly move Army forces across strategic lines of communication to theater operations. In emergency situations, the USAF may aerial deliver sustainment to forward areas or areas where terrain may be too restrictive for ground operations. The USAF through the Air Mobility Command (AMC) provides worldwide cargo and passenger airlift, air refueling, and aeromedical evacuation. AMC also provides Contingency Response Elements that provide enroute ground support for airlift operations.

The Naval Forces provide critical sustainment support to Army operations. Naval forces provide essential joint logistics over the shore (JLOTS) support ensuring sustainment is provided to land forces when ports may be austere, damaged, or non-existent. Naval forces may be responsible for removing sustainment from vessels and delivering them to port operations for release to Army forces. The Naval Construction Force provides port construction such as warehouses, storage facilities. The Navy also provides explosive ordnance disposal support to locate and dispose of mines along ports and channels.

A crucial role the Army plays as a joint interdependent force is opening and setting the theater. Setting the theater is described as all activities directed at establishing favorable conditions for conducting military operations in the theater, generally driven by the support requirements of specific operation plans and other requirements established in the geographic combatant commander's (GCC) theater campaign plan. Setting the theater includes whole-of-government initiatives such as bilateral or multilateral diplomatic agreements to allow U.S. forces to have access to ports, terminals, airfields, and bases within the area of responsibility (AOR) to support future military contingency operations. Setting the joint operations area (JOA) includes activities such as theater opening, establishing port and terminal operations, conducting reception, staging, onward movement, and integration, force modernization and theater-specific training, and providing Army support to other Services and common-user logistics to Army, joint, and multinational forces operating in the JOA (FM 3-93).

The U.S. Military Surface Deployment and Distribution Command (SDDC) is the Army service component command (ASCC) to U.S. Transportation Command (USTRANSCOM) and is responsible for port opening and operations. The theater sustainment command is responsible for theater opening and setting the theater. As a result of Title 10, United States Code (USC) and executive agent responsibility, the Army contributes a significant portion of sustainment to support joint operations.

Sustainment of Joint Forces

Sustainment of joint forces is the deliberate, mutual reliance by each Service component on the sustainment capabilities of two or more Service components. CCDRs and their staffs must consider a variety of sustainment factors including defining priorities for common sustainment functions and responsibilities.

Common sustainment consists of materiel, services, and/or support that is shared with or provided by two or more military Services, DOD agencies, or multinational partners to another Service, DOD agency, non-DOD agency, and/or multinational partner in an operation. It can be restricted by type of supply and/or service and to specific unit(s) times, missions, and/or geographic areas. Service component commands, DOD Agencies (such as Defense Logistics Agency [DLA]), and Army commands (such as U.S. Army Materiel Command (USAMC)/U.S. Army Medical Command (USAMEDCOM), provide common sustainment to other service components, multinational partners, and other organizations authorized to receive support.

Defense Logistics Agency provides support for joint forces during peace and war. DLA is the focal point for the industrial base and is the executive agent for all CL I, II, III (B) (P), IV, VIII and a majority of Class IX. Excluded supply items are munitions, missiles, and military Service unique items. DLA Disposition Services provides material reutilization, marketing, demilitarization and disposal services at sites throughout the world and is an active partner with deployed units in contingency environments. Authorized unserviceable and excess Department of Defense property destined for DLA Disposition Services sites is inspected and categorized upon receipt. An appropriate disposition is determined that may include reutilization, transfer, donation to approved organizations, demilitarization or disposal. As the theater matures, DLA-directed activities may expand to include theater storage and delivery. When the theater situation permits, DLA may use host nation or contractor support to assist in the storage, transportation or delivery of parts and material to the customer.

Department of Defense Directive (DODD) 5101.9 designates the DLA Troop Support as the executive agent (EA) for medical materiel. As the EA, DLA troop support is designated the DOD single point of contact to establish the strategic capabilities and systems integration necessary for effective and efficient Class VIII supply chain support to the combatant commander. As part of this directive, Army medical logistics units may be tasked to provide support to all Services and designated multinational partners (in accordance with applicable contracts and agreements) under the joint concepts of single integrated medical logistics manager (SIMLM), and theater lead agent for medical materiel (TLAMM). The TLAMM is designated by the combatant commander to provide the operational capability for medical supply chain management and distribution from strategic to tactical levels. In a land-based theater, the Army will normally be designated as the TLAMM, consistent with its traditional designation as SIMLM. Within the theater, these capabilities are provided by operational medical units that are task-organized under the control of the medical command (deployment support) (MEDCOM [DS]).

See page 6-20 and refer to JP 4-02 and FM 4-02.1 for further discussion.

III. Army Sustainment Responsibilities

Each Service retains responsibility for the sustainment of forces it allocates to a joint force. The Secretary of the Army exercises this responsibility through the Chief of Staff, United States Army (CSA) and the Theater Army assigned to each combatant command. The Theater Army is responsible for the preparation and administrative support of Army forces assigned or attached to the combatant command.

A. Army Title 10 Sustainment Requirements

Title 10, USC and DOD Directive 5100.1, Functions of the DOD and Its Major Components, describe the organization, roles, and responsibilities for the elements of the DOD to include the statutory requirements for each Military Department to provide support to assigned forces.

See following page (p. 6-9) for further discussion.

B. Executive Agent (EA)

Executive Agent (EA) is a term used to indicate a delegation of authority by the Secretary of Defense to a subordinate to act on behalf of the Secretary of Defense. An EA may be limited to providing only administration and support or coordinating common functions; or it may be delegated authority, direction, and control over specified resources for specified purposes (JP 1).

When designated as an EA, the Army is specifically tasked by the Secretary of Defense for certain responsibilities, sometimes limited by geography, sometimes for a particular operation, and sometimes for the entire DOD on a continuing basis. The list below (not all inclusive) is an example of some of the Army's sustainment EA responsibilities:

- DOD Combat Feeding Research and Engineering Program
- Management of Land-based Water Research in Support of Contingency Operations
- Law of War Program
- Defense Mortuary Affairs Program
- Military Postal Service
- Explosive Safety Management
- Armed Services Blood Program Office

C. Lead Service

The CCDR may choose to assign specific common support functions, to include both planning and execution to a lead Service. These assignments can be for single or multiple common user functions and may also be based on phases and/or locations within the AOR. The CCDR may augment the lead Service logistics organization with capabilities from another component's logistics organizations as appropriate. The lead Service must plan, issue procedures, and administer sustainment funding for all items issued to other Services as well as a method for collecting items from other Services.

D. Joint Command for Logistics

The CCDR may assign joint logistics responsibilities to a subordinate Service component and establish a joint command for logistics (JP 4-0). The senior logistics HQ of a designated Service component will normally serve as the basis for this command. In order for the joint command for logistics to succeed, the CCDR must augment it with the capabilities needed to integrate and control the delivery of theater support to meet the joint force requirements. When the Army is designated for establishing a joint command for logistics, the Theater Sustainment Command, Expeditionary Sustainment Command, or Sustainment Brigade might fulfill that role.

E. Directive Authority for Logistics (DAFL)

The Directive Authority for Logistics (DAFL) is the CCDR's authority to issue logistics directives to subordinate commanders, including peacetime measures, necessary to ensure the effective execution of approved operation plans (JP 1). The CCDRs may delegate directive authority for as many common support capabilities as required to accomplish the assigned mission. It includes peacetime measures to ensure the effective execution of approved OPLANs, effectiveness and economy of operation, prevention or elimination of unnecessary duplication of facilities, and overlapping of functions among the Service component commands.

When the CCDR gives a Service component common support responsibility, the responsibility must be specifically defined. When two or more Services have common commodities or support services, one Service may be given responsibility for management based on DOD designations or inter service support agreement. However, the CCDR must formally delineate this directive authority by function and scope. The Army, when directed to provide management of common sustainment functions which include other services, most often establishes and leads joint boards. These boards are ad hoc, and if directed by the Theater Army, the TSC may serve as the board lead.

IV. Generating Forces

The generating force consists of those Army organizations whose primary mission is to generate and sustain the operational Army's capabilities for employment by Joint Force Commanders (JFCs). The generating force activities include support of readiness, Army force generation, and the routine performance of functions specified and implied in Title 10 USC. As a consequence of its performance of functions specified and implied by law; the generating force also possesses operationally useful capabilities for employment by or in direct support of JFCs. The generating force organizations enable strategic reach by helping to project Army capabilities. Generating force capabilities include analyzing, understanding and adapting, and generating operational forces tailored to the specific context in which they will be employed.

The generating force's ability to develop and sustain potent land power capabilities is useful in developing partner security forces and governmental institutions, with its capability to develop, maintain, and manage infrastructure. Army generating forces train and advise partner generating force activities to build institutional capacity for professional education, force generation, and force sustainment (ADP 3-0). Army sustainment forces play a significant role in transitioning to HNS capacity. Sustainment forces assess host nation sustainment capacity, identify process improvements and then train and mentor the HN sustainment force in building its own capacity.

The generating force is responsible for moving Army forces to and from ports of embarkation. They also provide capabilities to assist in the management and operation of ports of embarkation and debarkation and provide capabilities to GCC to conduct reception, staging, onward movement, and integration (RSOI).

See following pages (pp. 6-10 to 6-11) for further discussion of generating forces.

Army Title 10 Sustainment Requirements

Ref: ADRP 4-0, Sustainment (Jul '12), p. 2-3.

There are 12 Army Title 10 responsibilities; of the 12, ten are sustainment related responsibilities:

- Recruiting
- Organizing
- Supplying
- Equipping (including research and development)
- Training
- Servicing
- Mobilizing
- Demobilizing
- Administering (including the morale and welfare of personnel)
- Maintaining
- Construction, outfitting, and repair of military equipment
- Construction, maintenance, repairs of building and structures, utilities, acquisition of real property and interests in real property necessary to carry out the responsibilities

The Secretary of the Army exercises this responsibility through the Chief of Staff the Army and the Theater Army assigned to each combatant command. The Theater Army is responsible for the preparation and administrative support of Army forces assigned or attached to the combatant command. However, the purposeful combination of complementary Service capabilities to create joint interdependent forces is often the most effective and efficient means by which to sustain a joint force. Therefore additional authorities to Title 10 have been developed to provide for inter-service and interagency mutual support. The options for executing sustainment of a joint force may include any combination of following:

- Executive Agent
- Lead Service
- Directive Authority for Logistics

All of these authorities may have the same possible impact on Army sustainment headquarters, which is that the Army may be required to provide support to other Services and agencies involved in an operation or that other Services or agencies may provide support to Army units which would normally receive such support from the Army. Army sustainment leaders must be prepared to plan and execute such operations as tasked.

Generating Forces

Ref: ADRP 4-0, Sustainment (Jul '12), pp. 2-6 to 2-8.

U.S. Army Materiel Command

The U.S. Army Materiel Command (USAMC) equips and sustains the Army, providing strategic impact at operational speed. The USAMC is the Army's materiel integrator. It provides national level sustainment, acquisition integration support, contracting services, and selected logistics support to Army forces. It also provides related common support to other Services, multinational, and interagency partners. The capabilities of USAMC are diverse and are accomplished through its various major subordinate commands and other subordinate organizations.

The USAMC is the lead for the Army's national-level maintenance and supply programs which are managed and executed by its subordinate Life Cycle Management Commands (LCMCs). These USAMC LCMCs coordinate with the USAMC staff as well as related Assistant Secretary of the Army, Acquisitions Logistic and Technology (ASA [ALT]), Program Executive Officers (PEOs) and Product/Project Managers offices. Together, USAMC LCMC and Assistant Secretary of the Army for Acquisition, Logistics and Technology elements work to ensure support for fielded weapon systems and equipment for their entire life cycle. The LCMCs support to deploying and deployed forces is coordinated through the Army Sustainment Command (ASC) and is executed under the control of the supporting Army Field Support Brigade (AFSB). LCMCs are discussed in more detail below.

In addition to the functions performed by the LCMCs, USAMC exercises overall responsibility of sustainment maintenance for the Army and managing secondary items through the National Maintenance Program, whose tenets are as follows:

- Managing sustainment maintenance unit workloads to meet national requirements
- Ensuring all component repairs are performed to a national standard
- Ensuring sustainment maintenance providers possess the facilities, tools, test, measurement, and diagnostic equipment, skills, and workforce required to meet national standards
- Facilitating quality assurance by ensuring that maintainers use documented quality systems and are technically certified to repair to standards

The USAMC is also the lead, but not sole, Army organization responsible for providing contracting services to the Army. USAMC contracting support includes the Logistics Civil Augmentation Program (LOGCAP). Through its subordinate contracting commands, USAMC provides both institutional and operational contract support planning assistance and contract execution support to Army forces (except the National Guard Bureau, U.S. Army Intelligence and Security Command, U.S. Army Corps of Engineers, U.S. Army Medical Command, U.S. Army Special Operations Command, U.S. Army Space and Missile Defense Command, and Program Executive Office for Simulation, Training, and Instrumentation).

The USAMC and its subordinate organizations work with forward deployed commands in executing sustainment support and synchronizing distribution and redistribution of materiel in and out of theaters. Keys to success are data accuracy, asset visibility, property accountability, and disposition instructions. Retrograde Property Assistance Teams (RPATs), an ad hoc organization, facilitate the turn-in of equipment for redistribution or retrograde:

- Aviation and Missile Life Cycle Management Command
- The CECOM Life Cycle Management Command
- The Tank-Automotive and Armaments Life Cycle Management Command
- The Joint Munitions and Lethality Life Cycle Management Command

U.S. Army Medical Command

The U.S. Army Medical Command (USAMEDCOM), commanded by the Army Surgeon General, provides AHS support for mobilization, deployment, sustainment, redeployment, and demobilization across a range of military operations. The USAMEDCOM integrates the capabilities of its subordinate operational Army medical units with generating force assets such as medical treatment facilities and research, development, and acquisition capabilities. The USAMEDCOM's generating force capabilities not only augment those of operating forces but also provide significant assistance in coping with unanticipated health threats.

The U.S. Army Medical Command maintains the capability to provide continuity of care for patients returning from theater. It also provides individual AMEDD training, medical materiel, and research and development activities to support the Army mobilization force. The USAMEDCOM's major subordinate commands include:

- U.S. Army Dental Command
- U.S. Army Medical Research and Materiel Command
- U.S. Army Medical Department Center and School
- U.S. Army Public Health Command
- U.S. Army Warrior Transition Command

The USAMEDCOM also has regional medical commands responsible for oversight of day-to-day operations in military treatment facilities, exercising mission command over the military treatment facilities (MTFs) in the supported region.

U. S. Army Financial Management Command

The U.S. Army Financial Management Command (USAFMCOM) is a strategic level command that serves as a field operating agency of the Assistant Secretary of the Army (Financial Management and Comptroller) (ASA [FM&C]). The USAFMCOM provides strategic financial management oversight and support to include: financial management operations policy; field coordination for national provider support; Army liaison with the Defense Finance and Accounting Service (DFAS); administering the Army banking program; strategic Electronic Commerce and financial management systems integration, deployment and training; contingency operations technical training and assessments for deploying units, G-8s, and Financial Management Support Operations (FM SPOs); systems deployment and support; coordination for Army financial management audits; classified financial management and accounting oversight; and Army/Joint staff coordination.

U.S. Army Installation Management Command

The U.S. Army Installation Management Command (IMCOM) supports unit commanders in the conduct of pre-deployment activities. Through its installation transportation offices, IMCOM plans and coordinates the movement of units from home station to ports of debarkation. IMCOM provides capabilities to operate and manage bases in support of Army and Joint Force commanders. It also provides capabilities to support the unit deployment, redeployment and reintegration. Operational Army organizations, headquarters and units, routinely rely on civilian specialists to execute the day-to-day tasks associated with the management of munitions in transportation and storage during peacetime. Most of these civilian specialists are not organic to these operational Army organizations. Instead, they are assigned to IMCOM installations or USAMC.

U. S. Army Space and Missile Defense Command

The Army is reliant on space-based capabilities and systems, such as global positioning, communication, weather satellites, and intelligence collection platforms. These systems are critical enablers for Army personnel to plan, communicate, navigate and maneuver, maintain situational awareness, engage the enemy, provide missile warning, and protect and sustain our forces.

V. Operating Forces

Ref: ADRP 4-0, Sustainment (Jul '12), pp. 2-8 to 2-12.

The operating forces are those forces whose primary missions are to participate in combat and the integral supporting elements thereof (FM 1-01). Operational Army units are typically assigned to CCDRs. The Army normally executes its responsibilities to organize, train, and equip operational Army units through ASCCs.

1. Army Service Component Command /Theater Army

When an Army Service component command (ASCC) is in support of a GCC, it is designated as a Theater Army (TA). The Theater Army is the primary vehicle for Army support to joint, interagency, intergovernmental, and multinational forces (MNFs). The TA HQ performs functions that include reception, staging, onward movement, and integration; logistics over-the-shore operations; and security coordination.

The Theater Army is responsible for providing support to Army forces and common sustainment/support to other Services as directed by the CCDR and other authoritative instructions. The Theater Sustainment Command (TSC) is assigned to the Theater Army and is the Army's senior logistics headquarters (HQ) within the theater of operations. When directed, the TSC provides lead Service and executive agency support for designated logistics and services to other government agencies, MNFs, and nongovernmental organizations (NGO). When directed, the MEDCOM (DS) provides AHS support to other services.

The TA exercises administrative control over all Army forces in the area of responsibility unless modified by DA. This includes Army forces assigned, attached, or OPCON to the combatant command. The TA coordinates with the TSC for operational sustainment planning and management. The TA defines theater policies and coordinates with the TSC for technical guidance and execution of force projection and sustainment.

2. Corps

The corps provides a HQ that specializes in operations as a land component command HQ and a joint task force for contingencies. When required, a corps may become an intermediate tactical HQ under the land component command, with OPCON of multiple divisions (including multinational or Marine Corps formations) or other large tactical formations. Its primary mission command is land combat operations. The corps HQ has the capability to provide the nucleus of a joint HQ.

3. Division

Divisions are the Army's primary tactical war fighting HQ. Their principal task is directing subordinate brigade operations. Divisions are not fixed formations. Therefore, they may not have all types of Brigade Combat Teams (BCT) in an operation or they may control more than one of a particular type of BCT. A division can control up to six BCTs with additional appropriate supporting brigades during major combat operations. The types of support brigades are combat aviation, fires, maneuver enhancement, battlefield surveillance, and sustainment. The sustainment brigade normally remains attached to the TSC or ESC but supports the division. The division may have OPCON of a SUSTAINMENT BRIGADE while conducting large-scale exploitation and pursuit operations.

4. Brigade Combat Team (BCT)

As combined arms organizations, Brigade Combat Teams (BCT) form the basic building block of the Army's tactical formations. They are the principal means of executing engagements. Three standardized BCT designs exist: armor, infantry, and Stryker. Battalion-sized maneuver, fires, reconnaissance, and Brigade Support Battalion (BSB) are organic to BCTs.

5. Theater Sustainment Command (TSC)

The Theater Sustainment Command (TSC) serves as the senior Army sustainment HQ (less medical) for the Theater Army. The TSC provides mission command of units assigned, attached, or OPCON. The mission of the TSC is to provide theater sustainment (less medical) (FM 4-94).

The Theater Sustainment Command is capable of planning, preparing, executing, and assessing logistics and human resource support for Army forces in theater. It provides support to unified land operations. As the distribution coordinator in theater, the TSC leverages strategic partnerships and joint capabilities to establish an integrated theater-level distribution system that is responsive to Theater Army requirements. It employs sustainment brigades to execute theater opening (TO), theater sustainment, and theater distribution operations.

The TSC includes units capable of providing multifunctional logistics: supply, maintenance, transportation, petroleum, port, and terminal operations. Other specialized capabilities, such as mortuary affairs (MA), aerial delivery, human resources, sustainment to internment/resettlement operations, and financial management, are available from the force pool. The combination of these capabilities gives the TSC commander the ability to organize and provide tailored support.

6. Expeditionary Sustainment Command (ESC)

Expeditionary Sustainment Commands (ESC) are force pooled assets. They are normally under the mission command of the TSC. The ESC provides mission command of sustainment units (less medical) in designated areas of a theater. The ESC plans, prepares, executes, and assesses sustainment, distribution, theater opening, and reception, staging, and onward movement operations for Army forces in theater. It may serve as a basis for an expeditionary command for joint logistics when directed by the GCC or designated multinational or joint task force commander. It normally deploys when the TSC determines that a forward command presence is required. This capability provides the TSC commander with the regional focus necessary to provide effective operational-level support to Army or JTF missions.

7. Theater Engineer Command (TEC)

The Theater Engineer Command (TEC) is designed to mission command engineer capabilities for all assigned or attached engineer brigades and other engineer units and missions for the joint force land component or Theater Army commander. It is the only organization designed to do so without augmentation and can provide the joint force commander with an operational engineer headquarters or augment an engineer staff for a JTF. The TEC is focused on operational-level engineer support across all three of the engineer disciplines and typically serves as the senior engineer headquarters for a Theater Army, land component headquarters, or potentially a JTF (see FM 3-34).

8. Human Resource Sustainment Center (HRSC)

The Human Resource Sustainment Center (HRSC) is a multifunctional, modular organization (staff element), and theater-level center assigned to a TSC that integrates and ensures execution of Personnel Accountability (PA), casualty, and postal functions throughout the theater as defined by the policies and priorities established by the ASCC G-1/AG. The HRSC, in coordination with the TSC, has a defined role to ensure that the theater HR support plan is developed and supported with available resources within the TSC. This includes collaborating with the ASCC G-1/AG and TSC to ensure appropriate HR support relationships are established and properly executed through the OPORD process.

Continued on next page

Operating Forces (Continued)

Ref: ADRP 4-0, Sustainment (Jul '12), pp. 2-8 to 2-12.

Continued from previous page

9. Financial Management Center (FMC)

The Financial Management Center (FMC) is a modular and tailorable operational financial management unit whose mission is inextricably linked to the TA G-8. In order to provide adequate theater and national-provider responsiveness and support, the FMC maintains visibility of all financial management operations and placement of all operational and tactical financial management units in theater. The primary mission of the FMC is to provide technical coordination of all theater finance operations and serve as the principal advisor to the TA G-8 and the TSC commander on all aspects of theater finance operations. Technical coordination of theater financial management units (financial management companies and their subordinate detachments) encompasses the provision of recommendations and advice to theater commanders and staff regarding the employment, integration, direction, and control of their financial management forces for the accomplishment of assigned missions. Other missions include but are not limited to: negotiations with host nation banking facilities, advising unit commanders on the use of local currency, and coordination with national providers (e.g., Department of the Treasury, DFAS, Assistant Secretary of the Army Financial Management & Comptroller, USAFMCOM) and the ECC to establish financial management support requirements (FM 1-06).

10. Army Field Support Brigade (AFSB)

The Army Field Support Brigade (AFSB) is assigned to the ASC-and when deployed, is placed OPCON to the supported theater Army. This OPCON relationship is normally delegated to the supporting TSC or ESC as appropriate. An AFSB provides materiel readiness focused support to include coordination of acquisition logistics and technology actions, less theater support contracting and medical, to Army operational forces. AFSBs serve as ASC's link between the generating force and the operational force. AFSBs are also responsible to integrate LOGCAP support into contract support integration plans, in coordination with the theater Army G-4 and the supporting CSB (ATP 4-91).

11. Sustainment Brigade

When deployed, the sustainment brigade is a subordinate command of the TSC, or by extension the ESC. The sustainment brigade is a flexible, multifunctional sustainment organization, tailored and task organized according to mission, enemy, terrain and weather, troops and support available, time available, and civil considerations (METT-TC). It plans, prepares, executes, and assesses sustainment operations within an area of operations. It provides mission command of sustainment operations and distribution management.

Continued from previous page

12. Combat Sustainment Support Battalion (CSSB)

The Combat Sustainment Support Battalion (CSSB) is a flexible and responsive unit that executes logistics throughout the depth of an area of operations including transportation, maintenance, ammunition, supply, MA, airdrop, field services, water, and petroleum. The CSSB is attached to a sustainment brigade and is the building block upon which the sustainment brigade capabilities are developed. The CSSB is tailored to meet specific mission requirements. Employed on an area basis, the CSSB plans, prepares, executes, and assesses logistics operations within an area of operations. The CSSB also supports units in or passing through its designated area.

13. Medical Command (Deployment Support)

The Medical Command (Deployment Support) (MEDCOM [DS]) serves as the senior medical command within the theater in support of the CCDR. The MEDCOM (DS) provides the mission command for medical units delivering health care in support of deployed forces. The MEDCOM (DS) is a regionally focused command and provides subordinate medical organizations to operate under the medical brigade (MEDBDE) and/or multifunctional medical battalion (MMB). The Medical Command (Deployment Support) is a versatile, modular mission command structure composed of a main command post (MCP) and an operational command post (OCP). *Refer to FM 4-02.12 for more information.*

14. Medical Brigade (MEDBDE)

The Medical Brigade (MEDBDE) provides a scalable expeditionary mission command capability for assigned and attached medical functional organizations task-organized for support of the BCTs and supported units at echelons above brigade (EAB). The MEDBDE provides all of the mission command and planning capabilities necessary to deliver responsive and effective AHS support. The MEDBDE ensures the right mixture of medical professional (operational, technical, and clinical) expertise to synchronize the complex system of medical functions.

The Medical Brigade has the capability to provide an early entry module, an expansion module, and the campaign module, thus enabling its capability to be tailored to METT-TC factors of a specific operation. As the supported forces grow in size and complexity, the MEDBDE can deploy additional modules that build upon one another to support unified land operations.

15. Multifunctional Medical Battalion (MMB)

The Multifunctional Medical Battalion (MMB) is designed as a multifunctional HQ. It can also be deployed to provide mission command to expeditionary forces in early entry operations and facilitate the RSOI of theater medical forces. All EAB medical companies, detachments, and teams in theater may be assigned, attached, or placed under the OPCON of an MMB. The MMB is under the mission command of the MEDBDE/MEDCOM (DS).

16. Sustainment Brigade (Special Operations) (Airborne)

The Sustainment Brigade (Special Operations) (Airborne) is a subordinate command of the U.S. Army Special Operations Command. Its mission is to provide limited sustainment, medical, and signal support to Army Special Operations Forces (ARSOF). ARSOF are not logistically self-sufficient. ARSOF units rely upon the GCC theater infrastructure for virtually all of their support above their organic capabilities. The planning and execution of logistics support to ARSOF must be nested within the GCC's concepts of operation and support, as well as tailored to interface with the theater logistics structures. *For further information on ARSOF logistics capabilities refer to FM 3-05.140.*

17. Brigade Support Battalion (BSB)

The Brigade Support Battalion (BSB) is an organic component of BCT, fires, and maneuver enhancement brigades. The BSB is tailored to support the particular brigade to which it is organic. For example, the BSB of an armor brigade combat team (HBCT) has more fuel distribution capabilities and maintenance than does a fires brigade BSB. The BSB provides supply, maintenance, motor transport, and medical support to the supported brigade. The BSB plans, prepares, and executes, logistics operations in support of brigade operations *(refer to FM 4-90).*

18. Aviation Support Battalion (ASB)

The Aviation Support Battalion is the primary aviation logistics organization organic to Combat Aviation Brigade and the Theater Aviation Brigade. The Aviation Support Battalion performs the BSB mission. It provides aviation and ground field maintenance, brigade-wide satellite signal support, replenishment of all supplies, and medical support to the aviation brigade. The Aviation Support Battalion has been optimized to support the Combat Aviation Brigade's forward support companies, aviation maintenance companies, and the brigade HQ and HQ company (FM 3-04.111).

VI. Intergovernmental and Interagency Coordination

Interagency coordination is the coordination that occurs between elements of DOD and U.S. Government agencies for the purpose of achieving an objective. It is an essential characteristic of unified action. The SECDEF may determine that it is in the national interest to task U.S. military forces with missions that bring them into close contact with (if not in support of) intergovernmental organizations (IGOs) and NGOs. In such circumstances, it is mutually beneficial to closely coordinate the activities of all participants.

In a national emergency or complex contingency operation, DOD and the U.S. military often serve in a supporting role to other agencies and organizations. Unified action partners normally provide for their own sustainment. However, when authorized by the SECDEF, U.S. military sustainment capabilities may be provided to these organizations. This support may include inter-theater and intra-theater airlift, ground transportation of personnel, equipment, and supplies, airfield control groups, and port and railhead operations groups.

The key to interagency coordination is in understanding the civil-military relationship as collaborative rather than competitive. The most productive way to look at this relationship is seeing the comparative advantages of each of the two communities—military and civilian. While the military normally focuses on reaching clearly defined and measurable objectives within given timelines under a command and control structure, civilian organizations are concerned with fulfilling changeable political, economic, social, and humanitarian interests using dialogue, bargaining, risk taking, and consensus building.

While the ways and means between military and civilian organizations may differ, they share many purposes and risks, and the ultimate overall goal may be shared. Unity of effort between IGOs, NGOs, and military forces should be the goal. However, this is not always the case. For instance, in a hostile or uncertain environment, the military's objective may be stabilization and security of its own force. NGOs, on the other hand, may be primarily interested in addressing humanitarian needs, an objective that does not always coincide with the military's objectives. Taskings to support IGOs and NGOs are normally for a short-term purpose due to extraordinary events. In most situations, sustainment, communications, mobility, and security are the capabilities most needed. Ideally, requests to support IGOs and NGOs should be coordinated through a Civil-Military Operations Center, which establishes a system in which the military and NGO/IGO communities can meet and work together in advancing common goals. Sustainment commanders should bear in mind the wide ranging size, capabilities, expertise, and purposes of the scores of NGOs they will likely encounter. Commanders should anticipate NGOs objecting to military actions they perceive as compromising to their impartiality, independence, humanitarianism, or neutrality, principles that NGOs vigorously protect. In all cases there must be specific legal authority authorizing DOD support to U.S. agencies, the United Nations, IGOs, NGOs, and MNFs.

The following basic steps support an orderly and systematic approach to building and maintaining coordination and collaboration:

- Forge a collective definition of the problem in clear and unambiguous terms. Appropriate representatives from relevant agencies, departments, and organizations, to include field offices, should be involved at the onset of the planning process and share their perspectives.
- Understand the objectives, end state, and transition criteria for each involved organization or agency. Commanders and decision makers should establish a clearly defined end state supported by attainable objectives and transition criteria. Not all agencies and organizations will necessarily understand or agree

to clearly define the objective with the same sense of urgency or specificity as military planners.

- Develop a common, agreed set of assumptions that will drive the planning among the supported and supporting agencies. Collectively amend the assumptions as necessary throughout the planning and execution of operations.

Operations such as stability and humanitarian support are often sustainment intensive particularly in logistics, financial management, medical and engineering capabilities. Therefore, the overall sustainment concept should be closely tied into the operational strategy and be mutually supporting. Planning also should consider the potential requirements to provide support to nonmilitary personnel (e.g., IGOs, NGOs, indigenous populations and institutions, and the private sector).

For some operations, sustainment forces may be employed in quantities disproportionate to their normal military roles and in nonstandard tasks. Furthermore, they may precede other military forces or may be the only forces deployed. They also may have continuing responsibility after the departure of combat forces in support of MNFs, OGAs, NGOs, and IGOs. In such cases, they must adhere to any applicable status-of-forces agreements and acquisition cross servicing agreements (ACSAs) to which the United States is a party.

U.S. sustainment capabilities are often tasked to support civilian populations. Sustainment support to the population occurs in stability tasks and defense support of civil authorities. Sustainment provided to civilians may very well determine the success or failure of the overall mission.

Refer to JFODS4: The Joint Forces Operations & Doctrine SMARTbook (Guide to Joint, Multinational & Interorganizational Operations) for further discussion. Topics and chapters include joint doctrine fundamentals, joint operations, joint operation planning, joint logistics, joint task forces, information operations, multinational operations, and interorganizational coordination.

VII. Sustainment in Multinational Operations

A major objective when Army forces participate in the sustainment of multinational deployments is to maximize operational effectiveness. Support provided and received in multinational operations must be in accordance with existing legal authorities. There are two types of multinational operations: alliances and coalitions.

In multinational operations, sustainment of forces is primarily a national responsibility. However, relations between the United States and its NATO allies have evolved to where sustainment is viewed as a collective responsibility (NATO Military Committee Decision 319/1). In multinational operations, the multinational commander must have sufficient authority and control mechanisms over assets, resources, and forces to effectively achieve the mission. For each nation to perform sustainment functions separately, it would be inefficient and expensive. It would also hinder the multinational commander's ability to influence and prioritize limited resources to support the operation and accomplish the mission.

See following pages (pp. 6-18 to 6-19) for further discussion.

Refer to TAA2: Military Engagement, Security Cooperation & Stability SMARTbook (Foreign Train, Advise, & Assist) for further discussion. Topics include the Range of Military Operations (JP 3-0), Security Cooperation & Security Assistance (Train, Advise, & Assist), Stability Operations (ADRP 3-07), Peace Operations (JP 3-07.3), Counterinsurgency Operations (JP & FM 3-24), Civil-Military Operations (JP 3-57), Multinational Operations (JP 3-16), Interorganizational Coordination (JP 3-08), and more.

Sustainment in Multinational Operations

Ref: ADRP 4-0, Sustainment (Jul '12), pp. 2-6 to 2-8.

NATO Logistics Options

NATO doctrine allows for the formation of a Combined Joint Force Land Component Command (CJFLCC). The CJFLCC HQ can be set at a sub-regional command level or formation level. The CJFLCC commander establishes requirements and sets priorities for support of forces in accordance with the overall direction given by the Joint Force Commander. The commander coordinates sustainment operations with all participating nations. See Allied Land Publication 4.2 (Standardization Agreement 2406) for additional details.

Merging national sustainment systems into multinational support systems requires the willingness to share the control of vital support functions with a NATO commander and requires technical interoperability of national support assets. Standardization Agreement (STANAGs) provide agreed policy and standards to NATO nations and contribute to the essential framework for specific support concepts, doctrine procedures, and technical designs. Non-NATO nations will be expected to comply with NATO publications while on NATO-led operations.

The basic sustainment support options for multinational operations may range from totally integrated multinational sustainment forces to purely national support. NATO Allied Publication 4.2 provides details on the following support options.

National Support Element

A National Support Element is any national organization or activity that supports national forces that are a part of a MNF. Their mission is nation-specific support to units and common support that is retained by the nation. It should also be noted that National Support Elements operating in the NATO commander's area of operation are subject to the status of forces agreement, memorandums of agreements, and other HN arrangements.

Host Nation Support

Host nation support (HNS) is civil and military assistance rendered by a nation to foreign forces within its territory during peacetime, crises or emergencies, or war based on agreements mutually concluded between nations. Many HNS agreements have already been negotiated between NATO nations. Potential HNS agreements may address labor support arrangements for port and terminal operations, using available transportation assets in country, using bulk petroleum distribution and storage facilities, possible supply of Class III (Bulk) and Class IV items, and developing and using field services. The U. S. initiates and continually evaluates agreements with multinational partners for improvement.

Multinational Integrated Logistics Units

A Multinational Integrated Logistics Unit is formed when two or more nations agree, under OPCON of a NATO commander, to provide logistics support to a MNF. Multinational Integrated Logistics Units are designed to provide specific logistics support where national forces cannot be provided, or could be better utilized to support the commander's overall sustainment plan.

Lead Nation

A lead nation for logistic support has agreed to assume overall responsibility for coordinating and/or providing an agreed range of sustainment for all or part of a MNF within a defined geographical area. This responsibility may also include procurement of goods and services with compensation and/or reimbursement subject to agreements between the parties involved.

Role Specialization

One nation may assume the responsibility for providing or procuring a particular class of supply or service for all or part of the MNF. A role specialization nation's responsibilities also include the provision of assets needed to deliver the supply or service. Compensation and/or reimbursement will then be subject to agreement between the parties involved.

Contracting Support to Multinational Operations

A deployed force may be required to set up contractual arrangements with contractors. These are normally negotiated individually with vendors to make use of available HN resources. Coordination between contributing nations and the NATO HQ in contractual arrangements is essential. Coordination should be accomplished at the highest appropriate level.

Third Party Logistics Support Services

Third party logistics support services is the use of preplanned civilian contracting to perform selected sustainment. The aim is to enable competent commercial partners to provide a proportion of deployed sustainment so that such support is assured for the commander and optimizes the most efficient and effective use of resources. Third party logistics support services is most likely to be used once the operational environment has become more benign. The third party logistics support services database, which NATO Maintenance & Supply Agency developed, contains details of potential contractors worldwide, capable of providing sustainment to NATO operations. NATO commands and nations may consider using the technical expertise of NATO Maintenance & Supply Agency for their contract activities.

Mutual Support Agreements

Participating nations have the option to develop mutual support arrangements (bi- and multilaterally) to ensure provision of logistics support to their forces. This is especially useful when nations have small force contingents collocated with the forces of another nation that have the capacity to support them. By working together and sharing resources (especially services capabilities), nations can achieve economies of scale in their sustainment operations. Mutual support arrangements have the advantage of being simple to set up and can take place on an ad hoc basis.

Acquisition Cross-Servicing Agreement

Under ACSA authority (Title 10 USC, sections 2341 and 2342), the SECDEF can enter into ACSA for logistics support, supplies, and services on a reimbursable, replacement-in-kind, or exchange-for-equal value basis. These agreements can be with eligible nations and international organizations of which the United States is a member. An ACSA is a broad overall agreement, which is generally supplemented with an implementing agreement. Implementing agreements contain points of contact and specific details of the transaction and payment procedures for orders for logistics support. Neither party is obligated until the order is accepted.

Under these agreements, common support may include food, billeting, transportation (including airlift), petroleum, oils, lubricants, clothing, communications services, medical services, ammunition, base operations, storage services, use of facilities, training services, spare parts and components, repair and maintenance services, calibration services, and port services. Items that may not be acquired or transferred under the ACSA authority include weapon systems, major end items of equipment, guided missiles, nuclear ammunition, and chemical ammunition (excluding riot control agents).

Other Sustainment Options

Chapter 138 of Title 10 USC authorizes exchanging support between U.S. services and those of other countries. It authorizes DOD acquisition from other countries by payment or replacement-in-kind, without establishing a cross-servicing agreement. Supplies and services authorized under Chapter 138 do not include major end items, missiles, or bombs. It does include food, billeting, petroleum, oils, transportation, communication services, medical services, ammunition, storage, spare parts, maintenance services, and training. Therefore, negotiations in advance of operations for sharing projection and sustainment resources are recommended.

VIII. Joint Logistics

Ref: JP 4-0, Joint Logistics (Oct '13), chap. I.

Sustainment is one of the six joint functions (command and control [C2], intelligence, fires, movement and maneuver, protection, and sustainment) described in Joint Publication (JP) 3-0, Joint Operations. Sustainment provides the joint force commanders (JFCs) freedom of action, endurance, and the ability to extend operational reach. Effective sustainment determines the depth to which the joint force can conduct decisive operations, allowing the JFC to seize, retain, and exploit the initiative. Sustainment is primarily the responsibility of the supported combatant commander (CCDR) and subordinate Service component commanders in close cooperation with the Services, combat support agencies (CSAs), and supporting commands. Sustainment is the provision of logistics and personnel services necessary to maintain and prolong operations until mission accomplishment and redeployment of the force. Joint logistics supports sustained readiness for joint forces.

Core Logistic Capabilities

Core Capabilities	Functional Capabilities
Supply	▪ Manage Supplies and Equipment ▪ Inventory Management ▪ Manage Supplier Networks
Maintenance Operations	▪ Depot Maintenance Operations ▪ Field Maintenance Operations ▪ Manage Life Cycle Systems Readiness
Deployment and Distribution	▪ Move the Force ▪ Sustain the Force ▪ Operate the Joint Deployment and ▪ Distribution Enterprise
Health Service Support	▪ Casualty Management ▪ Patient Movement ▪ Medical Logistics ▪ Preventive Medicine and Health Surveillance ▪ Theater Medical Information
Engineering	▪ Combat Engineering ▪ General Engineering ▪ Geospatial Engineering
Logistic Services	▪ Food Service ▪ Water and Ice Service ▪ Base Camp Services ▪ Hygiene Services
Operational Contract Support	▪ Contract Support Integration ▪ Contract Management

Logistics concerns the integration of strategic, operational, and tactical support efforts within the theater, while scheduling the mobilization and movement of forces and materiel to support the JFC's concept of operations (CONOPS). Joint logistics is the coordinated use, synchronization, and sharing of two or more Military Departments' logistics resources to support the joint force. The joint logistics enterprise (JLEnt) projects and sustains a logistically ready joint force by leveraging Department of Defense (DOD), interagency, nongovernmental agencies, multinational, and industrial resources.

Refer to JFODS4: The Joint Forces Operations & Doctrine SMARTbook (Guide to Joint, Multinational & Interagency Operations) for discussion of joint logistics to include sustainment as a joint function, core logistics capabilities, and planning/controlling/executing joint logistics (from a joint doctrine perspective).

Chap 1

II. Sustainment of Decisive Action

Ref: ADRP 4-0, Sustainment (Jul '12). chap. 3.

Decisive action is the continuous, simultaneous combinations of offensive, defensive, and stability or defense support of civil authorities tasks (ADRP 3-0). In unified land operations, commanders seek to seize, retain, and exploit the initiative while synchronizing their actions to achieve the best effects possible. Sustainment, through mission command, enables decisive action. Sustainment provides the operational commander with operational reach, freedom of action and endurance.

Operational Context

Any operational environment consists of many interrelated variables and sub-variables, as well as the relationships among those variables and sub-variables. How the many entities and conditions behave and interact with each other within an operational environment is difficult to discern and always results in differing circumstances. Different actor or audience types do not interpret a single message in the same way. Therefore, no two operational environments are the same (ADRP 3-0).

Unified Land Operations

Unified land operations require the integration of U.S. military operations with that of multinational partners and other government agencies and nongovernmental organizations. The Army's two core competencies—combined arms maneuver and wide area security— provide the means for balancing the application of the elements of combat power in unified action to defeat enemy ground forces; to seize, occupy, and defend land areas; and to achieve physical, temporal, and psychological advantages over the enemy to seize and exploit the initiative (ADP 3-0).

The sustainment warfighting function is essential for conducting operations and providing resources for generating and maintaining combat power. Sustainment provides the operational commander operational reach, freedom of action, and operational endurance. As mentioned previously, sustainment is inherently joint and requires a coordinated and collaborated effort between joint and multinational partners and other government agencies.

A sustaining operation is an operation at any echelon that enables the decisive operation or shaping operations by generating and maintaining combat power (ADRP 3-0). Sustaining operations are inseparable from decisive and shaping operations, though not decisive in and of itself. When executing sustainment operations, commanders, staffs, and subordinates ensure their decisions and actions comply with U.S., international, and in some cases host nation, laws and regulations. Commanders at all levels ensure their Soldiers operate in accordance with the law of war (see FM 27-10) and the rules of engagement.

Sustainment determines the depth and duration of Army operations. It is essential to retaining and exploiting the initiative and it provides the support necessary to maintain operations until mission accomplishment. Failure to provide sustainment could cause a pause or culmination of an operation resulting in the loss of the initiative.

Sustainment of Decisive Action

Ref: ADP 4-0, Sustainment (Jul '12), pp. 10 to 15.

Sustainment is one of the elements of sustaining operations. Sustaining operations, typically address important sustainment and protection actions essential to the success of decisive and shaping operations. A sustaining operation is an operation at any echelon that enables the decisive operation or shaping operations by generating and maintaining combat power and is inseparable from decisive and shaping operations.

Sustainment enables commanders with operational reach, freedom of action, and endurance. Operational reach is achieved by the ability to open theaters, deploy forces to support the combatant commander's mission. Mission command is the primary means by which sustainment headquarters plan, prepare, execute, and assess the sustainment of operations. An effective distribution system enables prolonged endurance by delivering sustainment in the right quantities to support decisive action.

I. Operational Reach

Operational reach is a necessity for successful operations. Operational reach is the distance and duration across which a unit can successfully employ military capabilities (JP 3-0). The limit of a unit's operational reach is its culminating point. Operational reach is facilitated by prepositioning stocks; capability to project Army forces and sustainment to an operational environment; to open theater ports; establish forward bases; and to close a theater upon conclusion of an operation.

See p. 6-26 for further discussion.

II. Freedom of Action

Freedom of action enables commanders with the will to act, to achieve operational initiative and control and maintain operational tempo. Enabling freedom of action requires that sustainment commanders synchronize the sustainment plan with the operations plan to ensure supported commanders can operate freely and unencumbered by limited resources. Sustainment commanders can enable freedom of action through preparing and putting in place sustainment capabilities.

Negotiating and agreements. Negotiating and establishing agreements with host nation resources is important for establishing freedom of action. Through negotiation and agreements, Army forces can reduce the military sustainment footprint and resources to focus on higher priority operations requiring greater military sustainment involvement. Host nation support agreements may include pre-positioning of supplies and equipment, OCONUS training programs, and humanitarian and civil assistance programs. These agreements are designed to enhance the development and cooperative solidarity of the host nation and provide infrastructure compensation should deployment of forces to the target country be required.

See p. 6-34 for further discussion.

III. Endurance

Endurance refers to the ability to employ combat power anywhere for protracted periods (ADRP 3-0). Endurance stems from the ability to maintain, protect, and sustain forces, regardless of how far away they are deployed, how austere the environment, or how long land power is required.

Distribution. Distribution is key for endurance. Endurance is enabled by an Army distribution system (referred to as theater distribution) that provides forces with a continuous flow of sustainment. The distribution system is a complex of facilities, installations, methods, and procedures designed to receive, store, maintain, distribute, and control the flow of military resources between point of receipt into the military system and point of issue to using activities and units (refer to ATTP 4-0.1). An important aspect of distribution is intransit visibility.

See p. 6-34 for further discussion.

Sustainment is a critical and essential enabler that allows the U.S. forces to deploy long distances (operational reach), conduct operations across the depth and breadth of the operational area (freedom of action), and maintain operations for extended durations (prolong endurance).

Sustainment Preparation

Preparation for the sustainment of operations consists of activities performed by units to improve their ability to execute an operation. Preparation includes but is not limited to plan refinement, rehearsals, information collection, coordination, inspections, and movements. For sustainment to be effective, several actions and activities are performed across the levels of war to properly prepare forces for operations.

See p. 6-35 for further discussion.

Sustainment Execution

Execution is putting a plan into action by applying combat power to accomplish the mission (ADP 5-0). It focuses on actions to seize, retain, and exploit the initiative. Sustainment determines the depth and duration of Army operations. It is essential to retaining and exploiting the initiative and it provides the support necessary to maintain operations until mission accomplishment. Failure to provide sustainment could cause a pause or culmination of an operation resulting in the loss of the initiative. It is essential that sustainment planners and operation planners work closely to synchronize all of the warfighting functions, in particular sustainment, to allow commanders the maximum freedom of action. Sustainment plays a key role in enabling decisive action.

- **Sustaining Offensive Tasks.** An offensive task is a task conducted to defeat and destroy enemy forces and seize terrain, resources, and population centers (ADRP 3-0). Sustainment operations in support of offensive tasks are high in intensity. Commanders and staffs plan for increased requirements and demands, anticipate where the greatest need might occur, and develop a priority of support. Sustainment planners may consider positioning sustainment units in close proximity to operations to reduce response times for critical support.
- **Sustaining Defensive Tasks.** A defensive task is conducted to defeat an enemy attack, gain time, economize forces, and develop conditions favorable for offensive or stability tasks (ADRP 3-0). For sustainment, the movement of materiel and troops within the area of operation has to be closely and continuously coordinated, controlled, and monitored. Distribution managers direct forecasted sustainment to designated units. Army health system support assets should be placed within supporting distance of maneuver forces but not close enough to impede ongoing operations.
- **Sustaining Stability Tasks.** Stability tasks are tasks conducted as part of operations outside the United States in coordination with other instruments of national power to maintain or reestablish a safe and secure environment, provide essential governmental services, emergency infrastructure reconstruction, and humanitarian relief. See Army doctrine on decisive action. Sustainment of stability tasks often involves supporting U.S. and unified action partners in a wide range of missions and tasks. It will almost always require interaction with other governmental agencies and nongovernmental organizations.
- **Sustaining Defense Support of Civil Authorities Tasks.** Defense Support of Civil Authorities is support provided by U.S. Federal military forces, DOD civilians, DOD contract personnel, DOD component assets, and National Guard forces (when the Secretary of Defense, in coordination with the Governors of the affected States, elects and requests to use those forces in Title 32, USC, status) in response to requests for assistance from civil authorities for domestic emergencies, law enforcement support, and other domestic activities, or from qualifying entities for special events.

Sustainment Planning

Sustainment planning begins with the operational commander's intent and concept of operations. This single, unifying idea provides direction for the entire operation. Based on a specific idea of how to accomplish the mission, commanders refine the concept of operations during planning. They adjust it throughout the operation as subordinates develop the situation or conditions change.

Sustainment planning indirectly focuses on a threat's ability to disrupt sustainment operations but more specifically on sustaining friendly forces to the degree that the Army as a whole accomplishes the desired end state. They must track developments and adjust plans as the operations unfold. Sustainment commanders must understand processes and procedures for how sustainment is provided, in relation to the operational environment and the resources available to them. Sustainment commanders build upon their understanding by collecting, processing, storing, displaying, and disseminating information that impacts the operation. As a result, the sustainment estimate and commanders' understanding have to be reviewed and re-evaluated throughout an operation.

Planning begins with analysis of the conditions in the operational environment with emphasis on the enemy and operational variables METT-TC. It involves understanding and framing the problem and envisioning the set of conditions that represent the desired end state (ADRP 3-0).

Planning sustainment support of an operation is vital to mission success. Sustainment commanders and their planning staffs must coordinate and synchronize every stage of the planning process with the operational staff. They must also coordinate, synchronize and integrate the sustainment plan with joint and multinational partners to ensure a continuous linkage with strategic level providers.

The sustainment staff's role in synchronizing sustainment planning with operations is necessary to assist operational commanders and staffs set the conditions for what is in the realm of the possibilities. To ensure maximum freedom of action sustainment planners must understand the commander's intent, be able to visualize the operation and articulate the operational risks. Limitations like, insufficient infrastructure or the availability of a key class of supply or replacement weapon systems has bearing on the commander's ability to execute the mission. Sustainment commanders and staffs must present credible courses of action commensurate with sustainment capabilities to allow as much freedom action as possible to accomplish the operational end state. While sustainment should not be an impediment to an operation, poor planning, lack of coordination, and understanding could severely impact the success of the operation.

Doctrinal Linkage

Sustainment commanders must ensure their staffs are conversant in operational and sustainment doctrine as well as joint and multinational doctrine. Doctrine is a guide and establishes a basic framework for the conduct of operations that facilitates the planning process. Sustainment is a complex operation with many branches and sequels. Doctrine, based on past experiences and knowledge, helps leaders understand the broad context of what is possible, and then allow commanders and staffs to use judgment in the application of doctrine to adjust to the unique circumstances facing them.

To effectively conduct sustainment in unified land operations, sustainment planners must fully understand the operations process—the major mission command activities performed during operations: planning, preparing, executing, and continuously assessing the operation; operational art; and mission command (see ADP 5-0, 3-0, and 6-0, respectively). Additionally, sustainment planners must understand the mechanics of Joint operations and the necessary links that ensure strategic level support (see JP 3.0, JP 4.0, and JP 5.0).

Mission Command of Sustainment Operations

Ref: ADRP 4-0, Sustainment (Jul '12), pp. 3-1 to 3-4.

Mission command is the exercise of authority and direction by the commander using mission orders to enable disciplined initiative within the commander's intent to empower agile and adaptive leaders in the conduct of unified land operations (ADP 6-0). It blends the art of command and the science of control while integrating the war fighting functions to accomplish decisive action. Mission command is a pivotal aspect of providing sustainment to operational forces.

1. Build Cohesive Teams Through Mutual Trust

Sustainment commanders and their staffs must be a part of the overall team and work as an integral team member. Sustainment must be synchronized and planned in conjunction with operational planning.

2. Create Shared Understanding

Understanding is fundamental to mission command. Sustainment commanders must understand the supported commanders' intents and concept of operations. They must understand how and what the supported commanders think.

3. Provide a Clear Commander's Intent

Sustainment commanders and staffs at every level must have a clear understanding of the supported commanders' intent. This is essential to making sure sustainment is synchronized and the appropriate level and type of sustainment is forecasted. Sustainment commanders then provide a clear supporting vision and intent to their subordinates to ensure required sustainment to supported commanders.

4. Exercise Disciplined Initiative

There are four sustainment principles that support the exercise disciplined initiative principle: anticipation, responsiveness, economy, and improvisation. Sustainment commanders cannot wait for a supported commander to request support.

5. Use Mission Orders

Mission orders are the most effective means for conducting mission command of sustainment units. Sustainment commanders must have confidence that subordinate sustainment leaders know and understand the maneuver commander's concept of operations and intent. Often sustainment leaders do not have the luxury of asking for permission.

6. Accept Prudent Risk

The ability of threats to disrupt the flow of sustainment could significantly degrade forces' ability to conduct operations as well as sustain them. Sustainment planners and staffs must consider risk factors when planning sustainment. Sustainment commanders may have to balance risk with survivability in considering redundant capabilities and alternative support plans.

7. Mission Command Headquarters and Staffs for Sustainment

Sustainment commanders facilitate confidence building by establishing and maintaining personal contact with supported commanders. Sustainment staffs must constantly coordinate with supported operational staffs to synchronize and assess support and make adjustments as METT-TC factors warrant.

The sustainment staffs are responsible for providing staff support activities for the commander. The sustainment staff integrator monitors and coordinates sustainment functions between the sustainment staffs and other war fighting function staffs and advises the commander on force readiness.

A firm doctrinal grasp enables sustainment staffs to use and apply the planning tools of the operations process (see ADP 5-0, 3-0 and 6-0, respectively). In addition to a firm foundation in planning, operational, and sustainment doctrine, planners must understand maneuver doctrine in order to arrange sustainment actions in a manner to effectively support the operation.

Planning Considerations

Sustainment staffs create viable plans that are well coordinated and synchronized, facilitate operational tempo and support the commander's priorities before, during, and after operations. Sustainment planners in an operational headquarters generally do not drive the planning process but must be fully integrated throughout the Army design methodology. Sustainment planners use the commander's intent, planning guidance, and the military decision making process to develop the sustainment concept of support.

The concept of support is derived from running estimates developed using a variety of planning tools. These running estimates project consumption rates for key classes of supply, casualty figures, maintenance requirements, and other sustainment requirements (see ADRP 5-0 for additional information). Sustainment planners participate in all aspects of the military decision making process to ensure synchronization and unity of effort.

Planning in a sustainment headquarters requires lead planners to take an active role in the planning process. They assist the development of the commander's understanding of the operational environment, identify the problems, and articulate the sustainment commander's vision. It requires they have regular access to the commander. Sustainment planners must have the most current products from the organizations they support as well as planning products from their higher headquarters to ensure proper nesting and synchronization. Developing effective plans facilitates well synchronized transitions between operational phases.

A comprehensive analysis of host nation capabilities and plans incorporating these resources, provides sustainment commanders with an array of options. For example, the availability of reliable contractible resources could reduce the burden on military resources and an already strained distribution system. Contracted resources could enable military resources to be focused on high priority operations that are unsuitable for civilian personnel. The use of contractors and host nation support are often directly tied to the level of violence and threat in the operational environment.

I. Operational Reach

Operational reach is a necessity in order to conduct decisive action. Operational reach is the distance and duration across which a unit can successfully employ military capabilities (JP 3-0). The limit of a unit's operational reach is its culminating point.

Sustainment enables operational reach. It provides Army forces with the lift, materiel, supplies, health services, and other support necessary to sustain operations for extended periods of time. Army forces require strategic sustainment capabilities and global distribution systems to deploy, maintain, and conduct operations over great distances. Army forces increase the joint force's ability to extend operational reach by securing and operating bases in the AOR. In many instances, land operations combine direct deployment with movements from intermediate staging bases located outside the operational area.

Extending operational reach is a paramount concern for commanders. To achieve the desired end state, forces must possess the necessary operational reach to establish and maintain conditions that define success. Commanders and staffs increase operational reach through deliberate, focused operational design, and the appropriate sustainment to facilitate endurance.

Army Prepositioned Stocks (APS)

Ref: ADRP 4-0, Sustainment (Jul '12), pp. 3-5 to 3-7.

Army Prepositioned Stocks (APS) is essential in facilitating strategic and operational reach. The APS program is a key Army strategic program. The USAMC manages and the ASC executes the APS program and provides accountability, storage, maintenance, and transfer (issue and receipt) of all equipment and stocks (except medical supplies and subsistence items) (ATTP 4-15). Medical APS stocks are managed by U.S. Army Medical Materiel Agency for the Office of the Surgeon General and subsistence items are managed for the Army by DLA.

Prepositioning of stocks in potential theaters provides the capability to rapidly resupply forces until air and sea lines of communication are established. Army pre-positioned stocks are located at or near the point of planned use or at other designated locations. This reduces the initial amount of strategic lift required for power projection, to sustain the war fight until the line of communication with CONUS is established, and industrial base surge capacity is achieved (FM 3-35.1). The four categories of APS are:

Prepositioned Unit Sets

Prepositioned unit sets consist of pre-positioned organizational equipment (end items, supplies, and secondary items) stored in unit configurations to reduce force deployment response time. Materiel is pre-positioned ashore and afloat to meet the Army's global prepositioning strategic requirements of more than one contingency in more than one theater of operations.

Operational Projects Stocks

Operational projects stocks are materiel above normal table of organization and equipment, table of distribution and allowances, and common table of allowance authorizations, tailored to key strategic capabilities essential to the Army's ability to execute force projection. They authorize supplies and equipment above normal modified table of organization and equipment authorizations to support one or more Army operation, plan, or contingency. They are primarily positioned in CONUS, with tailored portions or packages pre-positioned overseas and afloat. The operational projects stocks include aerial delivery, mortuary affairs, and Force Provider base camp modules.

Army War Reserve Sustainment Stocks

Army war reserve sustainment stocks are acquired in peacetime to meet increased wartime requirements. They consist of major and secondary materiel aligned and designated to satisfy wartime sustainment requirements. The major items replace battle losses and the secondary items provide minimum essential supply support to contingency operations. Stocks are pre-positioned in or near a theater of operations to reduce dependence on strategic lift in the initial stages of a contingency. They are intended to last until resupply at wartime rates or emergency rates are established.

War Reserve Stocks For Allies

War reserve stocks for allies is an Office of the Secretary of Defense –directed program that ensures U.S. preparedness to assist designated allies in case of war. The United States owns and finances war reserve stocks for allies and prepositions them in the appropriate theater.

Land-based APS in Korea, Europe, or Southwest Asia allows the early deployment of a BCT to those locations. These pre-positioned sets of equipment are essential to the timely support of the U.S. national military strategy in the areas of U.S. national interest and treaty obligations. Fixed land-based sites store Army pre-positioned sets of BCT equipment, operational projects stocks, and sustainment stocks. Land based sets can support a theater lodgment to allow the off-loading of Army pre-positioned afloat equipment and can be shipped to support any other theater worldwide.

Refer to FM 3-35.1.

Force Projection

Force projection is the ability to project the military instrument of national power from the United States or another theater, in response to requirements for military operations (JP 3-0). Force projection includes the processes of mobilization, deployment, employment, sustainment, and redeployment of forces. These processes are a continuous, overlapping, and repeating sequence of events throughout an operation. Force projection operations are inherently joint and require detailed planning and synchronization.

Sustainment of force projection operations is a complex process involving the GCC, strategic and joint partners such as USTRANSCOM, and transportation component commands like AMC, military sealift command, SDDC, USAMC, DLA, Service Component Commands, and Army generating forces.

- **Mobilization** is the process of bringing the armed forces to a state of readiness in response to a contingency. Upon alert for deployment generating force sustainment organizations, ensure Army forces are manned, equipped, and meet all Soldier readiness criteria.
- **Deployment** is the movement of forces to an operational area in response to an order. Sustainment is crucial to the deployment of forces. Joint transportation assets including air and sealift provide the movement capabilities for the Army.
- **Employment** encompasses a wide array of operations—including, but not limited to—entry operations, decisive action, conduct of operations, and post-conflict operations.
- **Sustainment** provides logistics, personnel services, and health service support to maintain forces until mission completion. It gives Army forces its operational reach, freedom of action and endurance.
- **Redeployment** is the return of forces and materiel to the home or mobilization station or to another theater. It requires retrograde of logistics, personnel services, and health service support and reuniting unit personnel and equipment at their home station.

See p. 3-10 for further discussion of deployment and redeployment operations.

A. Theater Opening

Theater opening (TO) is the ability to establish and operate ports of debarkation (air, sea, and rail) to establish a distribution system and sustainment bases, and to facilitate port throughput for the reception, staging, onward movement and integration of forces within a theater of operations (ADP 4-0). Preparing for TO operations requires unity of effort among the various commands and a seamless strategic-to-tactical interface. It is a complex joint process involving the GCC and strategic and joint partners such as USTRANSCOM and DLA. TO functions set the conditions for effective support and lay the groundwork for subsequent expansion of the theater distribution system.

See following pages (pp. 6-30 to 6-31) for further discussion.

B. Theater Closing

Theater closing is the process of redeploying Army forces and equipment from a theater, the drawdown and removal or disposition of Army non-unit equipment and materiel, and the transition of materiel and facilities back to host nation or civil authorities. Theater closing begins with the termination of joint operations.

See pp. 6-32 to 6-33 for further discussion.

Basing

Ref: ADRP 4-0, Sustainment (Jul '12), pp. 3-9 to 3-10.

Basing directly enables and extends operational reach, and involves the provision of sustainable facilities and protected locations from which units can conduct operations. Army forces typically rely on a mix of bases and/or base camps to deploy and employ combat power to operational depth. Options for basing range from permanent basing in CONUS to permanent or contingency (non-permanent) basing OCONUS. A base camp is an evolving military facility that supports military operations of a deployed unit and provides the necessary support and services for sustained operations.

Bases or base camps may be joint or single service and will routinely support both U.S. and multinational forces, as well as interagency partners, operating anywhere along the range of military operations. Commanders often designate a single commander as the base or base camp commander that is responsible for protection, terrain management, and day-to-day operations of the base or base camp. This allows other units to focus on their primary function. Units located within the base or base camp are under the tactical control of the base or base camp commander for base security and defense.

Within large echelon support areas, controlling commanders may designate base clusters for mutual protection and mission command. Within a support area, a designated unit such as a brigade combat team or maneuver enhancement brigade provides area security, terrain management, movement control, mobility support, clearance of fires, and required tactical combat forces.

1. Intermediate Staging Bases

An intermediate staging base (ISB) is a tailorable, temporary location used for staging forces, sustainment and/or extraction into and out of an operational area (JP 3-35). While not a requirement in all situations, the intermediate staging base may provide a secure, high-throughput facility when circumstances warrant. The commander may use an ISB as a temporary staging area en route to a joint operation, as a long-term secure forward support base, and/or secure staging areas for redeploying units, and noncombatant evacuation operations.

An intermediate staging base is task organized to perform staging, support, and distribution functions as specified or implied by the CCDR and the theater Army operations order. The intermediate staging base task organization is dependent on the operational situation and the factors of METT–TC. It may provide life support to staging forces in transit to operations or serve as a support base supporting the theater distribution plan.

As a support base, an intermediate staging base may serve as a transportation node that allows the switch from strategic to intratheater modes of transportation. Whenever possible an intermediate staging base takes advantage of existing capabilities, serving as a transfer point from commercial carriers to a range of tactical intratheater transport means that may serve smaller, more austere ports. Army forces may use an intermediate staging base in conjunction with other joint force elements to pre-position selected sustainment capabilities.

2. Forward Operating Bases

Forward operating bases extend and maintain the operational reach by providing secure locations from which to conduct and sustain operations. They not only enable extending operations in time and space; they also contribute to the overall endurance of the force. Forward operating bases allow forward deployed forces to reduce operational risk, maintain momentum, and avoid culmination.

Forward operating bases are generally located adjacent to a distribution hub. This facilitates movement into and out of the operational area while providing a secure location through which to distribute personnel, equipment, and supplies.

A. Theater Opening

Ref: ADRP 4-0, Sustainment (Jul '12), pp. 3-7 to 3-10.

Theater opening (TO) is the ability to establish and operate ports of debarkation (air, sea, and rail) to establish a distribution system and sustainment bases, and to facilitate port throughput for the reception, staging, onward movement and integration of forces within a theater of operations (ADP 4-0). Preparing for TO operations requires unity of effort among the various commands and a seamless strategic-to-tactical interface. It is a complex joint process involving the GCC and strategic and joint partners such as US-TRANSCOM and DLA. TO functions set the conditions for effective support and lay the groundwork for subsequent expansion of the theater distribution system.

When given the mission to conduct TO, a sustainment brigade, designated a sustainment brigade (TO), and a mix of functional battalions and multi-functional CSSBs are assigned based on mission requirements. The sustainment brigade HQ staff may be augmented with a Transportation Theater Opening Element to assist in managing the TO mission. The augmentation element provides the sustainment brigade with additional manpower and expertise to command and control TO functions, to conduct transportation planning, and provide additional staff management capability for oversight of RSOI operations, port operations, node and mode management, intermodal operations, and movement control. The sustainment brigade will participate in assessing and acquiring available HN infrastructure capabilities and contracted support and coordinating with military engineers for general engineering support (FMI 4-93.2 and ATTP 4-0.1)

Port Opening

Port opening is a subordinate function of theater opening. Port opening is the ability to establish, initially operate and facilitate throughput for ports of debarkation (POD) to support unified land operations. The port opening process is complete when the POD and supporting infrastructure is established to meet the desired operating capacity for that node. Supporting infrastructure can include the transportation needed to support port clearance of cargo and personnel, holding areas for all classes of supply, and the proper in-transit visibility systems established to facilitate force tracking and end to end distribution.

Port opening and port operations are critical components for preparing TO. Commanders and staffs coordinate with the HN to ensure sea ports and aerial ports possess sufficient capabilities to support arriving vessels and aircraft. USTRANSCOM is the port manager for deploying U.S. forces (ATTP 4-0.1).

Joint Task Force Port Opening

The Joint Task Force Port Opening (JTF-PO) is a joint capability designed to rapidly deploy and initially operate aerial and sea ports of debarkation, establish a distribution node, and facilitate port throughput within a theater of operations (JP 4-0). The JTF-PO is a standing task force that is a jointly trained, ready set of forces constituted as a joint task force at the time of need. The Army contribution to the JTF-PO is the Rapid Port Opening Element (RPOE) which deploys within hours to establish air and sea ports of debarkation in contingency response operations. The RPOE also provides in-transit visibility and cargo clearance.

The Joint Task Force Port Opening facilitates joint RSOI and theater distribution by providing an effective interface with the theater JDDOC and the sustainment brigade for initial aerial port of debarkation (APOD) operations. The JTF-PO is designed to deploy and operate for up to 60 days. As follow-on theater logistics capabilities arrive, the JTF-PO will begin the process of transferring mission responsibilities to arriving sustainment brigade forces or contracted capabilities to ensure the seamless continuation of airfield and distribution operations.

Seaports

Surface Deployment and Distribution Command is the single port manager (SPM) for all common user seaports of debarkation (SPOD) and as the SPM it develops policy and advises the GCC on port management, recommends ports to meet operational demands, and is primarily responsible for the planning, organizing, and directing the operations at the seaport. The TSC and its subordinate Sustainment Brigades, Terminal Battalions and Seaport Operating Companies perform the port operator functions at SPODs. These functions can include port preparations and improvement, cargo discharge and upload operations, harbor craft services, port clearance and cargo documentation activities. If the operational environment allows, SDDC may have the ability to contract locally for port operator support eliminating or decreasing the requirement for the TSC and its subordinate units.

The single port manager may have OPCON of a port support activity which is an ad hoc organization consisting of military and/or contracted personnel with specific skills to add in port operations. The TSC and SDDC will coordinate the PSA requirement. The PSA assists in moving unit equipment from the piers to the staging/marshaling/loading areas, assisting the aviation support element with movement of helicopters in preparation for flight from the port, providing limited maintenance support for equipment being offloaded from vessels, limited medical support, logistics support, and security for port operations.

Ideally, the SPOD will include berths capable of discharging large medium speed roll-on/roll-off ships. The SPOD can be a fixed facility capable of discharging a variety of vessels, an austere port requiring ships to be equipped with the capability to conduct their own offloading, or beaches requiring the conducting of JLOTS operations. Whatever the type of SPOD, it should be capable of accommodating a HBCT.

The Theater Gateway Personnel Accounting Team and supporting HR company and platoons will normally operate at the SPOD as well as movement control teams to facilitate port clearance of personnel and equipment. The movement control team that has responsibility for the SPOD, coordinates personnel accounting with the supporting CSSB or sustainment brigade for executing life support functions (billeting, feeding, transportation, and so forth) for personnel who are transiting into or out of the theater.

Aerial Ports

Airfields supporting strategic air movements for deployment, redeployment, and sustainment are designated aerial ports. Aerial ports are further designated as either an aerial port of embarkation (APOE) for departing forces and sustainment, or as an aerial port of debarkation (APOD) for arriving forces and sustainment. Reception at the APOD is coordinated by the senior logistics commander and executed by an Air Force contingency response group/element and an arrival/departure airfield control group (A/DACG). The A/DACG is an ad hoc organization established to control and support the arrival and departure of personnel, equipment, and sustainment cargo at airfields and must be a lead element when opening an APOD. Elements of a movement control team and an inland cargo transfer company typically operate the A/DACG however the mission can be performed by any unit with properly trained personnel and the appropriate equipment.

USTRANSCOM's Air Mobility Command (AMC) is the single port manager for all common user APODs. Ideally, the APOD will provide runways of varying capacity, cargo handling equipment, adequate staging areas, multiple links to the road and rail network, and a qualified work force. The single port manager works with the service provided A/ADACG of offload aircraft and assists in moving unit equipment to the staging/marshaling/loading areas. The A/ADCG also assists the aviation support element with movement of helicopters in preparation for flight from the APOD.

Basing

A base camp is an evolving military facility that supports military operations of a deployed unit and provides the necessary support and services for sustained operations.

See p. 6-29 for further discussion.

B. Theater Closing

Ref: ADRP 4-0, Sustainment (Jul '12), pp. 3-10 to 3-12.

Theater closing is the process of redeploying Army forces and equipment from a theater, the drawdown and removal or disposition of Army non-unit equipment and materiel, and the transition of materiel and facilities back to host nation or civil authorities. Theater closing begins with the termination of joint operations.

Terminating Joint Operations

Terminating joint operations is an aspect of the CCDR's functional or theater strategy that links to achievement of national strategic objectives (JP 5-0). Based on the President's strategic objectives that compose a desired national strategic end state, the supported CCDR can develop and propose termination criteria. The termination criteria describe the standards that must be met before conclusion of a joint operation. These criteria help define the desired military end state, which normally represents a period in time or set of conditions beyond which the President does not require the military instrument of national power as the primary means to achieve remaining national objectives. Termination criteria should account for a wide variety of operational tasks that the joint force may need to accomplish, to include disengagement, force protection (including force health protection support to conduct retrograde cargo inspections and pest management operations), transition to post-conflict operations, reconstitution, and redeployment. While there may be numerous terminating tasks the Army must achieve, the discussion below is deliberately broad and not all inclusive. The discussion focuses on redeployment, drawdown of non-unit materiel, and transitioning of materiel, facilities and capabilities to host nation or civil authorities.

Planning for the transition from sustained combat operations to the termination of joint operations, and then a complete handover to civil authority, must commence during plan development and be ongoing during all phases of a campaign or major operation. Planning for redeployment should be considered early and continued throughout the operation and is best accomplished in the same time-phased process in which deployment was accomplished.

Redeployment

Redeployment involves the return of personnel, equipment, and materiel to home and/or demobilization stations and is considered as an operational movement critical in reestablishing force readiness (FM 3-35). Under the ARFORGEN model deployment and redeployment of forces in support of extended operations is a cyclic process. However, for terminating joint operations, Army forces may be completely redeployed from the joint operational area. Many of the procedures used to in the initial deployment of forces to theater apply during redeployment. Unlike cyclic deployment where units fall in on positioned unit equipment and sets, termination redeployment efforts require movement of unit sets to APOEs/SPOEs for shipment to home station or other designated locations. After completion of military operations, redeploying forces move to designated assembly areas or directly to redeployment assembly areas. The same elements that operate and manage the theater distribution system during deployment and sustainment will usually perform support roles during redeployment.

Two critical aspects of equipment and materiel redeployment are property book accountability and asset visibility. Furthermore, the identification of how much equipment is on the ground, location of the equipment, type of equipment, condition of the equipment, and reporting procedures will allow for timely planning, as it will impact mode of transportation, resources, timeline, personnel, storage capabilities, and the like. Moreover, the accurate reporting of equipment on property books, locations, and condition will influence strategic-level decision making in terms of funding, field or sustainment reset, and disposal of equipment. *See also p. 3-10.*

Drawdown

Planning for drawdown of non-unit equipment and materiel should occur early in the operational and strategic planning process. Drawdown planning entails more than returning equipment to CONUS. At the strategic level, the requirement for specific types of equipment may necessitate the redistribution of equipment to another AOR.

Even though equipment drawdown is an important mission in the redeployment operation, it may not be the Army's or the GCC main priority; thus, prioritization of equipment redistribution/disposition must be established early on to maximize distribution capacity and velocity. A challenge is visibility of strategic level materiel requirements synthesized into the already established priority timeline. Overcoming this challenge is through strategic-level collaboration between partners including Service Headquarters, GCCs, USAMC, DLA, and USTRANSCOM to effectively and efficiently strategically reset both Joint and Army forces.

As planners begin the process of reducing forces in a theater of operations, they must develop a balance between operational capability and sustainment capability. There is a natural tendency to eliminate the sustainment and enabler forces first because they do not provide an inherent capability to engage with the population or enemy. However, as the sustainment and enabling force are withdrawn, there is a direct impact on the operational forces in the form of reduced operational reach and requirements for assumption of additional missions.

To provide unity of effort and ensure operational freedom of action through rapid return, repair, redistribution, and combat power regeneration for the Army, a USAMC Responsible Reset Task Force provides a comprehensive solution for drawdown. Reset is a coordinated effort to methodically plan and execute the timely, repair, redistribution, and/or disposal of non-unit equipment, non-consumable and materiel identified as excess to theater requirements, to home station, sources of repair, or storage or disposal facilities. Through the phased redeployment of forces, the Responsible Reset Task Force mission will reset the Army in the shortest time possible.

The TSC/ESC work closely with the DLA in the close out of materiel in the theater. Support Team serves as the single point of contact to the TSC/ESC. The DLA Support Teams are tasked to provide support to the theater closure plan and are focused on providing support to echelons at the theater level and below based on the priorities of effort. During theater closure, the DLA provides support in the form of adjusting the flow of CL I, II, III (B) (P), IV, VIII and IX to ensure support to the war fighter.

Closing Operational Contracts

The supporting contracting organization will be required to terminate and close out existing contracts and orders. Ratifications and claims must be processed to completion. Contracting for life support services and retrograde support may continue until the last element departs, but standards of support should be reduced as much as possible prior to final contract closeout. In some operations, the supporting contracting organization may be required to assist in the transition of contracted support (the contracts themselves are not transferable) to the Department of State, a multi-national partner or to the host nation. This transition of contract support may include limited continuation of existing contracts in support of high priority Department of State operations.

Port Closing

USTRANSCOM, through SDDC is responsible for providing and managing strategic common-user sealift, and terminal services in support GCC's drawdown or termination requirements. As the single port manager, it is Sod's responsibility to integrate and synchronize strategic and theater re-deployment execution and distribution operations within each CCDR's area of responsibility. It ensures termination/termination requirements are met through the use of both military and commercial transportation assets based on the supported commander business rules and JDDE best business practices.

II. Freedom of Action

Freedom of action enables commanders with the will to act, to achieve operational initiative and control, and maintain operational tempo. Enabling freedom of action requires that the sustainment commanders synchronize the sustainment plan with the operation plan to ensure supported commanders can operate freely and unencumbered due to limited resources. Sustainment commanders can enable freedom of action through preparing and putting in place sustainment activities.

A. Sustainment Preparation

Preparation for the sustainment of operations consists of activities performed by units to improve their ability to execute an operation. Preparation includes but is not limited to plan refinement, rehearsals, information collection, coordination, inspections, and movements. For sustainment to be effective, several actions and activities are performed across the levels of war to properly prepare forces for operations.

See facing page for further discussion to include negotiations and agreements, and sustainment preparation of the battlefield.

B. Sustainment Execution

Execution is putting a plan into action by applying combat power to accomplish the mission (ADP 5-0). It focuses on actions to seize, retain, and exploit the initiative.

Sustainment determines the depth and duration of Army operations. It is essential to retaining and exploiting the initiative and it provides the support necessary to maintain operations until mission accomplishment. Failure to provide sustainment could cause a pause or culmination of an operation resulting in the loss of the initiative. It is essential that sustainment planners and operation planners work closely to synchronize all of the war fighting functions, in particular sustainment, to allow commanders the maximum freedom of action.

Sustainment plays a key role in enabling decisive action. For example, general engineering support provides construction support to protect key assets such as personnel, infrastructure, and bases. Horizontal and vertical construction enables assured mobility of transportation networks and survivability operations to alter or improve cover and concealment to ensure freedom of action, extend operational reach, and endurance of the force. Legal personnel supporting rule of law activities may find themselves working closely with host nation judicial, law enforcement, and corrections systems personnel.

See page 6-23 for an overview of how sustainment supports decisive action tasks (offense, defense, stability and defense support to civil authorities)

III. Endurance

Endurance refers to the ability to employ combat power anywhere for protracted periods. Endurance stems from the ability to create, protect, and sustain a force, regardless of the distance from its base and the austerity of the environment (ADRP 3-0). Endurance involves anticipating requirements and continuity of integrated networks of interdependent sustainment organizations. Prolonged endurance is enabled by an effective distribution system and the ability to track sustainment from strategic to tactical level.

The sustainment principle continuity is paramount for ensuring endurance. Sustainment commanders must ensure the continuous link between strategic to tactical levels are maintained and free flowing. Commanders must be able to track sustainment in near real time and quickly make decisions resulting from changes to missions or operations.

See following page (p. 6-36) for an overview of distribution (primary means of prolonging endurance).

Sustainment Preparation

Ref: ADRP 4-0, Sustainment (Jul '12), pp. 3-12 to 3-13.

Preparation for the sustainment of operations consists of activities performed by units to improve their ability to execute an operation. Preparation includes but is not limited to plan refinement, rehearsals, information collection, coordination, inspections, and movements. For sustainment to be effective, several actions and activities are performed across the levels of war to properly prepare forces for operations.

Negotiations and Agreements

Negotiating and establishing agreements with host nation resources is an important task of mission command for sustainment operations. Through negotiation and agreements, Army forces are able to reduce its military sustainment footprint and/or all military sustainment resources to focus on higher priority operations that may not be conducive to civilian support functions.

Negotiation of agreements enables access to HNS resources identified in the requirements determination phase of planning. This negotiation process may facilitate force tailoring by identifying available resources (such as infrastructure, transportation, warehousing, and other requirements) which if not available would require deploying additional sustainment assets to support.

Host Nation Support agreements may include pre-positioning of supplies and equipment, OCONUS training programs, and humanitarian and civil assistance programs. These agreements are designed to enhance the development and cooperative solidarity of the host nation and provide infrastructure compensation should deployment of forces to the target country be required. The pre-arrangement of these agreements reduces planning times in relation to contingency plans and operations.

Sustainment Preparation of the Operational Environment

Sustainment preparation of the operational environment is the analysis to determine infrastructure, physical environment, and resources in the operational environment that will optimize or adversely impact friendly forces means for supporting and sustaining the commander's operations plan (ADP 4-0). The sustainment preparation of the operational environment assists planning staffs to refine the sustainment estimate and concept of support. It identifies friendly resources (HNS, contractible, or accessible assets) or environmental factors (endemic diseases, climate) that impact sustainment. Some of the factors considered (not all inclusive) are as follows:

- **Geography.** Information on climate, terrain, and endemic diseases in the AO to determine when and what types of equipment are needed. For example, water information determines the need for such things as early deployment of well-drilling assets and water production and distribution units.
- **Supplies and Services.** Information on the availability of supplies and services readily available in the AO. Supplies (such as subsistence items, bulk petroleum, and barrier materials) are the most common. Common services consist of bath and laundry, sanitation services, and water purification.
- **Facilities.** Information on the availability of warehousing, cold-storage facilities, production and manufacturing plants, reservoirs, administrative facilities, hospitals, sanitation capabilities, and hotels.
- **Transportation**. Information on road and rail networks, inland waterways, airfields, truck availability, bridges, ports, cargo handlers, petroleum pipelines, materials handling equipment, traffic flow, choke points, and control problems.
- **Maintenance**. Availability of host nation maintenance capabilities.
- **General Skills**. Information on the general skills such as translators and skilled and unskilled laborers.

Distribution

Ref: ADRP 4-0, Sustainment (Jul '12), pp. 3-16 to 3-18.

Distribution is the primary means to prolong endurance. Distribution is the operational process of synchronizing all elements of the logistic system to deliver the "right things" to the "right place" at the "right time" to support the geographic combatant commander. Additionally, it is also the process of assigning military personnel to activities, units, or billets (JP 4-0).

The distribution system consists of a complex of facilities, installations, methods, and procedures designed to receive, store, maintain, distribute, manage, and control the flow of military materiel between point of receipt into the military system and point of issue to using activities and units.

The Joint segment of the distribution system is referred to as global distribution. It is defined as the process that synchronizes and integrates the fulfillment of joint requirements with the employment of joint forces (JP 4-09). It provides national resources (personnel and materiel) to support the execution of joint operations.

The Army segment of the distribution system is referred to as theater distribution. Theater distribution is the flow of equipment, personnel, and materiel within theater to meet the CCDR's mission. The theater segment extends from the ports of debarkation or source of supply (in theater) to the points of need (Soldier). It is enabled by a distribution management system synchronizes and coordinates a complex of networks (physical, communications, information, and resources) and the sustainment war fighting function to achieve responsive support to operational requirements. Distribution management includes the management of transportation and movement control, warehousing, inventory control, order administration, site and location analysis, packaging, data processing, accountability for equipment (materiel management), people, and communications. See ATTP 4-0.1, Army Theater Distribution for details.

The distribution management of medical materiel is accomplished by a support team from the Medical Logistics Management Center (MLMC). The MLMC support team collocates with the DMC of the TSC/ESC to provide the MEDCOM (DS) with visibility and control of all Class VIII.

In-Transit Visibility

In-transit visibility is the ability to track the identity, status, and location of DOD units, and non-unit cargo (excluding bulk petroleum, oils, and lubricants) and passengers; patients and personal property from origin to consignee, or destination across the range of military operations (JP 3-35). This includes force tracking and visibility of convoys, containers/pallets, transportation assets, other cargo, and distribution resources within the activities of a distribution node.

Retrograde of Materiel

Another aspect of distribution is retrograde of materiel. Retrograde of materiel is the return of materiel from the owning/using unit back through the distribution system to the source of supply, directed ship-to location, and/or point of disposal (ATTP 4-0.1). Retrograde includes turn-in/classification, preparation, packing, transporting, and shipping. To ensure these functions are properly executed, commanders must enforce supply accountability and discipline and utilize the proper packing materials. Retrograde of materiel can take place as part of theater distribution operations and as part of redeployment operations. Retrograde of materiel must be continuous and not be allowed to build up at supply points/nodes.

Early retrograde planning is essential and necessary to preclude the loss of materiel assets, minimize environmental impact, and maximize use of transportation capabilities. Planners must consider environmental issues when retrograding hazardous materiel.

(III. Elements of Sustainment) A. Logistics

Ref: ADRP 4-0, Sustainment (Jul '12). pp. 4-1 to 4-6.

Logistics involves both military art and science. Knowing when and how to accept risk, prioritizing a myriad of requirements, and balancing limited resources all require military art while understanding equipment capabilities incorporates military science. Logistics integrates strategic, operational, and tactical support of deployed forces while scheduling the mobilization and deployment of additional forces and materiel. Logistics include; maintenance, transportation, supply, field services, distribution, operational contract support, and general engineering support. Distribution was previously discussed elsewhere.

A. Maintenance

Maintenance is all actions taken to retain materiel in a serviceable condition or to restore it to serviceability. The Army's two levels of maintenance are field maintenance and sustainment maintenance (see ATTP 4-33). Maintenance is necessary for endurance and performed at the tactical through strategic levels of war.

1. Field Maintenance

Field maintenance is repair and return to user and is generally characterized by on-(or near) system maintenance, often utilizing line replaceable unit, component replacement, battle damage assessment, repair, and recovery (see ATTP 4-33). It is focused on returning a system to an operational status. Field level maintenance is not limited to remove and replace, but also provides adjustment, alignment, and fault/failure diagnoses. Field maintenance also includes battlefield damage and repair tasks performed by either the crew or support personnel to maintain system in an operational state.

2. Sustainment Maintenance

Sustainment maintenance is generally characterized as "off system" and "repair rear" (see ATTP 4-33). The intent is to perform commodity-oriented repairs on all supported items to one standard that provides a consistent and measurable level of reliability. Off-system maintenance consists of overhaul and remanufacturing activities designed to return components, modules, assemblies, and end items to the supply system or to units, resulting in extended or improved operational life expectancies.

B. Transportation Operations

Army transportation units play a key role in facilitating endurance. Transportation units move sustainment from ports to points of need and retrograde materiel as required. Transportation operations encompass the wide range of capabilities needed to allow joint and Army commanders to conduct operations. Important transportation functions are movement control, intermodal operations (terminal and mode), and container management.

1. Movement Control

Movement control is the dual process of committing allocated transportation assets and regulating movements according to command priorities to synchronize distribution flow over lines of communications to sustain land forces. Movement control balances requirements against capabilities and requires continuous synchronization to integrate military, host nation, and commercial movements by all modes of trans-

Principles of Sustainment (and Logistics)

Ref: ADRP 4-0, Sustainment (Jul '12), pp. 1-2 to 1-4.

The principles of sustainment are essential to maintaining combat power, enabling strategic and operational reach, and providing Army forces with endurance. While these principles are independent, they are also interrelated. The principles of sustainment and the principles of logistics are the same.

Priniciples of Sustainment (and Logistics)

1. **Integration**
2. **Anticipation**
3. **Responsiveness**
4. **Simplicity**
5. **Economy**
6. **Survivability**
7. **Continuity**
8. **Improvisation**

1. Integration

Integration is combining all of the sustainment elements within operations assuring unity of command and effort. It requires deliberate coordination and synchronization of sustainment with operations across all levels of war. Army forces integrate sustainment with joint and multinational operations to maximize the complementary and reinforcing effects of each Service component's and national resources. One of the primary functions of the sustainment staff is to ensure the integration of sustainment with operations plans.

2. Anticipation

Anticipation is the ability to foresee operational requirements and initiate necessary actions that most appropriately satisfy a response without waiting for operations orders or fragmentary orders. It is shaped by professional judgment resulting from experience, knowledge, education, intelligence, and intuition. Commanders and staffs must understand and visualize future operations and identify appropriate required support. They must then start the process of acquiring the resources and capabilities that best support the operation. Anticipation is facilitated by automation systems that provide the common operational picture upon which judgments and decisions are based. Anticipation is also a principle of personnel services.

3. Responsiveness

Responsiveness is the ability to react to changing requirements and respond to meet the needs to maintain support. It is providing the right support in the right place at the right time. It includes the ability to anticipate operational requirements. Responsiveness involves identifying, accumulating, and maintaining sufficient resources, capabilities, and information necessary to meet rapidly changing requirements. Through responsive sustainment, commanders maintain operational focus and pressure, set the tempo of friendly operations to prevent exhaustion, replace ineffective units, and extend operational reach.

4. Simplicity

Simplicity relates to processes and procedures to minimize the complexity of sustainment. Unnecessary complexity of processes and procedures leads to the confusion. Clarity of tasks, standardized and interoperable procedures, and clearly defined command relationships contribute to simplicity. Simplicity enables economy and efficiency in the use of resources, while ensuring effective support of forces. Simplicity is also a principle of financial management (see FM 1-06).

5. Economy

Economy is providing sustainment resources in an efficient manner that enables the commander to employ all assets to the greatest effect possible. Economy is achieved through efficient management, discipline, prioritization, and allocation of resources. Economy is further achieved by eliminating redundancies and capitalizing on joint interdependencies. Disciplined sustainment assures greatest possible tactical endurance and constitutes an advantage to commanders. Economy may be achieved by contracting for support or using host nation resources that reduce or eliminate the use of limited military resources.

6. Survivability

Survivability is all aspects of protecting personnel, weapons, and supplies while simultaneously deceiving the enemy (JP 3-34). Survivability consists of a quality or capability of military forces to avoid or withstand hostile actions or environmental conditions while retaining the ability to fulfill their primary mission. This quality or capability of military forces is closely related to protection (the preservation of a military force's effectiveness) and to the protection/force protection warfighting function (the tasks or systems that preserve the force). Hostile actions and environmental conditions can disrupt the flow of sustainment and significantly degrade forces' ability to conduct and sustain operations. In mitigating risks to sustainment, commanders often must rely on the use of redundant sustainment capabilities and alternative support plans.

7. Continuity

Continuity is the uninterrupted provision of sustainment across all levels of war. Continuity is achieved through a system of integrated and focused networks linking sustainment to operations. Continuity is achieved through joint interdependence; linked sustainment organizations; a strategic to tactical level distribution system, and integrated information systems. Continuity assures confidence in sustainment allowing commanders freedom of action, operational reach, and endurance.

8. Improvisation

Improvisation is the ability to adapt sustainment operations to unexpected situations or circumstances affecting a mission. It includes creating, inventing, arranging, or fabricating resources to meet requirements. It may also involve changing or creating methods that adapt to a changing operational environment. Sustainment leaders must apply operational art to visualize complex operations and understand additional possibilities. These skills enable commanders to improvise operational and tactical actions when enemy actions or unexpected events disrupt sustainment operations. In regards to financial management, it includes task organizing units in non-traditional formations, submitting fiscal legislative proposals to acquire new fiscal authorities, applying existing financial and communication technologies (FM 1-06).

portation to ensure seamless transitions from the strategic through the tactical level of an operation. It is a means of providing commanders with situational awareness to control movements in their operational area. Movement control responsibilities are imbedded in an infrastructure that relies on coordination for the planning and execution to ensure transportation assets are utilized efficiently while ensuring LOCs are deconflicted to support freedom of access for military operations.

2. Intermodal Operations

Intermodal operations is the process of using multiple modes (air, sea, highway, rail) and conveyances (i.e. truck, barge, containers, pallets) to move troops, supplies and equipment through expeditionary entry points and the network of specialized transportation nodes to sustain land forces. It uses movement control to balance requirements against capabilities against capacities to synchronize terminal and mode operations ensuring an uninterrupted flow through the transportation system. It consists of facilities, transportation assets and material handling equipment required to support the deployment and distribution enterprise.

a. Terminal Operations

Terminal operations consist of the receiving, processing, and staging of passengers; the receipt, transit storage and marshalling of cargo; the loading and unloading of transport conveyances; and the manifesting and forwarding of cargo and passengers to a destination (JP 4-01.5). Terminal operations are a key element in supporting operational reach and endurance. They are essential in supporting deployment, redeployment and sustainment operations. There are three types of terminals: air, water, and land.

b. Mode Operations

Mode operations are the execution of movements using various conveyances (truck, lighterage, railcar, aircraft) to transport cargo. It includes the administrative, maintenance, and security tasks associated with the operation of the conveyances.

3. Container Management

Container management is the process of establishing and maintaining visibility and accountability of all cargo containers moving within the Defense Transportation System. In theater, container management is conducted by commanders at the operational and tactical levels.

The TSC distribution management center coordinates intermodal operations with the movement control battalion at transportation, storage, and distribution nodes. The TSC maintains information on the location and status of containers and flat racks in the theater. The movement control battalion provides essential information on container location, use, flow and condition. They assist with control of containers by identifying that they are ready for return to the distribution system. The distribution management center sets priorities for container shipment and diversion.

C. Supply

Supply is essential for enhancing Soldiers' quality of life. Supply provides the materiel required to accomplish the mission. Supply includes the following classes.

See facing page for a listing and further discussion of the classes of supply.

Classes of Supply

Ref: ADRP 1-02, Operational Terms and Military Symbols (Feb '15).

The Army divides supply into ten classes for administrative and management purposes.

Class	Symbol	Description
Class I		Subsistence, including health and welfare items.
Class II		Clothing, individual equipment, tentage, organizational tool sets and kits, hand tools, administrative and housekeeping supplies and equipment (including maps).
Class III		POL, petroleum and solid fuels, including bulk and packaged fuels, lubricating oils and lubricants, petroleum specialty products; solid fuels, coal, and related products.
Class IV		Construction materials, to include installed equipment and all fortification/barrier materials.
Class V		Ammunition of all types (including chemical, radiological, and special weapons), bombs, explosives, mines, fuses, detonators, pyrotechnics, missiles, rockets, propellants, associated items.
Class VI		Personal demand items (nonmilitary sales items).
Class VII		Major items: A final combination of end products which is ready for its intended use.
Class VIII		Medical material, including medical peculiar repair parts.
Class IX		Repair parts and components, including kits, assemblies and subassemblies, reparable and non repairable, required for maintenance support of all equipment.
Class X	CA	Material to support nonmilitary programs; such as, agricultural and economic development, not included in Class I through Class IX.

D. Field Services

Field services maintain combat strength of the force by providing for its basic needs and promoting its health, welfare, morale, and endurance. Field services provide life support functions.

1. Shower and Laundry

Shower and laundry capabilities provide Soldiers a minimum of one weekly shower and up to 15 pounds of laundered clothing each week (comprising two uniform sets, undergarments, socks, and two towels). The shower and laundry function does not include laundry decontamination support.

2. Field Feeding

Food preparation is a basic unit function and one of the most important factors in Soldiers' health, morale, and welfare. The standard is to provide Soldiers at all echelons three quality meals per day (AR 30-22). Proper refuse and waste disposal is important to avoid unit signature trails and maintain field sanitation standards.

3. Water Production and Distribution

Water production and distribution are essential for hydration, sanitation, food preparation, medical treatment, hygiene, construction, and decontamination. The water production is both a field service and a supply function. Quartermaster supply units normally perform purification in conjunction with storage and distribution of potable water.

4. Clothing and Light Textile Repair

Clothing and light textile repair is essential for hygiene, discipline, and morale purposes. Clean, serviceable clothing is provided as far forward as the brigade area.

5. Aerial Delivery

Aerial delivery includes parachute packing, air item maintenance, and rigging of supplies and equipment. This function supports airborne insertions, airdrop and air-land resupply. It is a vital link in the distribution system and provides the capability of supplying the force even when land LOCs have been disrupted or terrain is too hostile, thus adding flexibility to the distribution system. See FM 4-20.41 for details.

6. Mortuary Affairs

Mortuary affairs is a broadly based military program to provide for the necessary care and disposition of deceased personnel. The Army is designated as the Executive Agent for the Joint Mortuary Affairs Program (JP 4-06, Mortuary Affairs).

E. Operational Contract Support

Operational contract support is the integration of commercial sector support into military operations. Operational contract support consists of two complementary functions: contract support integration and contractor management. Operational contract support has three types of contract support: theater support, external support, and systems support.

Refer to ATTP 4-10 for full discussion on operational contract support.

Contract Support Integration

Contract support integration is the process of synchronizing operational planning, requirements development and contracting in support of deployed military forces and other designated organizations in the area of operations (ATTP 4-10). The desired end state of contract support integration actions include:

- Increased effectiveness, efficiencies, and cost savings of the contracting effort
- Increased visibility and control of contracting functions

Types of Operational Contract Support

Ref: ADRP 4-0, Sustainment (Jul '12), p..4-6

There are three types of operational contract support: theater support contracts, external support contracts, and system support contract.

Theater Support Contracts

Theater support contracts are a type of contingency contract awarded by contracting officers deployed to the AO serving under the direct contracting authority of the designated head of contracting activity for that particular contingency operation. These contracts, often executed under expedited contracting authority (e.g. reduced time frames for posting of contract solicitations; allowing for simplified acquisition procedures for higher dollar contracts, etc.), provide goods, services, and minor construction from commercial sources, normally within the AO. Also important from a contractor management perspective are local national employees that often make up the bulk of the theater support contract workforce.

External Support Contracts

External support contracts are awarded by contracting organizations whose contracting authority does not derive directly from the theater support contracting head(s) of contracting activity or from systems support contracting authorities. External support contracts provide a variety of logistics and other non combat related services and supply support. External support contracts normally include a mix of U. S. citizens, host nation, and local national contractor employees. Examples of external support contracts are:

- Service (Air Force, Army and Navy) civil augmentation programs
- Special skills contract (e.g. staff augmentation, linguists, etc.)
- Defense Logistics Agency prime vendor contract

The largest and most commonly known external support contract is the Army's LOGCAP. LOGCAP can provide a complete range of logistics services, including supply services (e.g. storage, warehousing, distribution, etc.) for the 9 classes of supplies, but the Services source the actual commodities. LOGCAP does not provide personal services type contracts.

System Support Contracts

System support contracts are prearranged contracts awarded by and funded by acquisition PEOs and project/product management officers. These contracts provide technical support, maintenance support and, in some cases, Class IX support for a variety of Army weapon and support systems. Systems support contracts are routinely put in place to provide support to newly fielded weapon systems, including aircraft, land combat vehicles and automated command and control information systems. Systems support contracting authority and contract management resides with the Army Contracting Command, while program management authority and responsibility for requirements development and validation resides with the system materiel acquisition program executive officers and project/product management offices. The AFSB assists in systems support integration. Systems support contractor employees, made up mostly of U.S. citizens, provide support both in garrison and in contingency operations. Operational commanders generally have less influence on the execution of systems support contracts than other types of contracted support.

For more information on operational contract support refer to ATTP 4-10 and JP 4-10.

- Minimized competition for scarce commercial resources
- Increased ability for the Army force commander to enforce priorities of support
- Decreased and/or mitigated contract fraud
- Limiting sole source (vice competitively awarded) and cost-plus contracts (vice fixed price) as much as practical
- Enhanced command operational flexibility through alternative sources of support

Contractor Management

Contractor management is the process of managing and integrating contractor personnel and their equipment into military operations (ATTP 4-10). Contractor management includes planning and deployment/redeployment preparation; in-theater management; force protection and security; and executing government support requirements. Integrating the two related operational contract support functions is a complex and challenging process. Multiple organizations are involved in this process including commanders, their primary/special staffs (at the ASCC down to, and including, battalion levels) and the supporting contracting organizations.

F. General Engineering Support

The Army has a broad range of diverse engineer capabilities, which commanders can use to perform various tasks for various purposes. One such purpose is to provide support that helps ground force commanders enable logistics. To accomplish this purpose, engineers combine and apply capabilities from all three engineer disciplines (combat, general, and geospatial engineering) to establish and maintain the infrastructure necessary for sustaining military operations in the AO. This involves primarily general engineering tasks that consist largely of building, repairing, and maintaining roads, bridges, airfields, and other structures and facilities needed for APODS, SPODS, main supply routes, and base camps. Depending on the range of military operations, other tasks include the planning, acquisition, management, remediation and disposition of real estate, supplying mobile electric power, utilities and waste management, environmental support and firefighting (see FM 3-34.400).

Although engineering tasks that help enable logistics are primarily considered general engineering tasks, engineers also use capabilities from the other engineer disciplines to enable logistics. Similarly, although general engineering tasks are often used to enable logistics, engineers also use capabilities from the general engineering discipline for other purposes and to support other war fighting functions. FM 3-34 provides additional information about all three engineer disciplines and how they are used for various purposes and to support all the war fighting functions. FM 3-34.400 provides additional information about general engineering. ATTP 3-34.80 and JP 2-03 provide additional information about geospatial engineering.

(III. Elements of Sustainment)
B. Personnel Services

Ref: ADRP 4-0, Sustainment (Jul '12), pp. 4-6 to 4-11.

Personnel services relate to personnel welfare (i.e. readiness, quality of life) and economic power. Personnel services facilitate the Army's capability to achieve endurance. Personnel services include: human resources, financial management, legal, religious, and band support.

A. Human Resources Support

Human resources support maximizes operational effectiveness and facilitates support to Soldiers, their families, Department of Defense civilians, and contractors authorized to accompany the force. Human resources support includes personnel readiness management; personnel accountability; strength reporting; personnel information management; casualty operations; essential personnel services, band support, postal operations; reception, replacement, return-to-duty, rest and recuperation, and redeployment operations; morale, welfare, and recreation (MWR); and human resource planning and staff operations (see FM 1-0).

1. Personnel Accountability

Personnel accountability is the process for recording by-name data on Soldiers, Department of the Army civilians, and contractors when they arrive and depart from units; when their location or duty status changes (such as from duty to hospital); or when their grade changes. These activities include the reception of personnel, the assignment and tracking of replacements, return-to-duty, rest and recuperation, and redeployment operations.

2. Strength Reporting

Strength reporting is a numerical end product of the accounting process, achieved by comparing the by-name data obtained during the personnel accountability process (faces) against specified authorizations (spaces or in some cases requirements) to determine a percentage of fill. Strength reporting relies on timely, accurate, and complete personnel information into the database of record. It is a command function conducted by the G1/S1 to enable them to provide a method of measuring the effectiveness of combat power.

3. Personnel Information Management

Personnel information management encompasses the collecting, processing, storing, displaying, and disseminating of information about Soldiers, units, and civilians. Personnel information management is the foundation for conducting or executing all human resources functions and tasks.

4. Personnel Readiness Management

Personnel readiness management involves analyzing personnel strength data to determine current combat capabilities, projecting future requirements, and assessing conditions of individual readiness. Personnel readiness management is directly interrelated and interdependent upon the functions of personnel accountability, strength reporting, and personnel information management.

5. Casualty Operations Management

The casualty operations management process includes the recording, reporting, verifying, and processing of information from unit level to HQ, Department of the

Army. The process collects casualty information from multiple sources and then collates, analyzes, and determines the appropriate action.

6. Essential Personnel Services

Essential personnel services provide Soldiers and units timely and accurate personnel services that efficiently update Soldier status, readiness and quality of life. It allows the Army leadership to effectively manage the force, including actions supporting individual career advancement and development, proper identification documents for security and benefits entitlements, recognition of achievements, and service. It also includes personal actions such as personnel support.

7. Personnel Support

Personnel support encompasses command interest/human resources programs, MWR, and retention functions. Personnel support also includes substance abuse and prevention programs, enhances unit cohesion, and sustains the morale of the force.

8. Postal Operations

The Military Postal Service serves as an extension of the U.S. Postal Services; therefore, its services are regulated by public law and federal regulation. Postal operations require significant logistics and planning for air and ground transportation, specialized equipment, secured facilities, palletization crews, and mail handlers.

9. Morale, Welfare, and Recreation and Community Support

Morale, welfare, and recreation (MWR) and community support provide Soldiers, Army civilians, and other authorized personnel with recreational and fitness activities, goods, and services. The morale, welfare, and recreation support network provides unit recreation, library books, sports programs, and rest areas for brigade-sized and larger units. Community support programs include the American Red Cross, Army Air Force Exchange System, and family support system. They capitalize on using cellular, e-mail, and video-teleconference technologies to provide links between Soldiers and their Families. Soldiers are also entertained through the latest in visual and audio entertainment over satellite, worldwide web, and virtual reality technologies.

B. Financial Management

The financial management mission is to ensure that proper financial resources are available to accomplish the mission in accordance with commander's priorities. The financial management mission generates economic power by providing banking and disbursing support, as well as resources to fund the force. This is accomplished by two mutually supporting core functions: finance and resource management operations. See FM 1-06 for additional information on financial management.

1. Finance Operations

The finance operations mission is to support the sustainment of Army, joint, and multinational operations through the execution of key finance operations tasks. These key finance operations tasks are to provide timely commercial vendor services and contractual payments, various pay and disbursing services, oversee and manage the Army's Banking Program and to implement financial management policies and guidance prescribed by the Office of the Under Secretary of Defense (Comptroller) and national providers (e.g., U.S. Treasury, Defense Finance and Accounting Service, Federal Reserve Bank).

- **Banking.** Banking support encompasses financial management activities ranging from currency support of U.S. military operations to liaison with host nation banking officials to strengthen local financial institutions. Other finan-

Principles of Personnel Services

Ref: ADRP 4-0, Sustainment (Jul '12), pp. 1-4 to 1-5.

The principles of personnel services guide the functions for maintaining Soldier and Family support, establishing morale and welfare, funding the force, and enforcing the rules of law. In addition to the principles of sustainment, the following principles are unique to personnel services:

Priniciples of Personnel Services

1. **Synchronization**
2. **Timeliness**
3. **Stewardship**
4. **Accuracy**
5. **Consistency**

1. Synchronization

Synchronization is ensuring personnel services are effectively aligned with military actions in time, space, and purpose to produce maximum relative readiness and operational capabilities at a decisive place and time. It includes ensuring that personnel services are synchronized with the operations process: plan, prepare, execute, and assess.

2. Timeliness

Timeliness ensures decision makers have an access to relevant personnel services information and analysis that support current and future operations. It also supports a near real-time common operational picture across all echelons of support.

3. Stewardship

Stewardship is the careful and responsible management of resources entrusted to the government in order to execute responsible governance. Stewardship most closely relates to financial management operations The Department of Defense (DOD) is entrusted by the American people as a steward of vital resources (funds, people, material, land, and facilities) provided to defend the nation (JP 1-06, Financial Management Support in Joint Operations). The Army operates under the mandate to use all available resources in the most effective and efficient means possible to support the CCDR. Good stewardship requires the availability of timely and accurate financial information to facilitate sound decision making and ensures that resources are used in compliance with existing statutory and regulatory guidance.

4. Accuracy

Accuracy of information impacts the decisions made by commanders and also Soldiers and their Families. For Soldiers, accurate information impacts their careers, retention, compensation, promotions, and general well being. For Family members, accuracy of information is critical for next of kin (NOK) notification. Personnel services providers must understand the dynamic nature of a system's architecture and the fact that data input at the lowest level has direct impact on decisions being made at the highest level.

5. Consistency

Consistency involves providing uniform and compatible guidance and support to forces across all levels of operations. Providers of personnel services must coordinate with the appropriate DOD organizations, governmental organizations and Services to ensure uniformity of support. For example, in financial management consistency is essential for making appropriate provisions for pay support and services, establishing banking and currency support, payment of travel entitlements and cash operations to support the procurement process (JP 1-06).

cial management activities within banking support include Limited Depositary selection and Limited Depositary Account establishment, coordination with U.S. embassies, USAFMCOM, DFAS, and Department of the Treasury in order to integrate all agencies in support of banking initiative.

- **Disbursing.** Disbursing is the arm within financial management that ensures all payments are made IAW DOD regulations. It is strongly recommended that all elements of the fiscal triad are co-located to facilitate fiscal communication, accuracy of documentation, and timely payment of goods and services. Disbursing is the paying of public funds to entities in which the U.S. Government is indebted; the collection and deposit of monies; the safeguarding of public funds; and the documenting, recording, and reporting of such transactions (FM 1-06). Disbursements are cash, check, electronic funds transfer, intra-governmental payment and collection system, or inter-fund payments that liquidate established obligations, disburse amounts previously collected into a deposit fund account, or provide payment in advance of performance.
- **Pay Support.** This competency provides for full U.S. pay (including civilian pay where not supported by DFAS); travel support; local and partial payments; check-cashing and currency exchange to Soldiers, civilians and U.S. contractors; and non-U.S. pay support (e.g., enemy prisoner of war, host nation employees, day laborers, civilian internee). Pay support also includes support to noncombatant evacuation operations in the form of travel advances. Financial management units providing pay support must ensure that all Soldiers, regardless of component, receive timely and accurate pay in accordance with existing statutes and regulations.

2. Resource Management

The resource management mission is to analyze resource requirements, ensure commanders are aware of existing resource implications in order for them to make resource informed decisions, and then obtain the necessary funding that allows them to accomplish their mission. Resource management is the critical capability within the financial management competency that matches legal and appropriate sources of funds with thoroughly vetted and validated requirements. Key resource management tasks are providing advice and recommendations to the commander, identifying sources of funds, forecasting, capturing, analyzing and managing costs; acquiring funds, distributing and controlling funds; tracking costs and obligations; establishing and managing reimbursement processes; and establishing and managing the Army Mangers' Internal Control Program.

C. Legal Support

The Office of the Staff Judge Advocate participates in actions related to mission command of its subordinates (ADP 6-0). The Office of the Staff Judge Advocate's command and staff functions include advising commanders, their staffs, and Soldiers on the six core legal disciplines: military justice, international and operational law, contracts and fiscal law, administrative and civil law, claims, and legal assistance. For additional information on legal support see FM 1-04.

See facing page for further discussion.

D. Religious Support

Religious support facilitates the Soldier's right to the free exercise of religion, provides religious activities that support resiliency efforts to sustain Soldiers, and advises commands on matters of religion, morals, morale, and their impact on military operations (see FM 1-05). As chaplain sections and unit ministry teams, Chaplains and Chaplain Assistants provide and perform religious support in the Army. Three core competencies provide the fundamental direction as the Chaplain

Legal Support

Ref: ADRP 4-0, Sustainment (Jul '12), pp. 4-9 to 4-10.

Military Justice. Military justice is the administration of the Uniform Code of Military Justice (FM 1-04). The purpose of military justice, as a part of military law, is "to promote justice, to assist in maintaining good order and discipline in the armed forces, to promote efficiency and effectiveness in the military establishment, and thereby to strengthen the national security of the United States." Military justice (Preamble, Manual for Courts-Martial, 2008). The Judge Advocate General is responsible for the overall supervision and administration of military justice within the Army. Commanders are responsible for the administration of military justice in their units and must communicate directly with their servicing Staff Judge Advocate about military justice matters (AR 27-10).

International and Operational Law. International law is the application of international agreements, U.S. and international law, and customs related to military operations and activities (FM 1-04). The practice of international law includes the interpretation and application of foreign law, comparative law, martial law, and domestic law affecting overseas activities, intelligence, security assistance, counter-drug, operations with a stability focus, and rule of law activities. Operational law is that body of domestic, foreign, and international law that directly affects the conduct of military operations (FM 1-04). Operational law encompasses the law of war, but goes beyond the traditional international law concerns to incorporate all relevant aspects of military law that affect the conduct of operations.

Administrative and Civil Law. Administrative and civil law is the body of law containing the statutes, regulations, and judicial decisions that govern the establishment, functioning, and command of military organizations (FM 1-04). The practice of administrative law includes advice to commanders and litigation on behalf of the Army involving many specialized legal areas, including military personnel law, government information practices, investigations, relationships with private organizations, labor relations, civilian employment law, military installations, regulatory law, intellectual property law, and government ethics.

Contract and Fiscal Law. Contract law is the application of domestic and international law to the acquisition of goods, services, and construction (FM 1-04). The practice of contract law includes battlefield acquisition, contingency contracting, bid protests and contract dispute litigation, procurement fraud oversight, commercial activities, and acquisition and cross-servicing agreements. Fiscal law is the application of domestic statutes and regulations to the funding of military operations and support to non-federal agencies and organizations (FM 1-04).

Claims. The Army claims program investigates, processes, adjudicates, and settles claims on behalf of and against the United States world-wide under the authority conferred by statutes, regulations, international and interagency agreements, and DOD Directives. The claims program supports commanders by preventing distractions to the operation from claimants, promoting the morale of Army personnel by compensating them for property damage suffered incident to service, and promoting good will with the local population by providing compensation for personal injury or property damage caused by Army or DOD personnel.

Legal Assistance. Legal assistance is the provision of personal civil legal services to Soldiers, their dependents, and other eligible personnel (FM 1-04). The mission of the Army Legal Assistance Program is to assist those eligible for legal assistance with their personal legal affairs quickly and professionally. The program assists eligible people by meeting their needs for help and information on legal matters and resolving their personal legal problems whenever possible. The legal assistance mission ensures that Soldiers have their personal legal affairs in order before deploying.

Corps executes its mission through nurturing the living, caring for the wounded, and honoring the dead.

1. Nurture the Living

In preparation for and during the execution of missions, unit ministry teams develop and execute a religious support plan that seeks to strengthen and sustain the resilience of Soldiers and Family members. Unit ministry teams also provide religious support, care, comfort, and hope to the living.

2. Caring for the Wounded

Unit ministry teams bring hope and strength to those who have been wounded and traumatized in body, mind, and spirit, by assisting in the healing process. Through prayer and presence, the unit ministry teams provide the Soldier with courage and comfort in the face of death.

3. Honoring the Dead

Our nation reveres those who have died in military service. Religious support honors the dead. Memorial ceremonies, services, and funerals reflect the emphasis the American people place on the worth and value of the individual. Chaplains conduct these services and ceremonies, fulfilling a vital role in rendering tribute to America's sons and daughters who paid the ultimate price serving the nation in the defense of freedom.

E. Band Support

Army bands provide support to the force by tailoring music support throughout military operations. Music instills in Soldiers the will to fight and win, fosters the support of our citizens, and promotes America's interests at home and abroad. Music serves as a useful tool to reinforce relations with host nation populations and favorably shapes the civil situation throughout the peace building process. Inherently capable of providing a climate for international relations, bands serve as ambassadors in multi-national operations or to the host nation population (see FM 1-0 and ATTP 1-19).

(III. Elements of Sustainment)
C. Health Service Support

Ref: ADRP 4-0, Sustainment (Jul '12). pp. 4-11 to 4-12.

Under the Army sustainment war fighting function, the health service support provides continual, flexible, and deployable medical support designed to sustain a force projection Army and its varied missions. The health service support mission includes— casualty care, medical evacuation, and medical logistics.

A. Casualty Care

Casualty care encompasses all issues pertaining to the provision of clinical services for the treatment of Soldiers from the point of injury to successive roles of care. Casualty care includes the following sub-functions: organic and area medical support, hospitalization, the treatment aspects of dental care and behavioral health/neuropsychiatric treatment, clinical laboratory services, and treatment of chemical, biological, radiological, and nuclear patients.

1. Organic and Area Medical Support

The medical treatment function encompasses Roles 1 and 2 medical treatment support. Role 1 medical treatment is provided by the combat medic or by the physician, the physician assistant, or the health care specialist in the battalion aid station/ Role 1 medical treatment facility. Role 2 medical care provides greater resuscitative capability than is available at Role 1 and is rendered by the medical company (brigade support battalion) or by the medical company (area support), which is an echelons above brigade asset. These roles of care are provided by organic assets or on an area support basis from supporting medical companies or detachments. The area support function encompasses emergency medical treatment, advanced trauma management, routine sick call, emergency dental care, preventive medicine, and combat and operational stress control support. See ATTP 4-02 for additional information on organic and area medical support and a full description of the roles of medical care.

2. Hospitalization

The Army's hospitalization capability consists of Role 3 combat support hospitals purposely positioned to provide support in the area of operations. At Role 3, the combat support hospital expands the support provided at Role 2 and is staffed and equipped to provide care for all categories of patients, to include resuscitation, initial wound surgery, damage control surgery, and postoperative treatment. Hospitalization capabilities deploy as modules or multiple individual capabilities that provide incrementally increased medical services in a progressively more robust area of operations. The hospitalization capability in the area of operations offers essential care to either return the patient to duty (within the theater patient movement policy) and/or stabilization to ensure the patient can tolerate evacuation to a definitive care facility outside the area of operations (this support is key to early identification and treatment of mild traumatic brain injuries).

3. Dental Care

Dental care provided as part of health service support includes far forward dental treatment, treatment of oral and dental disease, and early treatment of severe oral and maxillofacial injuries. Dental personnel may also be used to augment medical personnel (as necessary) during mass casualty operations.

Principles of the Army Health System

Ref: ADRP 4-0, Sustainment (Jul '12), pp. 1-5 to 1-6.

The principles of the Army health system (AHS) are the enduring tenets upon which the delivery of health care in a field environment is founded. The principles guide medical planners in developing operational plans which are effective, efficient, flexible, and executable. The AHS plans are designed to support the tactical commander's scheme of maneuver while still retaining a Soldier/patient focus. The AHS principles apply across all medical functions and are synchronized through medical mission command and close coordination and synchronization of all deployed medical assets though medical technical channels.

Priniciples of the Army Health System

1. Conformity
2. Proximity
3. Flexibility
4. Mobility
5. Continuity
6. Control

1. Conformity

Conformity with the tactical plan is the most basic element for effectively providing AHS support. In order to develop a comprehensive concept of operations, the medical commander must have direct access to the tactical commander. AHS planners must be involved early in the planning process and once the plan is established it must be rehearsed with the forces it supports.

2. Proximity

Proximity is to provide AHS support to sick, injured, and wounded Soldiers at the right time and to keep morbidity and mortality to a minimum. AHS support assets are placed within supporting distance of the maneuver forces which they are supporting, but not close enough to impede ongoing combat operations. As the battle rhythm of the medical commander is similar to the tactical commander's, it is essential that AHS assets are positioned to rapidly locate, acquire, stabilize, and evacuate combat casualties. Peak workloads for AHS resources occur during combat operations.

3. Flexibility

Flexibility is being prepared and empowered to shift AHS resources to meet changing requirements. Changes in tactical plans or operations make flexibility in AHS planning and execution essential. In addition to building flexibility into operation plans to support the tactical commander's scheme of maneuver, the medical commander must also ensure that he has the flexibility to rapidly transition from one level of violence to another across the range of military operations. As the current era is one characterized by conflict, the medical commander may be supporting simultaneous actions along the continuum from stable peace through general war (JP 3-0). The medical commander exercises command authority to effectively manage scarce medical resources to benefit the greatest number of Soldiers in the area of operations.

4. Mobility

Mobility is to ensure that AHS assets remain in supporting distance to support maneuvering forces. The mobility, survivability (such as armor plating and other force protection measures), and sustainability of medical units organic to maneuver elements must be equal to the forces being supported. Major AHS headquarters in echelons above brigade continually assess and forecast unit movement and redeployment. AHS support must be continually responsive to shifting medical requirements in the operational environment. In noncontiguous operations, the use of ground ambulances may be limited depending on the security threat and air ambulance use may be limited by environmental conditions and enemy air defense threat. Therefore, to facilitate a continuous evacuation flow, medical evacuation must be a synchronized effort to ensure timely, responsive, and effective support is provided to the tactical commander. The only means available to increase the mobility of medical units is to evacuate all patients they are holding. Medical units anticipating an influx of patients must medically evacuate patients on hand prior to the start of the engagement.

5. Continuity

Continuity in care and treatment is achieved by moving the patient through progressive, phased roles of care, extending from the point of injury or wounding to the continental United States (CONUS)-support base. Each type of AHS unit contributes a measured, logical increment in care appropriate to its location and capabilities. In current operations, lower casualty rates, availability of rotary-wing air ambulances, and other situational variables often times enables a patient to be evacuated from the point of injury directly to the supporting combat support hospital. In more traditional combat operations, higher casualty rates, extended distances, and patient condition may necessitate that a patient receive care at each role of care to maintain physiologic status and enhance chances of survival. The medical commander's depth of medical knowledge, ability to anticipate follow-on medical treatment requirements, and assessment of the availability of specialized medical resources can adjust the patient flow to ensure each Soldier receives the care required to optimize patient outcome. The medical commander can recommend changes in the theater evacuation policy to adjust patient flow within the deployed setting.

6. Control

Control is required to ensure that scarce AHS resources are efficiently employed and support the tactical through strategic plans. It also ensures that the scope and quality of medical treatment meet professional standards, policies, and U.S. and international law. As the Army Medical Department (AMEDD) is comprised of 10 medical functions which are interdependent and interrelated. Control of AHS support operations requires synchronization to ensure the complex interrelationships and interoperability of all medical assets remain in balance to optimize the effective functioning of the entire system.

4. Behavioral Health

The primary focus of behavioral health/neuropsychiatric treatment is to screen and evaluate Soldiers with maladaptive behaviors. The purpose of this function is to provide diagnosis, treatment, and disposition for Soldiers with neuropsychiatric/behavioral health-related issues.

5. Clinical Laboratory Services

Clinical laboratory services provide basic support within the theater, to include procedures in hematology, urinalysis, microbiology, and serology. Role 2 area support medical companies and brigade support medical companies receive, maintain, and transfuse blood products. The combat support hospital performs procedures in biochemistry, hematology, urinalysis, microbiology, and serology in support of clinical activities. The hospital also blood-banking services.

6. Treatment of Chemical, Biological, Radiological, and Nuclear Patients

Health service support operations in a chemical, biological, radiological, and nuclear (CBRN) environment are complex. Medical personnel may be required to treat CBRN injured and contaminated casualties in large numbers. Medical treatment must be provided in protected environments and protective clothing must be worn. Movement of CBRN casualties can spread contamination to clean areas. All casualties are decontaminated as far forward as the situation permits and must be decontaminated before they are admitted into a clean medical treatment facility. The admission of one contaminated casualty into a clean medical treatment facility will contaminate the facility; thereby, reducing treatment capabilities in the facility. See FM 4-02.7 for additional information.

B. Medical Evacuation

Medical evacuation provides en route medical care and emergency medical intervention. En route medical care enhances the Soldiers' prognosis, reduces long-term disability, and provides a vital linkage between the roles of care necessary to sustain the patient during transport.

C. Medical Logistics

Medical logistics encompasses planning and executing all Class VIII supply support to include medical materiel procurement and distribution, medical equipment maintenance and repair, blood management, optical fabrication and repair, and the centralized management of patient movement items. It also includes contracting support, medical hazardous waste management and disposal, and production and distribution of medical gases. The system is anticipatory with select units capable of operating in a split based mode.

Chap 7

Protection Warfighting Function

Ref: ADP 3-37, Protection (Aug '12) and ADRP 3-0, Operations (Nov '16), p. 5-7.

Commanders and staffs synchronize, integrate, and organize capabilities and resources throughout the operations process to preserve combat power and the freedom of action and to mitigate the effects of threats and hazards. Protection safeguards the force, personnel (combatants and noncombatants), systems, and physical assets of the United States and unified action partners. Survivability refers to the capacity, fitness, or tendency to remain alive or in existence. For the military, survivability is about much more than mere survival—it is also about remaining effective. Military forces are composed of personnel and physical assets, each having their own inherent survivability qualities or capabilities that permit them to avoid or withstand hostile actions or environmental conditions while retaining the ability to fulfill their primary mission. These inherent qualities or capabilities are affected by various factors (dispersion, redundancy, morale, leadership, discipline, mobility, situational understanding, terrain and weather conditions) and can be enhanced by tasks within the protection warfighting function.

I. The Protection Warfighting Function

The protection warfighting function is the related tasks and systems that preserve the force so the commander can apply maximum combat power to accomplish the mission. Preserving the force includes protecting personnel (combatants and noncombatants) and physical assets of the United States and unified action partners, including the host nation. The protection warfighting function enables the commander to maintain the force's integrity and combat power. Protection determines the degree to which potential threats can disrupt operations and then counters or mitigates those threats. Protection is a continuing activity; it integrates all protection capabilities to safeguard bases, secure routes, and protect forces. Protection activities ensure maintenance of the critical asset list and defended asset list.

The protection warfighting function includes the following tasks:

- Conduct survivability operations.
- Provide force health protection.
- Conduct chemical, biological, radiological, and nuclear operations.
- Provide explosive ordnance disposal support.
- Coordinate air and missile defense.
- Conduct personnel recovery.
- Conduct detention operations.
- Conduct risk management.
- Implement physical security procedures.
- Apply antiterrorism measures.
- Conduct police operations.
- Conduct populace and resource control.

In addition to the principles of protection described in ADRP 3-37, commanders consider the following when performing protection warfighting function tasks:

- Security of forces and means enhances force protection by identifying and reducing friendly vulnerability to hostile acts, influence, or surprise.
- Physical security measures, like any defensive measures, should be overlapping and deployed in depth.

II. The Role of Protection

Ref: ADP 3-37, Protection (Aug '12), pp. 1 to 2.

Protection is the preservation of the effectiveness and survivability of mission-related military and nonmilitary personnel, equipment, facilities, information, and infrastructure deployed or located within or outside the boundaries of a given operational area (Joint Publication [JP] 3-0). Commanders and staffs synchronize, integrate, and organize capabilities and resources throughout the operations process to preserve combat power and mitigate the effects of threats and hazards. Protection is a continuing activity; it integrates all protection capabilities to safeguard the force, personnel (combatants and noncombatants), systems, and physical assets of the United States and unified action partners.

Operational environments are uncertain, marked by rapid change and a wide range of threats and hazards. These evolving operational environments will provide significant challenges for commanders and staffs who are integrating protection capabilities. Protection preserves the combat power potential of the force by providing capabilities to identify and prevent threats and hazards and to mitigate their effects. Army units may also be required to provide protection for civilians in order to support mission objectives. This may include protecting civilians from widespread violence (such as mass atrocities), mitigating civilian casualties, and ensuring a secure environment for the population and nonmilitary partners.

Protection can be maximized by integrating the elements of combat power to reinforce protection or to achieve complementary protective effects. The goal of protection integration is to balance protection with the freedom of action throughout the duration of military operations. This is accomplished by integrating reinforcing or complementary protection capabilities into operations until all significant vulnerabilities have been mitigated, have been eliminated, or become assumed risks. The employment of synchronized and integrated reinforcing and complementary protection capabilities preserves combat power and provides flexibility across the range of military operations. The collaboration, integration, and synchronization between the warfighting functions assist in identifying and preventing threats and hazards and in mitigating their effects.

Army leaders are responsible for clearly articulating their visualization of operations in time, space, purpose, and resources. The commander's inherent responsibility to protect and preserve the force and secure the area of operations is vital in seizing, retaining, and exploiting the initiative. Protection must be considered throughout the operations process to—

- Identify threats and hazards
- Implement control measures to prevent or mitigate enemy or adversary actions
- Manage capabilities to mitigate the effects and time to react or maneuver on the adversary to gain superiority and retain the initiative

A shared understanding and purpose of the joint protection function (see JP 3-0) allows Army leaders to integrate actions within the unified action and to synchronize operations. The joint protection function focuses on preserving the joint force fighting potential in four primary ways:

- **Active defensive measures** to protect the joint force, its information, its bases/base camps, critical infrastructure, and lines of communications from an enemy or adversary attack
- **Passive defensive measures** to make friendly forces, systems, and facilities difficult to locate, strike, and destroy
- The application of technology and procedures to **reduce the risk of fratricide**
- **Emergency management and response** to reduce the loss of personnel and capabilities due to accidents, health threats, and natural disasters

Protection Logic Map

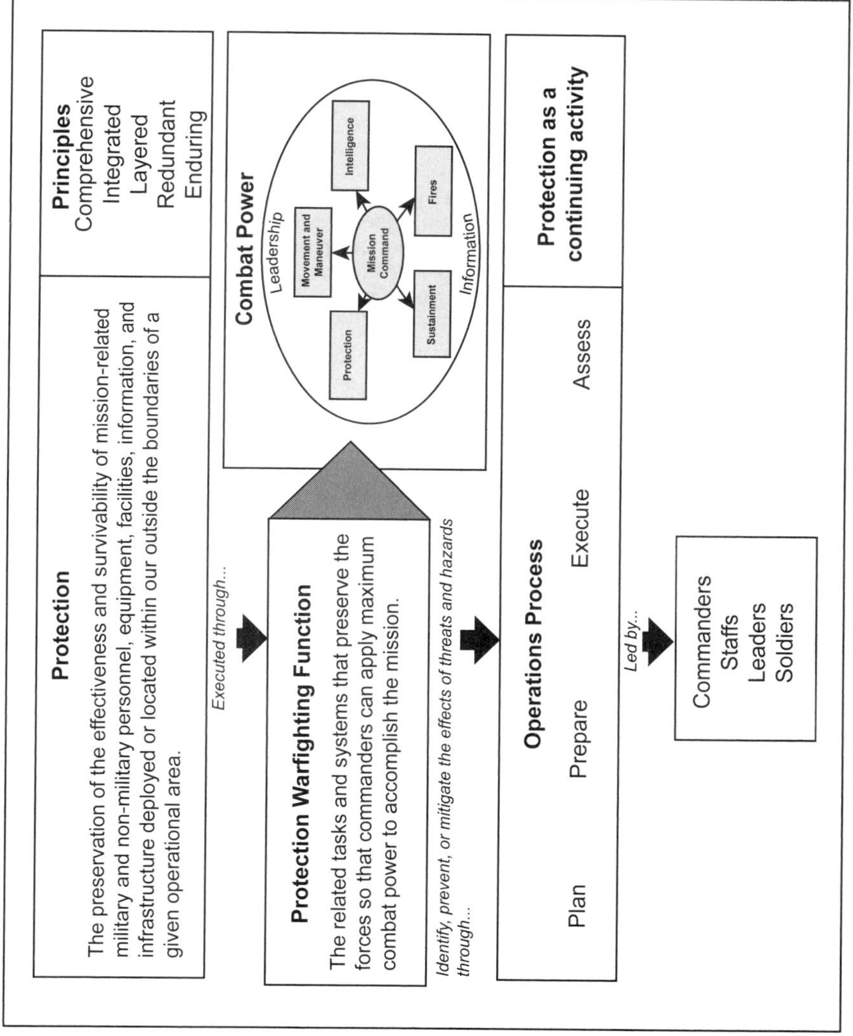

Protection
The preservation of the effectiveness and survivability of mission-related military and non-military personnel, equipment, facilities, information, and infrastructure deployed or located within our outside the boundaries of a given operational area.

Principles
Comprehensive
Integrated
Layered
Redundant
Enduring

Executed through...

Protection Warfighting Function
The related tasks and systems that preserve the forces so that commanders can apply maximum combat power to accomplish the mission.

Combat Power
Leadership
Movement and Maneuver
Intelligence
Mission Command
Fires
Protection
Sustainment
Information

Identify, prevent, or mitigate the effects of threats and hazards through...

Operations Process
Plan Prepare Execute Assess

Protection as a continuing activity

Led by...

Commanders
Staffs
Leaders
Soldiers

Ref: ADP 3-37, Protection, fig. 1, p. iii.

Army Doctrine Publication (ADP) 3-37 provides guidance on protection and the protection warfighting function. It also provides the guiding protection principles for commanders and staffs who are responsible for planning and executing protection in support of unified land operations. ADP 3-37 corresponds with the Army operations doctrine introduced in ADP 3-0.

III. ADRP 3-37, Protection (Aug '12)

Ref: ADRP 3-37, Protection (Aug '12), introduction.

ADRP 3-37 is a new publication that expands on the protection principles found in ADP 3-37. The doctrine described in this publication is nested within ADRP 3-0 and describes protection as a continuing activity and a warfighting function. It presents overarching doctrinal guidance and direction for conducting protection within unified land operations. ADRP 3-37 provides emphasis to Soldiers, leaders, and organizations on integrating protection capabilities into the operations process in order to identify, prevent, or mitigate the effects of threats and hazards. Overall, ADRP 3-37 remains consistent with previous doctrine. This manual modifies the current protection definition by aligning it with the joint definition of protection—preservation of the effectiveness and survivability of mission-related military and nonmilitary personnel, equipment, facilities, information, and infrastructure deployed or located within or outside the boundaries of a given operational area (Joint Publication [JP] 3-0). The protection principle of "full dimension" is replaced with "comprehensive," which expands on the definition of an all-inclusive utilization of complementary and reinforcing protection tasks and systems available to commanders, incorporated into the plan, to preserve the force. The protection warfighting function is updated to exemplify the tasks and systems that are synchronized and integrated throughout the operations process and with the other elements of combat power to preserve the force so that the commander can apply maximum combat power to accomplish the mission (introductory tables 1 through 3):

- "Air and missile defense" was modified. "Coordinate air and missile defense" is within the protection warfighting function, and "conduct air and missile defense" is within the fires warfighting function. Coordinating air and missile defense protects the force from missile attack, air attack, and aerial surveillance.
- The definition of fire support now includes air missile defense
- "Information protection" is now within the mission command warfighting function (see ADP 6-0 and ADRP 6-0)
- "Conduct operational area security," "implement physical security procedures," "conduct law and order," and "conduct internment and resettlement" were added to the protection warfighting function. "Conduct internment and resettlement" moved from the sustainment warfighting function.
- "Fratricide avoidance" was incorporated into "employ safety techniques (including fratricide avoidance)" within the protection warfighting function

The joint protection function tasks and key considerations in JP 3-0 are not directly aligned, one for one, with the Army protection warfighting function tasks, but they are integrated in how the Army operates. A shared understanding and purpose of the joint protection function allows Army leaders to integrate their actions within the unified action and to synchronize operations. The joint protection function focuses on preserving the joint force fighting potential in four primary ways:

- Active defensive measures to protect the joint force, its information, its bases/ base camps, critical infrastructure, and lines of communications from an enemy or adversary attack
- Passive defensive measures to make friendly forces, systems, and facilities difficult to locate, strike, and destroy
- The application of technology and procedures to reduce the risk of fratricide
- Emergency management and response to reduce the loss of personnel and capabilities due to accidents, health threats, and natural disasters

See JP 3-0 for additional information on the joint protection function, its tasks, and key considerations.

The protection warfighting function, the Army Protection Program, and force protection are important aspects of protection that reinforce each other:

Protection Warfighting Function

The protection warfighting function is the related tasks and systems that preserve the force so the commander can apply maximum combat power to accomplish the mission (ADRP 3-0).

The protection warfighting function supports unified action and unified land operations, whereas the Army Protection Program manages and executes the programs within the non-warfighting functional elements. Additionally, the protection warfighting function focuses on preserving the force and protecting personnel (friendly combatants and noncombatants) and physical assets of the United States and unified action partners.

- Emergency management
- Computer network defense
- Continuity of operations
- Critical infrastructure risk management
- Operations security (OPSEC)
- Antiterrorism (AT)
- Fire and emergency services
- Force health protection
- High-risk personnel
- Law enforcement
- Information assurance
- Physical security

The complementary capabilities within the non-warfighting elements and the protection warfighting function assist in reinforcing protection throughout.

Army Protection Program

The Army Protection Program is a management framework to synchronize, prioritize, and coordinate protection policies and resources (see Army Directive 2011-04). It includes the twelve non-warfighting functional elements:

Force Protection

Force protection is preventative measures taken to mitigate hostile actions against Department of Defense personnel (to include family members), resources, facilities, and critical information (JP 3-0).

The chapters within ADRP 3-37 expound on the protection framework (chapter 1) and protection integration and synchronization throughout the operations process (chapters 2 through 5). Chapter 1 provides a common understanding of protection principles, protection nesting within unified land operations, the protection warfighting function, and the integration of the protection warfighting function tasks and systems. Chapters 2 through 5 elaborate on protection within the operations process—plan, prepare, execute, and assess.

New and Modified Army terms

Introductory Table-1. New Army terms

Term	Remarks
critical asset security	New term and definition.
fratricide	New term and definition.

Introductory Table-2. Modified Army terms

Term	Remarks
protection	Adopts the joint definition.

IV. Protection Integration in the Operations Process

Ref: ADP 3-37, Protection (Aug '12), pp. 3 to 7.

Protection is integrated throughout the operations process to provide a synchronization of efforts and an integration of capabilities. The protection warfighting function tasks are incorporated into the process in a layered and redundant approach to complement and reinforce actions to achieve force protection.

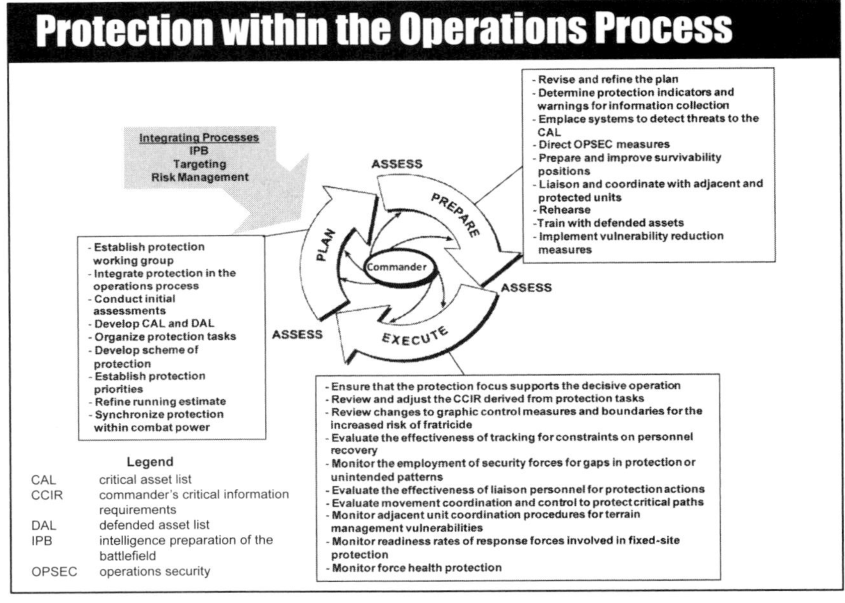

Ref: ADRP 3-37, Protection, introductory fig. 1, p. vi.

A. Plan *(pp. 7-25 to 7-38)*

Planning is the first step toward effective protection. Commanders consider the most likely threats and hazards and decide which personnel, physical assets, and information to protect. They set protection priorities for each phase or critical event of an operation. The military decisionmaking process and troop leading procedures provide a deliberate process to develop and examine information for use in the various continuing activities and integrating processes that comprise the operations process. Effective protection schemes and risk decisions are developed based on information that flows from mission analysis, allowing a thorough understanding of the environment (operational and mission variables). The integrating processes provide a context to identify and analyze threats and hazards, to develop a situational understanding of the operational environment, and to develop a scheme of protection. Staffs assess threats, hazards, criticality, vulnerability, and capability to help commanders determine protection priorities, task organizations, and protection task integration.

Commanders and staffs apply protection considerations in relation to the mission and the operational environment throughout the operations process. They discern hazards that are preventable and divide threats into those that may be deterred and those that may require the application of security or defensive measures to achieve protection. Commanders provide risk guidance, critical information requirements, essential elements of friendly information, and asset or capability criticality to help focus staffs and subordinate leaders. Commanders direct staffs to conduct the necessary tasks to protect the force, secure the area, and mitigate the effects of current and potential threats and hazards.

The keys to protection planning are identifying the threats and hazards, assessing the threats and hazards to determine the risks, developing preventive measures, and integrating protection tasks into a comprehensive scheme of protection that includes mitigating measures. The warfighting functions are synchronized throughout the operations process to assist in the development of an enduring scheme of protection. The critical asset list and the defended asset are developed and revised during this process.

During planning, the protection cell/working group—

- Establishes a protection working group
- Conducts initial assessments
- Develops a critical asset list and a defended asset list
- Integrates and layers protection tasks
- Develops a scheme of protection.
- Recommends protection priorities
- Refines the running estimate
- Synchronizes protection within the elements of combat power
- Identifies communication channels among key personnel within protection and leadership
- Develops and publishes personnel recovery guidance
- Establishes personnel recovery that is related to the commander's critical information requirements

Continued on next page

B. Prepare *(pp. 7-39 to 7-42)*

During the preparation phase, protection focuses on deterring and preventing the enemy or adversary from actions that would affect combat power and the freedom of action. The implementation of protection tasks with ongoing preparation activities assists in the prevention of negative effects. Commanders ensure the integration of protection warfighting function tasks and systems to safeguard bases/base camps, secure routes, and protect the force while it prepares for operations. Active defense measures assist in denying the initiative to the enemy or adversary, while the execution of passive defense measures prepares the force against the threat and hazard effects and speeds the mitigation of those effects.

Assessment occurs during preparation and includes activities required to maintain situational understanding; monitor and evaluate running estimates and tasks, methods of evaluation, and measures of performance; and identify variances for decision support. These assessments generally provide commanders with a composite estimate of preoperational force readiness or status in time to make adjustments.

Preparation includes increased application and emphasis on protection measures. During preparation, the protection cell/working group—

- Revises and refines the plan
- Determines protection indicators and warnings for information collection
- Emplaces systems to detect threats to the critical assets
- Directs operations security measures
- Prepares and improves survivability positions
- Conducts liaison and coordinates with adjacent and protected units
- Rehearses
- Trains with defended assets
- Reviews the personnel recovery readiness of subordinate units
- Establishes personnel recovery architecture
- Implements vulnerability reduction measures

Continued on next page

IV. Protection Integration in the Operations Process (Cont.)

Ref: ADP 3-37, Protection (Aug '12), pp. 3 to 7.

Continued from previous page

C. Execute *(pp. 7-43 to 7-44)*

Commanders who exercise mission command decide, direct, lead, access, and provide leadership to organizations and Soldiers during execution. As operations develop and progress, the commander interprets information that flows from systems for indicators and warnings that signal the need for execution or adjustment of decisions. Commanders may direct and redirect the way that combat power is applied or preserved, and they may adjust the tempo of operations through synchronization. The continuous and enduring character of protection makes the continuity of protection tasks and systems essential during execution. Commanders implement control measures and allocate resources that are sufficient to ensure protection continuity and restoration.

Commanders and leaders must be flexible and adaptive as they seek opportunities to seize, retain, and exploit the initiative. Leaders must have situational understanding during simultaneous operations due to the diversity of threats, the proximity to civilians, and the impact of information during operations. The changing nature of operations may require the surge of certain capabilities, such as protection, to effectively link decisive operations to shaping or stabilizing activities in the area of operations.

Commanders must accept prudent risk to exploit time-sensitive opportunities by acting before enemies or adversaries discover vulnerabilities, take evasive or defensive action, and implement countermeasures. Commanders and leaders can continue to act on operational and individual initiative by making prudent risk decisions faster than the enemy or adversary, ultimately breaking their will and morale through relentless pressure. Commanders can leverage technological advancements or processes that minimize fratricide and increase the probability of mission accomplishment.

An accurate assessment is essential for effective decisionmaking and the assignment of combat power to protection tasks. The staff monitors the conduct of operations during execution, looking for variances from the scheme of maneuver and protection. When variances exceed a threshold value, adjustments are made to prevent a developing vulnerability or to mitigate the effects of the unforecasted threat or hazard. The status of protection assets is tracked and evaluated on the effectiveness of the protection systems as they are employed. Commanders maintain protection by applying comprehensive protection capabilities, from main and supporting efforts to decisive and shaping operations. Protection can be derived as a by-product or a complementary result of some combat operations (such as security operations), or it can be deliberately applied as commanders integrate and synchronize tasks that comprise the protection warfighting function.

Continued from previous page

The force continues to identify and prevent threats and hazards. The effects of a threat or hazard are identified to verify the presence of adversary action, qualifying and quantifying the specific hazard and collecting and preserving forensic evidence. Once a threat or hazard is known, it is imperative that the force is warned and begins responding to the action. Response actions save lives, protect property, and continue essential services, mitigating the effects of the threat or hazard and allowing the force to retain the initiative and deny it to the enemy or adversary. Restoring mission readiness and implementing measures from assessments prepare the force to continue operations and prepare for future operations.

The protection cell/working group monitors and evaluates several critical ongoing functions associated with execution for operational actions or changes that impact protection cell proponents, which include—

- Ensuring that the protection focus supports the commander's intent and concept of the operation
- Reviewing and recommending adjustment to the commander's critical information requirements and essential elements of friendly information derived from protection tasks
- Reviewing changes to graphic control measures and boundaries for the increased risk of fratricide
- Monitoring and evaluating personnel recovery operations
- Monitoring the employment of security forces for gaps in protection or unintended patterns
- Evaluating the effectiveness of liaison personnel for protection activities
- Evaluating movement coordination and control to protect critical paths
- Monitoring adjacent unit coordination procedures for terrain management vulnerabilities
- Monitoring the readiness rates of response forces involved in fixed-site protection
- Monitoring force health protection
- Coordinating continuously with unified action partners
- Coordinating with the Mission Management Center, U.S. Army Space and Missile Defense Command, on personnel recovery operations

D. Assess *(pp. 7-45 to 7-46)*

Assessing protection is an essential, continuous activity that occurs throughout the operations process. While a failure in protection is typically easy to detect, the successful application of protection may be difficult to assess and quantify.

Assessment is the determination of the progress toward accomplishing a task, creating a condition, or achieving an objective (JP 3-0). Commanders typically base assessments on their situational understanding, which is generally a composite of several informational sources and intuition. Assessments help the commander determine progress toward attaining the desired end state, achieving objectives, and performing tasks. It also involves continuously monitoring and evaluating an operational environment to determine what changes might affect the conduct of operations.

Staff members develop running estimates that illustrate the significant aspects of a particular activity or function over time. These estimates are used by commanders to maintain situational understanding and direct adjustments. Significant changes or variances among or within running estimates can signal a threat or an opportunity, alerting commanders to take action.

Staffs monitor and evaluate variances in threats and hazards, vulnerabilities, capabilities, risks, and priorities. They also track the status of protection assets and evaluate the effectiveness of the protection systems as they are employed. If an action appears to be failing in its desired effect, it may be attributed to personnel or equipment system failure, insufficient resource allocation at vulnerable points, or a variance in anticipated threat combat power ratio. This can result in an increased risk or ineffective supporting efforts and can lead to a cumulative failure of more critical elements. The staff then recommends adjustments to protection priorities, posture, resource allocation, systems, or the scheme of protection.

V. Protection in Support of Unified Land Operations

Ref: ADRP 3-37 (FM 3-37), Protection (Aug '12), p. 1-1.

The synchronization, integration, and organization of capabilities and resources to preserve combat power from the effects of threats and hazards are essential. The ability to protect and preserve the force and secure the area of operations is vital in seizing, retaining, and exploiting the initiative. Protection emphasizes the importance of planning and expanding our protection priorities, to include protecting unified action partners, civilian populations, equipment, resources, infrastructure, and cultural landmarks across the range of military operations. It focuses on adapting our force to better leverage, integrate, and synchronize unified action capabilities and better understand operational environments. It emphasizes the need for Soldiers, leaders, and organizations to identify, prevent, or mitigate threats and hazards. Mutually supporting and overlapping protection capabilities through operational and tactical level actions better position forces to defend, respond, and recover from threat and hazard effects and to deter, counterattack, neutralize, and defeat the threats.

Protection Principles

The following principles of protection provide military professionals with a context for implementing protection efforts, developing schemes of protection, and allocating resources:

Comprehensive

Protection is an all-inclusive utilization of complementary and reinforcing protection tasks and systems available to commanders, incorporated into the plan, to preserve the force.

Integrated

Protection is integrated with other activities, systems, efforts, and capabilities associated with unified land operations to provide strength and structure to the overall effort. Integration must occur vertically and horizontally with unified action partners throughout the operations process.

Layered

Protection capabilities are arranged using a layered approach to provide strength and depth. Layering reduces the destructive effect of a threat or hazard through the dispersion of energy or the culmination of the force.

Redundant

Protection efforts are often redundant anywhere that a vulnerability or a critical point of failure is identified. Redundancy ensures that specific activities, systems, efforts, and capabilities that are critical for the success of the overall protection effort have a secondary or auxiliary effort of equal or greater capability.

Enduring

Protection capabilities are ongoing activities for maintaining the objectives of preserving combat power, populations, partners, essential equipment, resources, and critical infrastructure in every phase of an operation.

Chap 7

I. Protection Supporting Tasks

Ref: ADRP 3-37 (FM 3-37), Protection (Aug '12), chap. 1.

Commanders and staffs synchronize, integrate, and organize capabilities and resources throughout the operations process to preserve combat power and the freedom of action and to mitigate the effects of threats and hazards. Protection safeguards the force, personnel (combatants and noncombatants), systems, and physical assets of the United States and unified action partners. Survivability refers to the capacity, fitness, or tendency to remain alive or in existence. For the military, survivability is about much more than mere survival—it is also about remaining effective. Military forces are composed of personnel and physical assets, each having their own inherent survivability qualities or capabilities that permit them to avoid or withstand hostile actions or environmental conditions while retaining the ability to fulfill their primary mission. These inherent qualities or capabilities are affected by various factors (dispersion, redundancy, morale, leadership, discipline, mobility, situational understanding, terrain and weather conditions) and can be enhanced by tasks within the protection warfighting function.

I. Supporting Tasks

Supporting task of the protection warfighting function include:

Protection Supporting Tasks

- **A** Conduct Operational Area Security
- **B** Employ Safety Techniques (including Fratricide Avoidance)
- **C** Implement OPSEC
- **D** Provide Intelligence Support to Protection
- **E** Apply Antiterrorism Measures
- **F** Implement Physical Security Procedures
- **G** Conduct Law and Order
- **H** Conduct Survivability Operations
- **I** Provide Force Health Protection
- **J** Provide Explosive Ordnance Disposal (EOD) and Protection Support
- **K** Conduct Chemical, Biological, Radiological, and Nuclear (CBRN) Operations
- **L** Coordinate Air and Missile Defense
- **M** Conduct Personnel Recovery
- **N** Conduct Internment and Resettlement

Ref: ADRP 3-37 (FM 3-37), Protection (Aug '12), p. 1-3.

A. Conduct Operational Area Security

The task of conducting operational area security is a form of security operations conducted to protect friendly forces, installations, routes, and actions within an area of operations. Forces engaged in operational area security protect the force, installation, route, area, or asset. Although vital to the success of military operations, operational area security is normally an economy-of-force mission, often designed to ensure the continued conduct of sustainment operations and to support decisive and shaping operations by generating and maintaining combat power.

Operational area security may be the predominant method of protecting support areas that are necessary to facilitate the positioning, employment, and protection of resources required to sustain, enable, and control forces. Operational area security is often an effective method of providing civil security and control during some stability operations. Forces engaged in operational area security can saturate an area or position on key terrain to provide protection through early warning, reconnaissance, or surveillance and to guard against unexpected enemy or adversary attack with an active response. This early warning, reconnaissance or surveillance may come from ground- and space-based sensors. Operational area security often focuses on named areas of interest in an effort to answer commander's critical information requirements, aiding in tactical decisionmaking and confirming or denying threat intentions. Forces engaged in operational area security are typically organized in a manner that emphasizes their mobility, lethality, and communications capabilities. The maneuver enhancement brigade and some military police units are specifically equipped and trained to conduct operational area security and may constitute the only available force during some phases of an operation. However, operational area security takes advantage of the local security measures performed by all units, regardless of their location in the area of operations.

All commanders apportion combat power and dedicate assets to protection tasks and systems based on an analysis of the operational environment, the likelihood of threat action, and the relative value of friendly resources and populations. Based on their assessments, joint force commanders may designate the Army to provide a joint security coordinator to be responsible for designated joint security areas. Although all resources have value, the mission variables of METT-TC make some resources, assets, or locations more significant to successful mission accomplishment from enemy or adversary and friendly perspectives. Commanders rely on the risk management process and other specific assessment methods to facilitate decisionmaking, issue guidance, and allocate resources. Criticality, vulnerability, and recoverability are some of the most significant considerations in determining protection priorities that become the subject of commander guidance and the focus of operational area security.

See facing page for further discussion.

B. Employ Safety Techniques (Including Fratricide Avoidance)

Safety techniques are used to identify and assess hazards to the force and make recommendations on ways to prevent or mitigate the effects of those hazards. Commanders have the inherent responsibility to analyze the risks and implement control measures to mitigate them. All staffs understand and factor into their analysis how their execution recommendations could adversely affect Soldiers. Incorporating protection within the risk management integrating process is key. It ensures a thorough analysis of risks and implements controls to mitigate their effects. All commands develop and implement a command safety program that includes fratricide avoidance, occupational health, risk management, fire prevention and suppression, and accident prevention programs focused on minimizing safety risks.

Operational Area Security

Ref: ADRP 3-37 (FM 3-37), Protection (Aug '12), pp. 1-4 to 1-5.

Base/Base Camp Defense

Base defense is the local military measures, both normal and emergency, required to nullify or reduce the effectiveness of enemy attacks on, or sabotage of, a base to ensure that the maximum capacity of its facilities is available to U.S. forces (JP 3-10).

Critical Asset Security

Critical asset security is the protection and security of personnel and physical assets or information that is analyzed and deemed essential to the operation and success of the mission and to resources required for protection.

Node Protection

Command posts and operations centers are often protected through area security techniques that involve the employment of protection and security assets in a layered, integrated, and redundant manner.

High-Risk Personnel Security

High-risk personnel are personnel who, by their grade, assignment, symbolic value, or relative isolation, are likely to be attractive or accessible terrorist targets (JP 3-07.2).

Response Force Operations

Response force operations expediently reinforce unit organic protection capabilities or complement that protection with maneuver capabilities based on the threat. Response force operations include planning for the defeat of Level I and II threats and the shaping of Level III threats until a designated combined arms tactical combat force arrives for decisive operations. *Refer to FM 3-39 for more information.*

Lines of Communications Security

The security and protection of lines of communications and supply routes are critical to military operations since most support traffic moves along these routes. The security of lines of communications and supply routes (rail, pipeline, highway, and waterway) presents one of the greatest security challenges in an area of operations. Route security operations are defensive in nature and are terrain-oriented (see FM 3-90).

Checkpoints and Combat Outposts

It is often necessary to control the freedom of movement in an area of operations for a specific period of time or as a long-term operation. This may be accomplished by placing checkpoints and combat outposts along designated avenues and roadways or on key terrain identified through METT-TC. *Refer to ATTP 3-90.4 for more information.*

Convoy Security

A convoy security operation is a specialized kind of area security operations conducted to protect convoys (FM 3-90). Units conduct convoy security operations anytime there are insufficient friendly forces to continuously secure routes in an area of operations and there is a significant danger of enemy or adversary ground action directed against the convoy. *Refer to FM 4-01.45 for more information.*

Port Area and Pier Security

Ground forces may typically provide area security for port and pier areas. The joint force commander and subordinate joint force commanders ensure that port security plans and responsibilities are clearly delineated and assigned. *Refer to JP 3-10 for more information.*

Area Damage Control

Commanders conduct area damage control when the damage and scope of the attack are limited and they can respond and recover with local assets and resources. Optimally, commanders aim to recover immediately.

Fratricide

Fratricide is the unintentional killing or wounding of friendly or neutral personnel by friendly firepower. The destructive power and range of modern weapons, coupled with the high intensity and rapid tempo of combat, increase the potential for fratricide. Tactical maneuvers, terrain, and weather conditions may also increase the danger of fratricide.

Fratricide is accidental and is usually the end product of an error by a leader or Soldier. Accurate information about locations and activities of friendly and hostile forces and an aggressive airspace management plan help commanders avoid fratricide. The U.S. Army Space and Missile Defense Command Mission Management Center provides joint friendly force tracking for all Services, which improves the situational awareness and helps prevent fratricide. Liaison officers increase situational understanding and enhance interoperability. Commanders, leaders, and Soldiers must know the range and blast characteristics of their weapons systems and munitions to prevent ricochet, penetration, and other unintended effects.

Commanders, leaders and Soldiers are responsible for preventing fratricide. They must lower the probability of fratricide without discouraging boldness and audacity. Good leadership that results in positive weapons control, the control of troop movements, and disciplined operational procedures contribute to achieving this goal. Situational understanding, friendly personnel identification methods, and combat identification methods also help. Soldiers must be confident that the probability of misdirected friendly fire is low. Contractors authorized to accompany the force; local, national day laborers; and nongovernmental organization personnel who support Army operations face the same risks as U.S. forces. Since these personnel work and often live in and among U.S. forces, commanders must include them in protection and combat identification plans.

The potential for fratricide may increase with the fluid nature of the noncontiguous battlefield and the changing disposition of attacking and defending forces. The presence of noncombatants in the area of operations further complicates the scheme of maneuver. Simplicity and clarity are often more important than a complex, detailed plan when developing fratricide avoidance methods.

The effects of fratricide can be devastating to unit moral and confidence and can quickly diminish the mission effectiveness of a unit. Known postfratricide events have resulted in the following unit behavior:

- Hesitation to conduct limited visibility operations
- Loss of confidence in unit leadership
- Increase of leadership self-doubt
- Hesitation to use supporting combat systems
- Oversupervision of units
- Loss of initiative
- General degradation of unit cohesiveness, morale, and combat power

See facing page for discussion of risk mitigation and fratricide prevention.

Fire Prevention, Fire suppression, and Firefighting

Fire prevention, fire suppression, and firefighting encompass all efforts aimed at preventing or stopping fires. Fire prevention programs exist at all levels, and all levels of command are responsible for the Army fire protection plan. Commanders and supervisors are responsible for fire safety policies and plans in their organizations. Army firefighting capabilities consist of general firefighting and tactical firefighting:

Refer to FM 5-415 for more information on fire prevention, fire suppression, and firefighting.

Risk Mitigation and Fratricide Prevention

Ref: ADRP 3-37 (FM 3-37), Protection (Aug '12), p. 1-8.

Commanders ensure that risk mitigation strategies and fratricide prevention methods are employed and trained to lessen the risk of fratricide on the battlefield. Prevention methods include fratricide prevention training, weapons control measures, rules of engagement training, assembly area procedures, reconnaissance, rehearsals, back-briefs, unexploded ordnance training and reporting procedures, field discipline, friendly troop marking procedures and, most importantly, awareness at all levels.

In any situation involving the risk of fratricide due to friendly fire, leaders must be prepared to take immediate actions to prevent casualties, equipment damage, and equipment destruction. The recommended actions in fratricide situations include—

- Identify the incident and order all parties involved to cease fire
- Conduct an in-stride risk assessment
- Identify and implement controls to prevent the incident from recurring

Fratricide may be more prevalent during joint and multinational operations when communications and interoperability challenges are not fully resolved. Fratricide avoidance is normally accomplished through a scheme of protection that emphasizes prevention and is centered on awareness and target identification:

Awareness

Awareness is the immediate knowledge of the conditions of the operation, constrained geographically and in time. It includes the real-time, accurate knowledge of one's own location and orientation and the locations, activities, and intentions of other friendly, enemy, adversary, neutral, or noncombatant elements in the area of operations, sector, zone, or immediate vicinity. As previously mentioned, the U.S. Army Space and Missile Defense Command joint friendly force tracking mission aids in the overall awareness of personnel location in the operational environment.

Target Identification

Target identification is the accurate and timely characterization of a detected object on the battlefield as friend, neutral, or enemy. This aspect of combat identification is time sensitive and directly supports a combatant's shoot or don't-shoot decision for detected objects on the battlefield (FM 3-20.15). Unknown objects should not be engaged; rather, the target identification process should continue until positive identification is made. An exception to this is a weapons-free zone where units can fire at anything that is not positively identified as friendly.

C. Implement Operations Security

Operations security is a process of identifying critical information and subsequently analyzing friendly actions attendant to military operations and other activities (JP 3-13.3). OPSEC may also be used to—

- Identify actions that can be observed by enemy or adversary intelligence systems
- Determine indicators of hostile intelligence that systems might obtain which could be interpreted or pieced together to derive critical information in time to be useful to adversaries or enemies
- Execute measures that eliminate or reduce (to an acceptable level) the vulnerabilities of friendly actions to enemy or adversary exploitation

OPSEC applies to all operations. All units conduct OPSEC to preserve essential secrecy. Commanders establish routine OPSEC measures in unit standing operating procedures. The unit OPSEC officer coordinates additional OPSEC measures with other staff and command elements and synchronizes with adjacent units. The OPSEC officer develops OPSEC measures during the military decisionmaking process. The assistant chief of staff, intelligence, assists the OPSEC process by comparing friendly OPSEC indicators with enemy or adversary intelligence collection capabilities.

Refer to JP 3-13.3 for additional OPSEC information.

D. Provide Intelligence Support to Protection

This is an intelligence warfighting function task that supports the protection warfighting function. It includes providing intelligence that supports measures which the command takes to remain viable and functional by protecting itself from the effects of threat activities. It also provides intelligence that supports recovery from threat actions. It includes analyzing the threats, hazards, and other aspects of an operational environment and utilizing the intelligence preparation of the battlefield process to describe the operational environment and identify threats and hazards that may impact protection. Intelligence support develops and sustains an understanding of the enemy, terrain and weather, and civil considerations that affect the operational environment. Information collection is an activity that synchronizes and integrates the planning and employment of sensors and assets as well as the processing, exploitation, and dissemination of systems in direct support of current and future operations (FM 3-55). Information collection can complement or supplement protection tasks. Through information collection, commanders and staffs continuously plan, task, and employ collection assets and forces. These forces collect, process, and disseminate timely and accurate information to satisfy the commander's critical information requirements and other intelligence requirements. When necessary, information collection assets (ground- and space-based reconnaissance and surveillance activities) focus on special requirements, such as personnel recovery.

Refer to ADRP 2-0 for additional intelligence information.

E. Apply Antiterrorism (AT) Measures

AT consists of defensive measures that are used to reduce the vulnerability of individuals and property to terrorist acts, including limited response and containment by local military and civilian forces. AT is a consideration for all forces during all military operations.

AT is an integral part of Army efforts to defeat terrorism. Terrorists can target Army elements at any time and in any location. By effectively preventing and, if necessary, responding to terrorist attacks, commanders protect all activities and people so that Army missions can proceed unimpeded. AT is neither a discrete task nor the sole responsibility of a single branch; all bear responsibility. AT must be integrated into all Army operations and considered at all times. Awareness must be built into every mission, every Soldier, and every leader. Integrating AT represents the foundation that is crucial for Army success.

F. Implement Physical Security Procedures

Ref: ADRP 3-37 (FM 3-37), Protection (Aug '12), p. 1-8.

Physical security consists of physical measures that are designed to safeguard personnel; to prevent unauthorized access to equipment, installations, material, and documents; and to safeguard them against espionage, sabotage, damage, and theft. The Army employs physical security measures in depth to protect personnel, information, and critical resources in all locations and situations against various threats through effective security policies and procedures. This total system approach is based on the continuing analysis and employment of protective measures, including physical barriers, clear zones, lighting, access and key control, intrusion detection devices, defensive positions, and nonlethal capabilities.

The goal of physical security systems is to employ security in depth to preclude or reduce the potential for sabotage, theft, trespass, terrorism, espionage, or other criminal activity. To achieve this goal, each security system component has a function and related measures that provide an integrated capability for—

Deterrence

A potential aggressor who perceives a risk of being caught may be deterred from attacking an asset. The effectiveness of deterrence varies with the aggressor's sophistication, the attractiveness of the asset, and the aggressor's objective.

Detection

A detection measure senses an act of aggression, assesses the validity of the detection, and communicates the appropriate information to a response force.

Assessment

Assessment—through the use of alarm systems, video surveillance systems, other types of detection systems, patrols, or fixed posts—assists in localizing and determining the size and intent of an unauthorized intrusion or activity.

Delay

Delay measures protect an asset from aggression by delaying or preventing an aggressor's movement toward the asset or by shielding the asset from weapons and explosives.

Response

Most protective measures depend on response personnel to assess unauthorized acts, report detailed information, and defeat an aggressor.

Refer to ATTP 3-39.32 for additional information on physical security procedures.

Typical Army AT programs are composed of several adjunct and information programs, including tasks for specialized, nonprotection military occupational specialties. AT includes the following areas at a minimum:

- Risk management (threat, critical asset, and vulnerability assessments of units, installations, facilities, and bases/base camps)
- AT planning (units, installations, facilities, and bases)
- AT awareness training and command information programs
- The integration of various vulnerability assessments of units, installations, facilities, bases/base camps, personnel, and activities
- AT protection measures to protect individual personnel, high-risk personnel, physical assets (physical security), and designated critical assets and information
- Resource application
- Civil and military partnerships
- Force protection condition systems to support terrorist threat and incident response plans
- Comprehensive AT program review

Refer to FM 3-37.2 for additional information on AT measures.

G. Conduct Law and Order Operations

Law and order operations encompass policing and the associated law enforcement activities to control and protect populations and resources and to facilitate the existence of a lawful and orderly environment. Law and order operations and the associated skills and capabilities inherent in that function provide the fundamental base on which all other military police functions are framed and conducted.

Law and order operations are conducted across the range of military operations. As the operation transitions and the operational environment stabilizes, civil control efforts are implemented and the rule of law is established. The closer the operational environment moves toward stability and full implementation of host nation governance under the rule of law, the more general policing activities transition to law enforcement activities.

The ultimate goal is to maintain order while protecting personnel and assets. Military police Soldiers and leaders apply this policing approach when conducting all operations. The military police view shares a common general understanding of the operational environment, while adding a degree of focus on those aspects that are necessary to maintain order and enforce laws. Care should be taken to eliminate jurisdictional overlap and under lap. Law and order operations include—

- Performing law enforcement
- Conducting criminal investigations
- Conducting traffic management and enforcement
- Employing forensics capabilities
- Conducting police engagement
- Providing customs support
- Providing host nation police development
- Supporting civil law enforcement
- Supporting border control, boundary security, and the freedom of movement

Refer to ATTP 3-39.10 for additional information on law and order operations.

H. Conduct Survivability Operations

Personnel and physical assets have inherent survivability qualities or capabilities that can be enhanced through various means and methods. When existing terrain features offer insufficient cover and concealment, survivability can be enhanced by altering the physical environment to provide or improve cover and concealment. Similarly, natural or artificial materials may be used as camouflage to confuse, mislead, or evade the enemy or adversary. Together, these are called survivability operations—those military activities that alter the physical environment to provide or improve cover, concealment, and camouflage. By providing or improving cover, concealment, and camouflage, survivability operations help military forces avoid or withstand hostile actions. Although such activities often have the added benefit of providing shelter from the elements, survivability operations focus on providing cover, concealment, and camouflage. All units conduct survivability operations within the limits of their capabilities. Engineer and CBRN personnel and units have additional capabilities to support survivability operations.

Survivability operations enhance the ability to avoid or withstand hostile actions by altering the physical environment. They accomplish this by providing or improving cover, concealment, and camouflage in four areas:

- Fighting positions
- Protective positions
- Hardened facilities
- Camouflage and concealment

The first three areas focus on providing cover (although not excluding camouflage and concealment). The fourth area focuses on providing protection from observation and surveillance. All four areas, but especially the first three, often have the added benefit of providing some degree of shelter from the elements. The areas of survivability operations are often addressed in combination. For example, fighting positions and protective positions usually require camouflage and concealment also. Camouflage and concealment activities often accompany activities to harden facilities.

Refer to FM 5-103 for more information on survivability and survivability operations. Refer to FM 3-34.400 for information on base camps. Refer to FM 90-7 for information on obstacle integration.

I. Provide Force Health Protection

Force health protection encompasses measures to promote, improve, or conserve the mental and physical well-being of Soldiers. These measures enable a healthy and fit force, prevent injury and illness, protect the force from health hazards, and include the prevention aspects of a number of Army Medical Department functions:

- Preventive medicine (medical surveillance, occupational and environmental health surveillance)
- Veterinary services (food inspection, animal care missions, prevention of zoonotic disease transmissible to man)
- Combat and operational stress control
- Dental services (preventive dentistry)
- Laboratory services (area medical laboratory support)

Army personnel must be physically and behaviorally fit. This requirement demands programs that promote and improve the capacity of personnel to perform military tasks at high levels, under extreme conditions, and for extended periods of time. These preventive and protective capabilities include physical exercise, nutritional diets, dental hygiene and restorative treatment, combat and operational stress management, rest, recreation, and relaxation that are geared to individuals and organizations.

Methods to prevent disease are best applied synergistically. Sanitation practices, waste management, and pest and vector control are crucial to disease prevention. Regional spraying and insect repellent application to guard against hazardous flora and fauna are examples of prevention methods. Prophylactic measures can encompass human and animal immunizations, dental chemoprophylaxis and treatment, epidemiology, optometry, counseling on specific health threats, and protective clothing and equipment.

The key to preventive and protective care is information—the capacity to anticipate the current and true health environment and the proper delivery of information to the affected human population. Derived from robust health surveillance and medical intelligence, this information addresses occupational, local environmental, and enemy- or adversary-induced threats from industrial hazards, air and water pollution, endemic or epidemic disease, CBRN threats or hazards, and directed-energy device weapons (highpowered microwaves, particle beams, lasers). Health service support must be capable of acquiring, storing, moving, and providing information that is timely, relevant, accurate, concise, and applicable to individuals. In summary, this information capability is crucial to force health protection. Force health protection includes—

- Preventing and controlling diseases
- Assessing environmental and occupational health
- Determining force health activities protection
- Employing preventive medicine toxicology and laboratory services
- Performing health risk assessments
- Disseminating health information

Refer to ATTP 4-02 for more information on force health protection.

J. Provide Explosive Ordnance Disposal and Protection Support

The role of EOD is to eliminate or reduce the effects of explosive ordnance and hazards to protect combat power and the freedom of action. Explosive ordnance and hazards are ever-present dangers in most areas of operation. They limit mobility, deny the use of critical assets, and potentially injure or kill Soldiers and civilians. The U.S. and multinational use of munitions that disperse submunitions across a wide area has led to increased amounts of unexploded ordnance on the battlefield. EOD forces have the capability to render-safe and destroy explosive ordnance and hazards across the range of military operations. EOD units are specifically trained in render-safe procedures and the disposal of explosive ordnance, explosive hazards, and CBRN munitions. While other forces may have the ability to destroy limited explosive ordnance by detonation, they are not properly equipped, trained, or authorized to perform render-safe procedures or other disposal procedures. EOD elements normally—

- Identify and collect information on explosive ordnance and hazards.
  - Perform an initial assessment of found munitions, which include single munitions discovered or captured during military operations (patrols, raids, maneuvers) and those obtained through buyback or amnesty programs
  - Assist commanders with AT, including intelligence support, electronic-warfare defense plans, bomb threat and search procedures, facility site surveys, and the development and implementation of EOD emergency response and AT plans
  - Collect weapons technical intelligence on explosive ordnance and hazards, including first-seen items of interest

K. Conduct Chemical, Biological, Radiological, and Nuclear (CBRN) Operations

Ref: ADRP 3-37 (FM 3-37), Protection (Aug '12), pp. 1-11 to 1-12.

CBRN threats and hazards include WMD, improvised weapons and devices, and toxic industrial material. All of these can potentially cause mass casualties and large-scale destruction. Many state and nonstate actors (including terrorists and criminals) possess or have the capability to possess, develop, or proliferate WMD.

CBRN operations include the employment of tactical capabilities that counter the entire range of CBRN threats and hazards through—

- Weapons of mass destruction (WMD) proliferation prevention (security cooperation and partner activities and threat reduction cooperation)
- WMD counterforce (interdiction, offensive operations, and elimination)
- CBRN defense (active and passive defense)
- CBRN consequence management

CBRN operations support operational and strategic objectives to combat WMD and operate safely in a CBRN environment. They include—

- **Providing WMD security cooperation and partner activities support.** WMD security cooperation and partner activities improve or promote defense relationships and the capacity of allied and partner nations to execute or support other military mission areas to combat WMD through military-to-military contact, burden-sharing arrangements, combined military activities, and support to international activities.
- **Providing WMD threat reduction cooperation support.** WMD threat reduction cooperation activities are undertaken with the consent and cooperation of host nation authorities in a permissive environment to enhance physical security and to reduce, dismantle, redirect, and/or improve the protection of an existing state WMD program, stockpiles, and capabilities.
- **Conducting WMD interdiction operations**. WMD interdiction operations track, intercept, search, divert, seize, or otherwise stop the transit of WMD, WMD delivery systems, or WMD-related materials, technologies, and expertise.
- **Conducting WMD offensive operations.** WMD offensive operations disrupt, neutralize, or destroy a WMD threat before it can be used; or they deter the subsequent use of a WMD.
- **Conducting WMD elimination operations.** WMD elimination operations are conducted in a hostile or uncertain environment to systematically locate, characterize, secure, disable, or destroy WMD programs and related capabilities. *Refer to ATTP 3-11.23 for more information.*
- **Conducting CBRN active defense.** CBRN active defense includes measures to defeat an attack with CBRN weapons by employing actions to divert, neutralize, or destroy those weapons or their means of delivery while en route to their target.
- **Conducting CBRN passive defense.** CBRN passive defense includes measures taken to minimize or negate the vulnerability to, and effects of, CBRN incidents. This mission area focuses on maintaining force ability to continue military operations in a CBRN environment. *Refer to FM 3-11.3 to 3-11.5.*
- **Conducting CBRN consequence management operations.** CBRN consequence management consists of actions taken to plan, prepare, respond to, and recover from a CBRN incident that requires forces and resource allocation beyond passive defense capabilities. *Refer to FM 3-11 and FM 3-11.21*

Refer to FM 3-11 for additional information on CBRN operations.

- Render-safe and dispose of explosive ordnance and hazards
  - Assist commanders with the implementation of protective works and consequence management
  - Provide technical advice and assistance to combat engineers during route, area, and minefield clearance operations
  - Support responses to nuclear and chemical accidents and incidents, including technical advice and procedures to mitigate hazards associated with such items
  - Provide EOD Soldiers in support of humanitarian assistance efforts that involve explosive ordnance and hazards

EOD is the only force equipped, manned, and trained to positively identify, render-safe, and dispose of U.S. and foreign explosive ordnance and improvised explosive devices.

Refer to ATTP 4-32 for additional EOD information.

L. Coordinate Air and Missile Defense

Air and missile defense protects the force from missile attack, air attack, and aerial surveillance by ballistic missiles, cruise missiles, conventional fixed- and rotary-wing aircraft, and unmanned aerial systems. It prevents enemies from interdicting friendly forces, while freeing commanders to synchronize movement and firepower. All members of the combined arms team perform air defense tasks; however, ground-based air defense artillery units execute most Army air and missile defense operations. Air and missile defense elements coordinate and synchronize defensive fires to protect installations and personnel from over-the-horizon strikes. Army air and missile defense capabilities increase airspace situational understanding and complement the area air defense commander.

Indirect-fire protection systems protect forces from threats that are largely immune to air defense artillery systems. The indirect-fire protection intercept capability is designed to detect and destroy incoming rocket, artillery, and mortar fires. This capability assesses the threat to maintain friendly protection and destroys the incoming projectile at a safe distance from the intended target.

The air and missile defense task consists of active and passive measures that protect personnel and physical assets from an air or missile attack. Passive measures include camouflage, cover, concealment, hardening, and OPSEC. Active measures are taken to destroy, neutralize, or reduce the effectiveness of hostile air and missile threats. The early warning of in-bound missile threats is provided in theater by the globally located, joint tactical ground stations.

Protection cell planners coordinate with the Air Defense Airspace Management Cell for Air and Missile Defense for the protection of the critical asset list (CAL) and defended asset list (DAL) and for other air and missile defense protection as required. There is continuous coordination to refine the CAL and DAL throughout operations, ensuring the protection of critical assets and forces from air and missile attack and surveillance.

The air and missile defense assets integrate protective systems by using the six employment guidelines—mutual support, overlapping fires, balanced fires, weighted coverage, early engagement, and defense in depth—and additional considerations necessary to mass and mix air and missile defense capabilities. These employment guidelines enable air defense artillery forces to successfully accomplish combat missions and support overall force objectives.

Refer to ADRP 3-09 for more information on air and missile defense operations and airspace control.

M. Conduct Personnel Recovery

Ref: ADRP 3-37 (FM 3-37), Protection (Aug '12), pp. 1-13 to 1-14.

Army personnel recovery is the sum of military, diplomatic, and civil efforts to prevent isolation incidents and to return isolated persons to safety or friendly control. Personnel recovery is the overarching term for operations that focus on recovering isolated or missing personnel before they become detained or captured.

Personnel recovery operations are conducted to recover and return personnel who are isolated, missing, detained, or captured in an operational area. These personnel consist of U.S. forces, Army civilians, or other personnel (as designated by the President or the Secretary of Defense) who are in an operational environment beyond the Army's positive or procedural control, requiring them to survive, evade, resist, or escape. Every unit must have procedures in place to recover personnel.

Commanders must understand the operational environment and the impact of PMESII-PT to ensure that personnel recovery is incorporated into and supports each mission. This includes the characteristics of the particular operational environment and how aspects of the environment become essential elements in shaping the way that Army forces conduct operations. Threats to isolated Soldiers will vary based on the operational environment.

Personnel recovery is not a stand-alone mission; it is incorporated into mission planning. Personnel recovery operations are supported through joint friendly force tracking activities. Personnel recovery guidance must synchronize the actions of commanders and staffs, recovery forces, and isolated individuals. In order to synchronize the actions of all three, commanders develop personnel recovery guidance based on command capabilities to conduct recovery operations. By knowing what actions they have dictated to potential isolated Soldiers, commanders develop situational understanding and provide guidance to their staffs and recovery forces to synchronize their actions with those of isolated personnel.

Commanders must integrate personnel recovery throughout operations. This requires an understanding of the complex, dynamic relationships between friendly forces and enemies and the other aspects of the operational environment (including the populace). This understanding helps commanders visualize and describe their intent for personnel recovery and helps them develop focused planning guidance. As commanders develop personnel recovery guidance for subordinate units, they must ensure that subordinates have adequate combat power for personnel recovery. Commanders must also provide resources and define command relationships with the requisite flexibility to plan and execute personnel recovery operations.

Commanders provide personnel recovery planning guidance within their initial guidance. Personnel recovery guidance provides a framework for how the unit and subordinates will synchronize the actions of isolated personnel and the recovery force. Effective personnel recovery planning guidance accounts for the operational environment and the execution of operations. Personnel recovery guidance is addressed in the synchronization of each warfighting function. It broadly describes how the commander intends to employ combat power to accomplish personnel recovery tasks within the higher commander's intent.

Refer to FM 3-50.1 for additional information on personnel recovery operations.

CBRN threats and hazards include WMD, improvised weapons and devices, and toxic industrial material. All of these can potentially cause mass casualties and large-scale destruction. Many state and nonstate actors (including terrorists and criminals) possess or have the capability to possess, develop, or proliferate WMD.

Refer to FM 3-11 for additional information on CBRN operations.

N. Conduct Internment and Resettlement

Internment and resettlement operations are conducted by military police to shelter, sustain, guard, protect, and account for populations (detainees, dislocated civilians, and U.S. military prisoners) as a result of military or civil conflict and natural or man-made disasters or to facilitate criminal prosecution:

Internment

Internment involves the detainment of a population or group that pose some level of threat to military operations.

Resettlement

Resettlement involves the quartering of a population or group for their protection.

These operations inherently control the movement and activities of their specific population for imperative reasons of security, safety, or intelligence gathering. The Army is the DOD executive agent for all detainee operations and for the long-term confinement of U.S. military prisoners.

Internment and resettlement operations include—

- Performing internment
- Interning U.S. military prisoners
- Supporting host nation corrections reform
- Conducting resettlement operations
- Conducting enemy prisoner of war operations
- Conducting detainee operations

Refer to FM 3-39.40 for more information on internment and resettlement operations.

II. Tasks and Systems Integration

In order to achieve protection and preserve combat power across the range of military operations, the scheme of protection must be comprehensive, integrated, layered, redundant, and enduring.

The protection warfighting function tasks and systems, when integrated throughout the operations process, help establish control measures against potential threats and hazards. The layering of protection tasks and systems, some even redundant, ensures a comprehensive scheme of protection. The layered approach of protection provides strength and depth. Units utilize their available capabilities to defend the protection priorities and a layering of capabilities reduces the destructive effect of threats and hazards.

Individuals are protected at the lowest level by awareness, personal protective equipment, an understanding of the rules of engagement, and fratricide avoidance measures. By implementing additional protection measures in the area surrounding an individual (fighting positions, vehicles, collective protection, and force health protection measures taken against accidents and disease), the force then provides a layering of protection. Implementing AT and physical security measures, enhancing survivability measures, and applying active and passive defense operations add to the next layer of a comprehensive, integrated, layered scheme of protection. Implementing the protection tasks and utilizing protection systems in a comprehensive, layered scheme of protection preserves the critical assets throughout the range of military operations in any operational environment.

Chap 7

II. Protection Planning

Ref: ADRP 3-37 (FM 3-37), Protection (Aug '12), chap. 2. See also p. 7-6.

Planning is the first step toward effective protection. Commanders consider the most likely threats and hazards and then decide which personnel, physical assets, and information to protect. They set protection priorities for each phase or critical event of an operation. The military decisionmaking process or troop leading procedures provide a deliberate process and context to develop and examine information for use in the various continuing activities and integrating processes that comprise the operations process. An effective scheme of protection and risk decisions are developed based on the information that flows from mission analysis, allowing a thorough understanding of the situation, mission, and environment. Mission analysis provides a context to identify and analyze threats and hazards, the situational understanding of the operational environment, and the development of the scheme of protection.

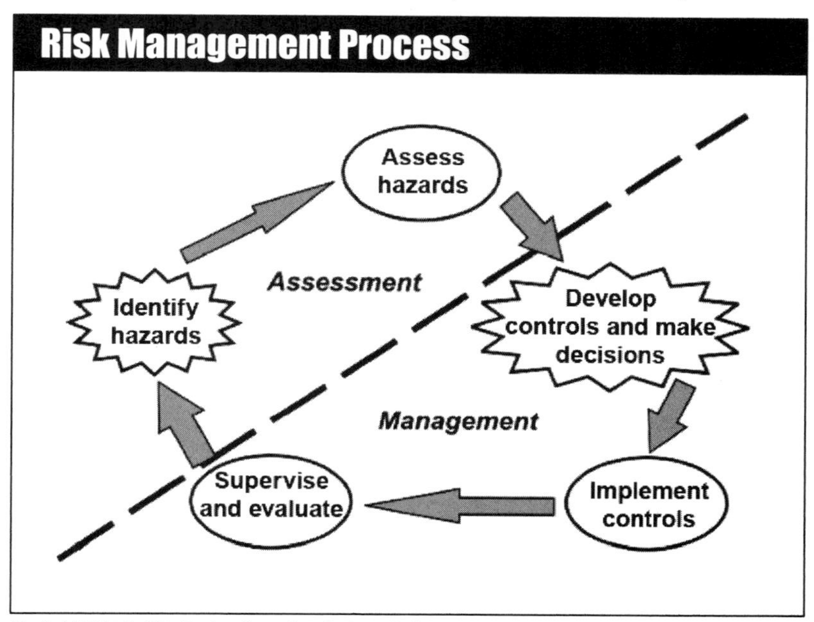

Ref: ADRP 3-37, Protection, fig. 2-1, p. 2-2.

I. Initial Assessments

Initial protection planning requires various assessments to support protection prioritization; namely, threat, hazard, vulnerability, criticality, and capability. These assessments are used to determine which assets can be protected given no constraints (critical assets) and which assets can be protected with available resources (defended assets). Commanders make decisions on acceptable risks and provide guidance to the staff so that they can employ protection capabilities based on the CAL and DAL. All forms of protection are utilized and employed during preparation and continue through execution to reduce friendly vulnerability.

Protection (ADRP 3-37)

Initial Assessments (Protection)

Threat and Hazard Assessment *(p. 7-28)*

Vulnerability Assessment *(p. 7-30)*

Criticality Assessement *(p. 7-30)*

Capability Assessement *(p. 7-31)*

Ref: ADRP 3-37, Protection, chap. 2.

II. Integrating Processes

The integrating processes of intelligence preparation of the battlefield, targeting, and risk management are essential in providing assessments or key information to assessments. They are a vital part of integrating protection within the other warfighting functions and throughout the operations process.

Intelligence Preparation of the Battlefield (IPB)

The intelligence preparation of the battlefield is a systematic process of analyzing and visualizing the mission variables of threat, terrain, weather, and civil considerations in a specific area of interest and for a specific mission. By applying the intelligence preparation of the battlefield, commanders gain the information necessary to selectively apply and maximize operation effectiveness at critical points in time and space.

Targeting

The targeting process integrates commander guidance and priorities to determine which targets to engage and how, when, and where to engage them in order to assign friendly capabilities to achieve the desired effect. The staff then assigns friendly capabilities that are best suited to produce the desired effect on each target. An important part of targeting is identifying possibilities for fratricide and collateral damage. Commanders establish control measures, including the consideration for restraint, that are necessary to minimize the chance of these events. The protection priorities must be integrated within the targeting process to achieve the desired effects while ensuring the preservation of combat power.

Risk Management

Risk management is the process of identifying, assessing, and controlling risks that arise from operational factors and making decisions that balance risk cost with mission benefits. Threat, hazard, capability, vulnerability, and criticality assessments are utilized to evaluate the risk to the force, determine the critical assets, ascertain available resources, and apply security or defensive measures to achieve protection. Risk management helps commanders preserve lives and resources, avoid or mitigate unnecessary risk, identify and implement feasible and effective control measures where specific standards do not exist, and develop valid courses of action (COAs). Risk management integration during operations process activities is the primary responsibility of the unit protection officer or operations officer.

See fig. 2-1 on previous page for an overview of the risk management process.

Threats and Hazards

Ref: ADRP 3-37 (FM 3-37), Protection (Aug '12), p. 2-3.

The protection warfighting function preserves the combat power potential and survivability of the force by providing protection from threats and hazards. Threats and hazards have the potential to cause personal injury, illness, or death; equipment or property damage or loss; or mission degradation.

- **Hostile actions.** Threats from hostile actions include any capability that forces or criminal elements have to inflict damage upon personnel, physical assets, or information. These threats may include improvised explosive devices, suicide bombings, network attacks, mortars, asset theft, air attacks, or CBRN weapons.
- **Nonhostile activities.** Nonhostile activities include hazards associated with Soldier duties within their occupational specialty, Soldier activity while off duty, and unintentional actions that cause harm. Examples include on- and off-duty accidents, OPSEC violations, network compromises, equipment malfunctions, or accidental CBRN incidents.
- **Environmental conditions.** Environmental hazards associated with the surrounding environment could potentially degrade readiness or mission accomplishment. Weather, natural disasters, and diseases are common examples. The staff also considers how military operations may affect noncombatants in the area of operations. Such considerations prevent unnecessary collateral damage and regard how civilians will affect the mission. Heavy civilian vehicle or pedestrian traffic adversely affects convoys and other operations.

Threats

The various actors in any area of operations can qualify as a threat, enemy, adversary, neutral, or friendly. Land operations often prove complex because actors intermix, often with no easy means to distinguish one from another.

- A **threat** is any combination of actors, entities, or forces that have the capability and intent to harm United States forces, United States national interests, or the homeland (ADRP 3-0). Threats may include individuals, groups of individuals (organized or not organized), paramilitary or military forces, nation-states, or national alliances.
- An **enemy** is a party identified as hostile against which the use of force is authorized (ADRP 3-0). An enemy is also called a **combatant** and is treated as such under the law of war.
- An **adversary** is a party acknowledged as potentially hostile to a friendly party and against which the use of force may be envisaged (JP 3-0)
- A **neutral** is a party identified as neither supporting nor opposing friendly or enemy forces (ADRP 3-0)
- A **friendly** is a contact positively identified as friendly (JP 3-01)
- A **hybrid threat** is the diverse and dynamic combination of regular forces, irregular forces, terrorist forces, and/or criminal elements unified to achieve mutually benefitting effects (ADRP 3-0).

Hazards

A hazard is a condition with the potential to cause injury, illness, or death of personnel; damage to or loss of equipment or property; or mission degradation (JP 3-33). Hazards are usually predictable and preventable and can be reduced through effective risk management efforts. Commanders differentiate hazards from threats and develop focused schemes of protection and priorities that match protection capabilities with the corresponding threat or hazard, while synchronizing those efforts in space and time. However, hazards can be enabled by the tempo or friction or by the complacency that sometimes develops during extended military operations.

A. Threat and Hazard Assessment

Ref: ADRP 3-37 (FM 3-37), Protection (Aug '12), pp. 2-3 to 2-5.

Personnel from all staff sections and warfighting functions help conduct threat and hazard analysis. This analysis comprises a thorough, in-depth compilation and examination of information and intelligence that address potential threats and hazards in the area of operations. The integrating processes (intelligence preparation of the battlefield, targeting, and risk management) provide an avenue to obtain the threats and hazards that are reviewed and refined. Threat and hazard assessments are continuously reviewed and updated as the operational environment changes.

Considerations for the threat and hazard assessment include—

- Enemy and adversary threats
  - Operational capabilities
  - Intentions
  - Activities
- Foreign intelligence and security service threats
- Crimes
- Civil disturbances
- Medical and safety hazards
- CBRN weapons and toxic industrial material
- Other relevant aspects of the operational environment
- Incident reporting and feedback points of contact

The threat and hazard assessment results in a comprehensive list of threats and hazards and determines the likelihood or probability of occurrence of each threat or hazard. Table 2-1 shows examples of potential threats and hazards in an area of operations. In the context of assessing risk, the higher the probability or likelihood of a threat or hazard occurring, the higher the risk of asset loss.

Potential Threats and Hazards

Area of Concern	Potential Threats and Hazards
Area security	• Assassination of, or attacks on, important personnel • Enemy, adversary or terrorist attacks on facilities • Ambushes or attacks on convoys • Enemy or adversary attacks on convoy routes
Safety	• Hazards associated with enemy or adversary activity • Accident potential • Weather or environmental conditions • Equipment
Fratricide avoidance	• Poor or reduced awareness • Inexperienced or poorly equipped or disciplined personnel • Complex or poorly defined mission against an experienced enemy or adversary
OPSEC	• Accidental friendly release of essential elements of friendly information • Enemy or adversary collection and exploitation of essential elements of friendly information • Enemy or adversary capture of unclassified friendly information • Physical security violations • Enemy or adversary intelligence gathering
AT	• Improvised explosive devices • Suicide bombs • Mail bombs • Snipers • Standoff weapons • WMD • Active shooters • Insider threats
Survivability	• Environmental conditions • Capabilities of threat weapons and sensors
Force health protection	• Endemic and epidemic diseases • Environmental factors • Diseases from animal bites, poisonous plants, animals, or insects • Risks associated with the health, sanitation, or behavior of the local populace
CBRN	• CBRN weapons • Toxic industrial materials
EOD	• Explosive ordnance and hazards (friendly and enemy) • Adversary attacks on personnel, vehicles, or infrastructure
Air and missile defense	• Artillery • Mortars • Rockets • Ballistic and cruise missiles • Fixed- and rotary-wing aircraft • Unmanned aerial systems
Personnel recovery	• Events that separate or isolate individuals or small groups of friendly forces from the main force

Legend:

AT	antiterrorism
CBRN	chemical, biological, radiological, and nuclear
EOD	explosive ordnance disposal
OPSEC	operations security
WMD	weapons of mass destruction

Ref: ADRP 3-37, Protection, table 2-1, pp. 2-4 to 2-5.

B. Vulnerability Assessment

A vulnerability assessment is an evaluation (assessment) to determine the magnitude of a threat or hazard effect against an installation, personnel, unit, exercise, port, ship, residence, facility, or other site. It identifies the areas of improvement necessary to withstand, mitigate, or deter acts of violence or terrorism. The staff addresses who or what is vulnerable and how it is vulnerable. The vulnerability assessment identifies physical characteristics or procedures that render critical assets, areas, infrastructures, or special events vulnerable to known or potential threats and hazards. Vulnerability is the component of risk over which the commander has the most control and greatest influence. The general sequence of a vulnerability assessment is—

Step 1. List assets and capabilities and the threats against them

Step 2. Determine the common criteria for assessing vulnerabilities

Step 3. Evaluate the vulnerability of assets and capabilities

Vulnerability evaluation criteria may include the degree to which an asset may be disrupted, quantity available (if replacement is required due to loss), dispersion (geographic proximity), and key physical characteristics.

DOD has created several decision support tools to perform criticality assessments in support of the vulnerability assessment process, including—

MSHARPP (mission, symbolism, history, accessibility, recognizability, population, and proximity)

MSHARPP is a targeting analysis tool that is geared toward assessing personnel vulnerabilities, but it also has application in conducting a broader analysis. The purpose of the MSHARPP matrix is to analyze likely terrorist targets and to assess their vulnerabilities from the inside out.

CARVER (criticality, accessibility, recuperability, vulnerability, effect, and recognizability)

The CARVER matrix is a valuable tool in determining criticality and vulnerability. For criticality purposes, CARVER helps assessment teams and commanders (and the assets that they are responsible for) determine assets that are more critical to the success of the mission. *Refer to FM 3-37.2 for more information on MSHARPP and CARVER.*

C. Criticality Assessment

A criticality assessment identifies key assets that are required to accomplish a mission. It addresses the impact of a temporary or permanent loss of key assets or the unit ability to conduct a mission. A criticality assessment should also include high-population facilities (recreational centers, theaters, sports venues) which may not be mission-essential. It examines the costs of recovery and reconstitution, including time, expense, capability, and infrastructure support. The staff gauges how quickly a lost capability can be replaced before giving an accurate status to the commander. The general sequence for a criticality assessment is—

Step 1. List the key assets and capabilities.

Step 2. Determine if critical functions or combat power can be substantially duplicated with other elements of the command or an external resource.

Step 3. Determine the time required to substantially duplicate key assets and capabilities in the event of temporary or permanent loss.

Step 4. Set priorities for the response to threats toward personnel, physical assets, and information.

The protection cell staff continuously updates the criticality assessment during the operations process. As the staff develops or modifies a friendly COA, information collection efforts confirm or deny information requirements. As the mission or threat changes, initial criticality assessments may also change, increasing or decreasing the subsequent force vulnerability. The protection cell monitors and evaluates these changes and begins coordination among the staff to implement modifications to the protection concept or recommends new protection priorities. Priority intelligence requirements, running estimates, measures of effectiveness (MOEs), and measures of performance (MOPs) are continually updated and adjusted to reflect the current and anticipated risks associated with the operational environment.

D. Capability Assessment

A capability assessment of an organization determines its current capacity to perform protection tasks based on the integrated material and nonmaterial readiness of the assets. A capability assessment considers the mitigating effects of existing manpower, procedures, and equipment. It is especially important in identifying capability gaps, which may be addressed to reduce the consequences of a specific threat or hazard. A capability assessment—

- Considers the range of identified and projected response capabilities necessary for responding to any type of hazard or threat
- Lists force resources, by type, and corresponding protection tasks
- Determines which assets are necessary to defend key areas

III. Protection Priorities

Criticality, vulnerability, and recuperability are some of the most significant considerations in determining protection priorities that become the subject of commander guidance and the focus of area security operations. The scheme of protection is based on the mission variables and should include protection priorities by area, unit, activity, or resource.

Although all military assets are important and all resources have value, the capabilities they represent are not equal in their contribution to decisive operations or overall mission accomplishment. Determining and directing protection priorities may be the most important decisions that commanders make and that staffs support. There are seldom sufficient resources to simultaneously provide the same level of protection to all assets. For this reason, commanders use risk management to identify increasingly risky activities and events, while other decision support tools assist in prioritizing protection resources.

Most prioritization methodologies assist in differentiating what is important from what is urgent. In protection planning, the challenge is to differentiate between critical assets and important assets and to further determine what protection is possible with available protection capabilities. Event-driven operations may be short in duration, enabling a formidable protection posture for a short time; while condition-driven operations may be open-ended and long-term, requiring an enduring and sustainable scheme of protection. In either situation, commanders must provide guidance on prioritizing protection capabilities and categorizing important assets.

IV. Critical and Defended Asset Lists

Ref: ADRP 3-37 (FM 3-37), Protection (Aug '12), pp. 2-6 to 2-7.

Initial assessments identify threats and hazards to the force, determine the criticality of systems and assets, and assess protection capabilities to mitigate vulnerabilities. The CAL and DAL are key protection products developed during initial assessments; they are dynamic lists that are continuously revised.

A. Critical Asset List (CAL)

The critical asset list is a prioritized list of assets, normally identified by phase of the operation and approved by the joint force commander, that should be defended against air and missile threats (JP 3-01). Once the threat, criticality, and vulnerability assessments are complete, the staff presents the prioritized CAL to the commander for approval. Commanders typically operate in a resource-constrained environment and have a finite amount of combat power for protecting assets. The protection cell/working group determines which assets are critical for mission success and recommends protection priorities based on the available resources. The CAL will vary depending on the mission variables.

During threat assessment, members of the protection cell/working group identify and prioritize the commander's critical assets using the vulnerability assessment, criticality assessment, and plan or order. Critical assets are generally specific assets of such extraordinary importance that their loss or degradation would have a significant and debilitating effect on operations or the mission. They represent what should be protected. The protection cell/working group uses information derived from command guidance, the intelligence preparation of the battlefield, targeting, risk management, warning orders, and the restated mission to nominate critical assets from their particular protection functional area. Vulnerability and criticality assessments are generally intended to be sequential. However, the criticality assessment can be conducted before, after, or concurrent with threat assessments. The vulnerability assessment should be conducted after the threat and criticality assessments to orient protection efforts on the most important assets. These assessments provide the staff with data to develop benchmarks, running estimates, commander's critical information requirements, change indicators, variances, MOEs, and MOPs.

CAL development may require the establishment of evaluation criteria, such as—

- Value (impact of loss)
- Depth (proximity in distance and time)
- Replacement impact (degree of effort, cost, or time)
- Capability (function and capacity for current and future operations)

The lack of a replacement may cause a critical asset to become the first priority for protection. Not all assets listed on the CAL will receive protection from continuously applied combat power. Critical assets with some protection from applied combat power become part of the DAL.

B. Defended Asset List (DAL)

The defended asset list is a listing of those assets from the critical asset list prioritized by the joint force commander to be defended with the resources available (JP 3-01). Critical assets that are reinforced with additional protection capabilities or capabilities from other combat power elements become part of the DAL. It represents what can be protected, by priority. The DAL allows commanders to apply finite protection capabilities to the most valuable assets. The combat power applied may be a weapons system, electronic sensor, obstacle, or combination.

V. Scheme of Protection Development

Ref: ADRP 3-37 (FM 3-37), Protection (Aug '12), pp. 2-7 to 2.8.

The scheme of protection describes how protection tasks support the commander's intent and concept of operations, and it uses the commander's guidance to establish the priorities of support to units for each phase of the operation. A commander's initial protection guidance may include protection priorities, civil considerations, protection task considerations, potential protection decisive points, high-risk considerations, and prudent risk.

Planners receive guidance as commanders describe their visualization of the operational concept and intent. This guidance generally focuses on the COA development by identifying decisive and supporting efforts, massing effects, and stating priorities. Effective planning guidance provides a broad perspective of the commander's visualization, with the latitude to explore additional options.

The scheme of protection is developed after receiving guidance and considering the principles of protection in relation to the mission variables, the incorporation of efforts, and the tasks that comprise the protection warfighting function. The scheme of protection is based on the mission variables, thus includes protection priorities by area, unit, activity, or resource. It addresses how protection is applied and derived during the conduct of operations. For example, the security for routes, bases/base camps, and critical infrastructure is accomplished by applying protection assets in dedicated, fixed, or local security roles; or it may be derived from economy-of-force protection measures such as area security techniques. It also identifies areas and conditions where forces may become fixed or static and unable to derive protection from their ability to maneuver and press the offensive. These conditions, areas, or situations are anticipated; and the associated risks are mitigated by describing and planning for the use of response forces. The staff considers the following items, at a minimum:

- Protection priorities
- Work priorities for survivability assets
- Air and missile defense positioning guidance
- Specific terrain and weather factors
- Intelligence focus and limitations for security efforts
- Areas or events where risk is acceptable
- Protected targets and areas
- Civilians and noncombatants in the area of operations
- Vehicle and equipment safety or security constraints
- Personnel recovery actions and control measures
- Force protection condition status
- Force health protection measures
- Mission-oriented protective posture guidance
- Environmental guidance
- Information operations condition
- Explosive ordnance and hazard guidance
- Ordnance order of battle
- OPSEC risk tolerance
- Fratricide avoidance measures
- Rules of engagement, standing rules for the use of force, and rules of interaction
- Escalation of force and nonlethal weapons guidance
- Operational scheme of maneuver
- Military deception
- Obscuration

VI. Protection Cell and Working Group

Commands utilize a protection cell and protection working group to integrate and synchronize protection tasks and systems for each phase of an operation or major activity.

A. Protection Cell

At division level and higher, the integration of the protection function and tasks is conducted by a designated protection cell and the chief of protection. At brigade level and below, the integration occurs more informally, with the designation of a protection coordinator from among the brigade staff or as an integrating staff function assigned to a senior leader. Chiefs of protection and protection coordinators participate in various forums to facilitate the continuous integration of protection tasks into the operations process. This occurs through protection working groups, staff planning teams, and staffs conducting integrating processes.

B. Protection Working Group

The protection cell forms the core membership of the protection working group, which includes other agencies as required. Protection cell and protection working group members differ in that additional staff officers are brought into the working group. These additional officers meet operational requirements for threat assessments, vulnerability assessments, and protection priority recommendations. The protection working group calls upon existing resources across the staff.

The protection working group is led by the chief of protection and normally consists of the following:

- Air and missile defense officer
- AT officer
- CBRN officer
- Engineer officer
- Electronic warfare element representative
- EOD officer
- Fire support representative
- OPSEC officer
- Provost marshal
- Safety officer
- Intelligence representative
- Civil affairs officer
- Personnel recovery officer
- Public affairs officer
- Staff judge advocate
- Surgeon
- Medical representative
- Veterinary representative
- Subordinate unit liaison officers
- Operations representative
- Area contracting officer

Commanders augment the team with other unit specialties and unified action partners depending on the operational environment and the unit mission. The protection officer determines the working group agenda, meeting frequency, composition, input, and expected output.

Running Estimate

Ref: ADRP 3-37 (FM 3-37), Protection (Aug '12), pp. 2-7 to 2.8.

The protection cell develops and refines the protection running estimate. The protection estimate provides a picture to the command on the protection warfighting function. It is developed from information (including the facts, assumptions, constraints, limitations, risks, and issues) pertaining to the protection mission and the scheme of protection. It includes the essential tasks from a higher order. Integrating process data and continuing activities (assets available, civil considerations, threat and hazard assessments, criticality assessments, vulnerability assessments, capability assessments, MOEs, MOPs, essential elements of friendly information, CALs, DALs, protection priorities, risk decision points, supporting tasks) feed updates to the running estimate.

Sample Protection Running Estimate

<table>
<tr><td>Past 24 hours</td><td colspan="3" rowspan="4">Protection common operational picture
• CAL/DAL
• Threats/hazards
• Incidents
• Asset locations</td><td>CAL/DAL</td></tr>
<tr><td>Next 48 to 72 hours</td><td>Asset task organization</td></tr>
<tr><td>Issues/risk decision points</td><td>Provost marshal officer</td></tr>
<tr><td>Protection priorities</td><td>Engineer</td></tr>
<tr><td>Essential elements of friendly information</td><td>FPCON
INFOCON</td><td>AMD
ADW
WCS</td><td>CBRN and EOD
MOPP
OEG</td><td>Safety</td></tr>
<tr><td colspan="5">Security banner</td></tr>
</table>

Legend:

ADW	air defense warning
AMD	air and missile defense
CAL	critical asset list
CBRN	chemical, biological, radiological, and nuclear
DAL	defended asset list
EOD	explosive ordnance disposal
FPCON	force protection condition
INFOCON	information operations condition
MOPP	mission-oriented protective posture
OEG	operation exposure guide
WCS	weapons control status

Ref: ADRP 3-37, Protection, fig. 2-2, p. 2-9.

C. Coordination and Relationships

Ref: ADRP 3-37 (FM 3-37), Protection (Aug '12), pp. 2-11 to 2-14.

The protection cell/working group ensures the integration of protection equities throughout the operations process via integrating processes, continuing activities, the military decisionmaking process, working groups, planning sessions, and coordination between warfighting functions. This develops and refines a scheme of protection and a protection plan that are comprehensive, integrated, layered, redundant, and enduring. All members of the protection cell/working group provide input and conduct actions that have beneficial output, which develops the scheme of protection and enhances the overall protection plan. The agenda, frequency, composition, input, and expected output for the working group are determined by the protection officer based on mission variables and military decisionmaking process integration.

Protection Working Group Actions

Key Input	Protection Actions	Steps	Protection Output	Key Output
• Higher HQ plan or order • New mission anticipated by the commander	• Consolidate protection-related running estimates from staffs • Review consolidated protection array of assets • Determine protection working group members • Ensure protection planner integration within the unit planning team	Step 1: Receipt of Mission Warning Order	• Protection working group • Warning and reporting systems • Protection running estimate	• Commander's initial guidance • Initial allocation of time
• Higher HQ plan or order • Higher HQ knowledge and intelligence products • Knowledge products from other organizations • Design concept (if developed)	• Provide input on critical networks or nodes that can be influenced • Identify requests for information • Determine available assets • Conduct and consolidate initial assessments • Conduct protection working group • Recommend and coordinate information collection assets for protection • Develop OPSEC indicators • Identify EEFI, and establish how long they need to be protected • Develop essential survivability and other engineering tasks • Identify available information on routes and key facilities • Analyze protection considerations of civilians in the area of operations • Determine available unified action partner capabilities • Determine funding sources, as required • Determine availability of construction and other engineering materials	Step 2: Mission Analysis Warning Order	• Consolidated HVT list • RFIs • Initial assessments • Recommended CAL • Recommended EEFI • Initial protection priorities • Input into information collection plan	• Problem statement • Mission statement • Initial commander's intent • Initial planning guidance • Initial CCIRs and EEFIs • Updated IPB and running estimates • Assumptions

Ref: ADRP 3-37, Protection, table 2-2, pp. 2-12 to 2-14.

Protection (ADRP 3-37)

Protection Working Group Actions (Cont.)

Key Input	Protection Actions	Steps	Protection Output	Key Output
• Mission statement • Initial commander's intent, planning guidance, CCIRs, and EEFIs • Updated IPB and running estimates • Assumptions	• Determine array of protection assets • Integrate protection tasks into COA • Determine initial scheme of protection • Coordinate health support requirements • Ensure that link architecture meets requirements and have been allocated from respective agents • Recommend appropriate level of survivability effort for each COA based on the expected threat • Determine alternate construction location, methods, means, materials, and timelines to give the commander options • Determine real-property and real estate requirements	Step 3: COA Development	• Recommended updates to CAL • Recommended updates to EEFIs • Initial scheme of protection	• COA statements and sketches • Tentative task organization • Broad concept of operations • Revised planning guidance • Updated assumptions
• Updated running estimates • Revised planning guidance • COA statements and sketches • Updated assumptions	• Identify limitations and shortfalls of protection tasks for each COA • Determine branches, sequels, decision points, unintended consequences, and second and third order effects • Develop risk management and decision points for risk tolerance • Develop MOE and MOP	Step 4: COA Analysis (War Game)	• Initial DAL • Refined EEFI • Refined information collection plan • Initial risk management and risk tolerance decision point matrix • Refined scheme of protection	• Refined COAs • Potential decision points • War-game results • Initial assessment measures • Updated assumptions
• Updated running estimates • Refined COAs • Evaluation criteria • War-game results • Updated assumptions	• Compare economy-of-force and risk reduction measures	Step 5: COA Comparison	• Refined protection priorities • Refined CAL/DAL • Refined EEFI • Refined scheme of protection	• Evaluated COAs • Recommended COAs • Updated running estimates • Updated assumptions
• Updated running estimates • Evaluated COAs • Recommended COA • Updated assumptions	• Brief scheme of protection • Brief protection task specifics, as required	Step 6: COA Approval Warning Order	• Refined protection priorities • Refined EEFI • Refined CAL/DAL • Refined scheme of protection	• Commander-selected COA and modifications • Refined commander's intent, CCIRs, and EEFIs • Updated assumptions
• Commander-selected COA with any modifications • Refined commander's intent, CCIRs, and EEFIs • Updated assumptions	• Refine and develop protection annex and supporting appendixes	Step 7: Orders Production, Dissemination, and Transition	• Protection annex and supporting appendixes	• Approved operation plan or order • Subordinate understanding of plan or order

D. Roles and Responsibilities

The protection cell/working group is responsible for integrating, coordinating, and synchronizing protection tasks and activities. The protection cell advises commanders on the priorities for protection and coordinates the implementation and sustainment of protective measures to protect assets according to the commander's priorities. The protection cell/working group helps develop a concept of protection tailored to the type of operation the unit is conducting.

During the planning process, the protection cell/working group provides input to the commander's military decisionmaking process by integrating the threat and hazard assessment with the commander's essential elements of friendly information, the CAL, and the DAL. While the planning cell develops plans, the protection cell/working group attempts to minimize vulnerability based on the developing COA. The intent is to identify and recommend refinements to the COA that are necessary to reduce vulnerability and ensure mission success. The protection cell/working group provides vulnerability mitigation measures to help reduce risks associated with a particular COA and conducts planning and oversight for unified land operations.

The protection working group—

- Determines likely threats and hazards from updated enemy or adversary tactics, the environment, and accidents
- Determines vulnerabilities as assessed by the vulnerability assessment team
- Establishes and recommends protection priorities, such as the CAL
- Provides recommendations for the CAL and DAL
- Reviews and coordinates unit protection measures
- Recommends force protection conditions and random AT measures
- Determines required resources and makes recommendations for funding and equipment fielding
- Provides input and recommendations on protection-related training
- Makes recommendations to cdrs on protection issues that require a decision
- Performs tasks required for a force protection working group and a threat protection working group according to DODI 2000.16
- Assesses assets and infrastructure that are designated as critical by HHQ
- Analyzes and provides recommendations for the protection of civilians in the AO
- Develops and refines the running estimate
- Develops a scheme of protection, nested with the operational concept
- Establishes the personnel recovery coordination center
- Develops personnel recovery guidance

The approved vulnerability reduction mitigation measures, commander's decisions for acceptable risks, CAL, and DAL represent running estimates that are incorporated into appropriate plans and orders. Based on these estimates, the protection cell develops the scheme of protection in the base order and appropriate annexes.

Planners integrate protection actions and information throughout specific plans and orders. Some significant, protection-related products that are often produced in the planning process include the—

- Scheme of protection that supports and nests with the operational concept
- Running estimate that reflects protection tasks and systems
- Quantifiable level of risk for specific events and activities
- Protection MOE and MOP and the threshold variances
- Recommendations for the commander's critical information requirements that reflect decision criteria from protection tasks and systems
- CAL and DAL
- Decision points based on the commander's risk tolerance level

III. Protection in Preparation

Ref: ADRP 3-37 (FM 3-37), Protection (Aug '12), chap. 3. See also p. 7-7.

The force is often most vulnerable to an enemy or adversary surprise attack during preparation. Preparation creates conditions that improve friendly force opportunities for success. It requires commander, staff, unit, and Soldier actions to ensure that the force is trained, equipped, and ready to execute operations. Preparation activities help commanders, staffs, and Soldiers understand a situation and their roles in upcoming operations. During the preparation phase, the protection focus is on deterring and preventing the enemy or adversaries from actions that would affect the combat power. The implementation of protection tasks with ongoing preparation activities helps prevent negative effects. Commanders ensure the integration of protection warfighting function tasks to safeguard bases, secure routes, and protect the force while it prepares for operations. Active defense measures help deny the initiative to the enemy or adversary, while the execution of passive defense measures prepares the force against the threat and hazard effects and accelerates the mitigation of those effects.

I. Protection Cell and Working Group

Preparation includes increased application and emphasis on protection measures. During preparation, the protection cell/working group—

- Provides recommendations to refine the scheme of protection
- Recommends systems to detect threats to the critical assets
- Proposes the refinement of OPSEC measures
- Monitors quick-reaction force or tactical and troop movements
- Provides recommendations on survivability position improvement
- Liaisons and coordinates with adjacent and protected units
- Determines protection indicators and warnings for information collection operations
- Monitors defended asset training
- Confirms backbriefs
- Analyzes and proposes vulnerability reduction measures
- Provides recommended revisions to tactical standing operating procedures
- Disseminates personnel recovery guidance

During preparation, the protection cell/working group ensures that the controls and risk reduction measures developed during planning have been implemented and are reflected in plans, standing operating procedures, and running estimates, even as the threat assessment is continuously updated. New threats and hazards are identified or anticipated based on newly assessed threat capabilities or changes in environmental conditions as compared with known friendly vulnerabilities and weaknesses. Commanders conduct after action reviews and war-game to identify changes to the threat. The protection cell and working group maintain a list of prioritized threats, adverse conditions, and hazard causes. The challenge is to find the root cause or nature of a threat or hazard so that the most effective protection solution can be implemented and disseminated.

II. Protection Considerations (Preparation)

Ref: ADRP 3-37 (FM 3-37), Protection (Aug '12), pp 3-1 to 3-3.

As the staff monitors and evaluates the performance or effectiveness of friendly COA, ground- and space-based information collection operations collect information that may confirm or deny forecasted threat COAs. As the threat changes, the risk to the force changes. Some changes may require a different protection posture or the implementation or cessation of specific protection measures and activities or restraints. The protection cell analyzes changes or variances that may require modifications to protection priorities and obtains guidance when necessary. Threat assessment is a dynamic and continually changing process. Chiefs of protection and planners stay alert for changing indicators and warnings in the operational environment that would signal new or fluctuating threats and hazards.

Detailed intelligence is used to develop threat assessments, and changes in the situation often dictate adjustments or changes to the plan when they exceed variance thresholds established during planning. During preparation, the staff continues to monitor and evaluate the overall situation because variable threat assessment information may generate new priority intelligence requirements, while changes in asset criticality could lead to new friendly force information requirements. Updated critical information requirements could be required based on changes to asset vulnerability and criticality when conjoined with the threat assessment.

Commanders who are exercising mission command direct and lead throughout the entire operations process as they provide supervision in concert with the process. Commanders' actions during preparation may also include—

- Reconciling the threat assessment with professional military judgment and experience
- Providing guidance on risk tolerance and making risk decisions
- Emphasizing protection tasks during rehearsals
- Minimizing unnecessary interference with subunits to allow maximum preparatory time
- Circulating throughout the environment to observe precombat inspections
- Directing control measures to reduce risks associated with preparatory movement
- Expediting the procurement and availability of resources needed for protection implementation
- Requesting higher headquarters support to reinforce logistical preparations and replenishment

Depending on the situation and the threat, some protection tasks may be conducted for short or long durations, covering the course of several missions or an entire operation. The staff coordinates the commander's protection priorities with vulnerability mitigation measures and clearly communicates them to—

- Superior, subordinate, and adjacent units
- Civilian agencies and personnel that are part of the force or those that may be impacted by the task or control

Subordinate leaders also conduct integration processes and provide supervision to ensure that Soldiers understand their responsibilities and the significance of protection measures and tasks. This is normally accomplished during mission preparation through training, rehearsals, task organization, and resource allocation. Rehearsals, especially those using opposing force personnel, can provide a measure of protection plan effectiveness.

Preparation Activities

Commanders, units, and Soldiers conduct preparation activities (as described in ADRP 5-0) to help ensure that the force is protected and prepared for execution. Protection is incorporated throughout all preparation activities, to include:

- Conduct rehearsals
- Conduct plans-to-operations transitions
- Refine the plan
- Initiate troop movement
- Integrate new Soldiers and units
- Initiate sustainment preparations
- Complete task organization
- Initiate network preparations
- Train
- Perform preoperation checks
- Prepare terrain
- Continue to build partnerships/teams
- Conduct confirmation briefs

Continue to Coordinate and Conduct Liaison

Coordination and liaison help ensure that leaders, internal and external to the headquarters, understand the unit role in upcoming operations and ensure that they are prepared to perform that role.

Initiate Information Collection

Throughout the operations process, commanders take every opportunity to improve their situational understanding. This requires aggressive and continuous information collection. Commanders and staffs continuously plan, task, and employ collection assets and forces to collect timely and accurate information that helps satisfy the commander's critical information requirements and other information requirements. For example, the protection working group uses staff analysis and coordination with higher headquarters to determine which critical assets or locations are likely to be attractive targets and require surveillance.

Initiate Security Operations

Commanders and staffs continuously plan and coordinate security operations throughout the conduct of operations. Security operations are those operations undertaken to—

- Provide an early and accurate warning of enemy or adversary operations
- Provide the force with the time and maneuver space necessary to react to the enemy or adversary
- Develop the situation so that commanders can effectively use the protected force

Security operations reflect increasing levels of combat power that can be applied to protect an asset or force from a directed threat, and they are typically conducted by operating forces designed to gain and exploit the initiative. The primary purpose of a **screen** operation is to provide early warning, thereby preventing surprise. **Guard and cover** operations involve combined arms units in combat, fighting to gain time with differing levels of capability and autonomy for independent action. **Operational area security** focuses on the protected force, installation, route, or area. **Local security** protection ranges from echelon headquarters to reserves and sustainment forces.

Manage and Prepare Terrain

Terrain management is the process of allocating terrain by establishing areas of operation, designating assembly areas, and specifying locations for units and activities to deconflict activities that might interfere with each other. Staffs deconflict operations, control movements, and deter the fratricide of units and unified action partners as they maneuver through the area of operations. The secure movement of theater resources is essential to ensure that commanders receive the forces, supplies, and equipment needed to support the operation plan and changing tactical situations; and it is an essential part of terrain management. Modifying the physical environment involves shaping the terrain to gain an advantage, such as improving cover, concealment, observation, fields of fire, obstacle effects through reinforcing obstacles, or mobility operations for the initial positioning of forces.

Subworking groups feed information to the protection working group and incorporate elements from the other warfighting functions. Commanders augment the working groups with other unit specialties and unified action partners depending on the operational environment and the unit mission. The lead for each working group determines the agenda, meeting frequency, composition, input, and expected output. Ultimately, the output from the working groups helps refine protection priorities, protection running estimates, assessments, essential elements of friendly information, CALs, DALs, and the scheme of protection.

A. Antiterrorism Working Group

The AT working group is led by the AT officer and includes members from the protection working group, subordinate commands, host nation agencies, and other unified action partners. It—

- Oversees the implementation of the AT program
- Develops and refines AT plans
- Addresses emergent and emergency AT program issues

B. Counter Improvised Explosive Device Working Group

The counter improvised explosive device working group is led by the EOD officer and includes members from the protection working group, subordinate commands, host nation agencies, and other unified action partners. It—

- Disseminates improvised explosive device information (including best practices), improvised explosive device trend analysis, and improvised explosive device defeat equipment and training issues
- Determines operational tactics to analyze and defeat the area of operations improvised explosive device networks
- Recommends the protection working group improvised explosive device defeat initiatives relating to equipment, intelligence, and operations.
- Identifies improvised explosive device defeat requirements and issues throughout the unit, including separate and subordinate units

C. Chemical, Biological, Radiological, and Nuclear Working Group

The CBRN working group is led by the CBRN officer and includes members from the protection working group, subordinate commands, host nation agencies, and other unified action partners. It—

- Disseminates CBRN operations information, including trend analysis, defense best practices and mitigating measures, operations, the status of equipment and training issues, CBRN logistics, and consequence management and remediation efforts
- Refines the CBRN threat, hazard, and vulnerability assessments

IV. Protection in Execution

Ref: ADRP 3-37 (FM 3-37), Protection (Aug '12), chap. 4. See also p. 7-8.

Commanders exercising mission command direct, lead, and assess organizations and Soldiers during execution. As operations develop and progress, commanders interpret information that flows from systems for indicators and warnings that signal the need for execution or adjustment decisions. Indicators and warnings can come from products such as space-based imagery and signals intelligence, among others. Commanders may direct and redirect the way combat power is applied or preserved and may adjust the tempo of operations through synchronization. Effective execution is aided by seizing the initiative through action and accepting prudent risk to exploit opportunities. The continuous and enduring character of protection activities makes the continuity of protection actions essential during execution. Commanders implement control measures and allocate resources that are sufficient to ensure protection continuity and restoration. The synchronization of applied protection warfighting function tasks and systems helps prevent threat and hazard effects to the force. Employed mitigation measures, planned and prepared for, allow the force to quickly respond and recover from the threat or hazard effects, ensuring a force that remains effective and continues the mission. Control measures may include restraint after careful and disciplined balancing decisions regarding the need for security and protection in the conduct of military operations.

I. Protection in Unified Land Operations

Unified land operations is the Army operating concept and its contribution to unified action. The central idea of unified land operations is how the Army seizes, retains, and exploits the initiative to gain and maintain a position of relative advantage in sustained land operations through simultaneous offensive, defensive, and stability or defense support of civil authorities operations to prevent or deter conflict, prevail in war, and create conditions for a favorable conflict resolution.

Commanders and leaders must be flexible and adaptive as they seek opportunities to seize, retain, and exploit the initiative. Leaders must have a situational understanding in simultaneous operations due to the diversity of threats, the proximity to civilians, and the impact of information during operations. The changing nature of operations may require a surge of certain capabilities, such as protection, to effectively link decisive operations to shaping or stabilizing activities in the area of operations. Commanders must accept risk to exploit time-sensitive opportunities by acting before adversaries discover vulnerabilities, take evasive or defensive action, and implement countermeasures.

Accurate assessment is essential for effective decisionmaking and the apportionment of combat power to protection tasks. Commanders fulfill protection requirements by applying comprehensive protection capabilities from main and supporting efforts to decisive and shaping operations. Protection can be derived as a by-product or a complementary result of some combat operations (such as security operations), or it can be deliberately applied as commanders integrate and synchronize tasks and systems that comprise the protection warfighting function.

ADRP 3-37, chap 4, provides discussion of protection considerations in support of the following types of operations: offense, defense, stability, civil security, civil control, and defense support of civil authorities.

II. Protection Considerations (Execution)

Ref: ADRP 3-37 (FM 3-37), Protection (Aug '12), pp. 4-13 to 4-14.

The protection cell/working group monitors and evaluates several critical ongoing functions associated with the execution of operational actions or changes that impact protection cell proponents. The protection cell/working group—

- Ensures that the protection focus supports the decisive operation
- Reviews and adjusts the commander's critical information requirements derived from protection tasks
- Reviews changes to graphic control measures and boundaries for the increased risk of fratricide
- Evaluates the effectiveness of battle tracking for constraints on personnel recovery
- Monitors the employment of security forces for gaps in protection or unintended patterns
- Evaluates the effectiveness of liaison personnel for protection activities
- Evaluates movement coordination and control to protect critical paths
- Monitors adjacent unit coordination procedures for terrain management vulnerabilities
- Monitors readiness rates of response forces involved in fixed-site protection
- Monitors force health protection
- Coordinates with the U.S. Army Space and Missile Defense Command for issues regarding personnel recovery operations

Staff members are also particularly alert for reports and events that meet the commander's critical information requirements. Once a threat to a critical or defended asset is detected by monitoring and evaluating running estimates and MOEs for indicators and warnings, the protection cell alerts the unit responsible for protecting the asset or recommends additional protective action. Unit commanders respond to the assessment of the threat or deliberate warning and then execute contingency or response plans.

Events frequently occur that prompt commanders to reevaluate assessed threats and their vulnerabilities. The protection cell/working group determines—

- Where protection assets can best help mission accomplishment with acceptable risk
- If protection assets should be committed to the mission immediately or be held in reserve
- If assets should be moved due to a change in the DAL
- If the commander needs to request assistance and, if so, for what purpose

There may be a change in the rules of engagement or the political, civil, or environmental situation. A failure to understand and comply with established rules of engagement can result in fratricide, mission failure, or national embarrassment. Commanders and Soldiers must limit collateral damage and apply force precisely to accomplish the mission without causing the unnecessary loss of life, suffering, or damage to property and infrastructure. The unanticipated changes may not require immediate action. However, commanders must consider how changes relate to the mission as they mitigate the vulnerability to civilians and the environment. They must—

- Determine if immediate actions will minimize damage
- Decide if actions will affect mission accomplishment
- Determine if the staff balance requires protective actions
- Ensure overall mission accomplishment

Chap 7

V. Protection Assessment

Ref: ADRP 3-37 (FM 3-37), Protection (Aug '12), chap. 5. See also p. 7-9.

Protection assessment is an essential, continuing activity that occurs throughout the operations process. While a failure in protection is typically easy to detect, the successful application of protection may be difficult to assess and quantify. Commanders and staffs monitor the current situation to—

- Collect data to assess
- Evaluate the progress toward attaining the end-state conditions, achieving objectives, and performing tasks
- Recommend or direct actions for improvement

I. Continuous Assessment

Assessment is the determination of the progress toward accomplishing a task, creating an effect, or achieving an objective (JP 3-0). Throughout the operations process, commanders integrate their assessments with those of the staff, subordinate commanders, and other unified action partners. The assessment plan is enabled by monitoring and evaluating criteria derived from the protection warfighting function tasks. Criteria used to monitor and evaluate the situation or operation may be represented as a MOE or a MOP.

Refer to The Battle Staff SMARTbook for discussion of the operations process, to include assessment (and MOEs/MOPs).

II. Lessons-Learned Integration

The way that organizations and Soldiers learn from mistakes is key in protecting the force. Although the evaluation process occurs throughout the operations process, it also occurs as part of the after action review and assessment following the mission. Leaders at all levels ensure that Soldiers and equipment are combat-ready. Leaders demonstrate their responsibility to sound stewardship practices and risk management principles required to ensure the minimal losses of resources and military assets due to hostile, nonhostile, and environmental threats and hazards. Key lessons learned are immediately applied and shared with other commands. Postoperational evaluations typically—

- Identify threats that were not identified as part of the initial assessment or identify new threats that evolved during the operation or activity.
- Assess the effectiveness of supporting operational goals and objectives. For example, determine if the controls positively or negatively impacted training or mission accomplishment and determine if they supported existing doctrine and TTP.
- Assess the implementation, execution, and communication of controls.
- Assess the accuracy of residual risk and the effectiveness of controls in eliminating hazards and controlling risks.
- Ensure coordination throughout the integration processes.

– Was the process integrated throughout all phases of the operation?
– Were risk decisions accurate?
– Were risk decisions made at the appropriate level?
– Did any unnecessary risks or benefits outweigh the cost in terms of expense, training benefit, or time?
– Was the process cyclic and continuous throughout the operation?

III. Protection Considerations (Assessment)

Ref: ADRP 3-37 (FM 3-37), Protection (Aug '12), pp. 5-1 to 5-2.

Assessment During Planning

The staff conducts analysis to assess threats, hazards, criticality, vulnerability, and capability to assist commanders in determining protection priorities, task organization decisions, and the integration of protection tasks.

The protection cell evaluates the COA during the military decisionmaking process against evaluation criteria derived from the protection warfighting function to determine if each COA is feasible, acceptable, and suitable in relation to its ability to protect or preserve the force.

Assessment During Preparation

Assessment occurs during preparation and includes activities required to maintain situational understanding; monitor and evaluate running estimates, tasks, MOEs, and MOPs; and identify variances for decision support. These assessments generally provide commanders with a composite estimate of preoperational force readiness or status in time to make adjustments.

During preparation, the protection cell focuses on threats and hazards that can influence preparatory activities, which includes monitoring new Soldier integration programs and movement schedules and evaluating live-fire requirements for precombat checks and inspections. The protection cell may evaluate training and rehearsals or provide coordination and liaison to facilitate effectiveness in high-risk or complex preparatory activities, such as movement and logistics preparation.

Assessment During Execution

The protection cell monitors and evaluates the progress of current operations to validate assumptions made in planning and to continually update changes to the situation. The protection cell and protection working group continually meet to monitor threats to the CAL and DAL, and they recommend changes to the protection plan as required. They also monitor the conduct of operations, looking for variances from the operations order that affect their areas of expertise. When variances exceed a threshold value developed or directed in planning, the protection cell may recommend an adjustment to counter an unforecasted threat or hazard or to mitigate a developing vulnerability. They also track the status of protection assets and evaluate the effectiveness of the protection systems as they are employed. Additionally, the protection cell and protection working group monitor the actions of other staff sections by periodically reviewing plans, orders, and risk assessments to determine if those areas require a change in protection priorities, posture, or resource allocation.

The protection cell and protection working group monitor and evaluate—

- Changes to threat and hazard assessments
- Changes in force vulnerabilities
- Changes to unit capabilities
- Relevancy of facts
- Validity of assumptions
- Reasons that new conditions affect the operation
- Running estimates
- Protection tasks
- System failures
- Resource allocations
- Increased risks
- Supporting efforts
- Force protection implementation measures, including site-specific AT measures

(AODS5) Index